Gravity

by

Trevor G. Underwood

What is gravity?

But is this such a simple law? What about the machinery of it? All we have done is to describe how the Earth moves around the Sun, but we have not said what makes it go. Newton made no hypothesis about this; was satisfied to find what it did without getting into the machinery of it. No one has since given any machinery. ... Why can we use mathematics to describe nature without a mechanism behind it? No one knows. We have to keep going because we find out more that way.

—Richard Feynman. (1963). *The Feynman Lectures on Physics*, Volume I, p. 7-9.

Published by Trevor G. Underwood
18 SE 10th Ave.
Fort Lauderdale, FL 33301

ISBN: 979-8-218-39654-1 (hardcover)
Library of Congress Control Number: 2024905518

Printed and distributed by Lulu Press, Inc.

627 Davis Dr.
Ste. 300
Morrisville, NC 27560
http://www.lulu.com/shop

CONTENTS

Page no.

39 **Kepler's laws of planetary motion.**

47 **Sir Isaac Newton (December 25, 1642 – March 20, 1727).**

50 **Newton, I. (July, 1687). *Philosophiæ Naturalis Principia Mathematica.*** (The Mathematical Principles of Natural Philosophy.) **Book I: The Motion of Bodies.** 1st Edition, London; 2nd Edition, Cambridge, 1713; 3rd Edition, London, 1726. (In Latin); translation below of 3rd Edition by A. Motte, (1729). London.); https://en.wikisource.org/ wiki/The_Mathematical_Principles _of_Natural_ Philosophy_(1729); also https://ia601604.us.archive.org/1/items/ newtonspmathema00newtrich/newtonspmathema00newtrich_bw.pdf. *Philosophiæ Naturalis Principia Mathematica* (Mathematical Principles of Natural Philosophy) is a work in three books written in Latin, first published July 5, 1687, with encouragement and financial help from Edmond Halley. After annotating and correcting his personal copy of the first edition, Newton published two further editions, during 1713 with errors of the 1687 corrected, and an improved version in 1726. The *Principia* includes Newton's three laws of motion, laying the foundation for classical mechanics; Newton's law of universal gravitation; and a derivation of Johannes Kepler's laws of planetary motion (which Kepler had first obtained empirically). In Book I: The Motion of Bodies, Newton addresses the motion of bodies attracted to each other by centripetal forces. Of the invention of *centripetal forces*, he propose that if two bodies S and P, attracting each other with forces reciprocally proportional to the squares of their distance, revolve about their common center of gravity; I say that the principal axis of the ellipse which either of the bodies as P describes by this motion about the other S, will be to the principal axis of the ellipse, which the same body P may describe in the same periodical time about the other body S quiescent, as the sum of the two bodies S+P to the first of two mean proportionals between that sum and the other body S. Of the motions of bodies tending to each other with centripetal forces, he proposes that two bodies attracting each other mutually, describe similar figures about their common center of gravity, and about each other mutually. Of the attractive forces of sphaerical bodies, he proposes that the same things supposed (if to the several points of a given sphere there tend equal centripetal forces decreasing in a duplicate ratio of the distances from the points), I say, that a corpuscle situate without the sphere is attracted with a force reciprocally proportional to the square of its distance from the center. He also proposes that if to the several points of a given sphere there tend equal centripetal forces decreasing in a duplicate ratio of the distances from the points; I say that another similar sphere will be attracted to it with a force reciprocally proportional to the square of the distance of the centers.

54 **Newton, I. (July, 1687). *Philosophiæ Naturalis Principia Mathematica.* Book III: Of the System of the World.** 1st Edition, London; 2nd Edition, Cambridge, 1713; 3rd Edition, London, 1726. (In Latin); translation below of 3rd Edition by A. Motte, (1729). London.); https://en.wikisource.org/ wiki/The_Mathematical _ Principles_of_Natural_Philosophy_(1729); also https://ia601604.us.archive.org/ 1/items/newtonspmathema00newtrich/newtonspmathema00newtrich_bw.pdf. In Book III, Newton notes that the centripetal force which arises between planets is the same as the gravitational force attracting matter to the Earth and focusses on gravitational attraction. He observes that the circumjovial planets, by radii drawn to Jupiter's center, describe areas proportional to the times of description; and that their periodic times, the fixed stars being at rest, are in the sesquiplicate proportion of their distances from, its center, etc. That the fixed Stars being at rest, the periodic times of the five primary Planets, and (whether of the Sun about the Earth, or) of the Earth about the Sun, are in the sesquiplicate proportion of their mean distances from the Sun. He noted that this proportion, first observed by Kepler, is now accepted by all astronomers. [The *sesquiplicate* ratio of given terms is the ratio between the square roots of the cubes of those terms.] He further observes that then the primary Planets, by radii drawn to the Earth, describe areas no wise proportional to the times: But that the areas, which they describe by radii drawn to the Sun, are proportional to the times of description; and that the Moon by a radius drawn to the Earth's center, describes an area proportional to the time of description. From this he proposes that the forces by which the circumjovial planets are continually drawn off from rectilinear motions, and retained in their proper orbits, tend to Jupiter's center; and are reciprocally as the squares of the distances of the places of those planets from that center; that the forces by which the primary planets are continually drawn off from rectilinear motions, and retained in their proper orbits, tend to the Sun; and are reciprocally as the squares of the distances of the places of those planets from the Sun's center; and that the force by which the Moon is retained in its orbit tends to the Earth; and is reciprocally as the square of the distance of its place from the Earth's center. He then proposes that that the moon gravitates towards the Earth, and by the force of gravity is continually drawn off from a rectilinear motion, and retained in its orbit; that the circumjovial planets gravitate towards Jupiter; the circumsaturnal towards Saturn; the circumsolar towards the Sun; and by the forces of their gravity are drawn off from rectilinear motions, and retained in curvilinear orbits; and concludes that the force which retains the celestial bodies in their orbits has been hitherto called centripetal force; but it being now made plain that it can be no other than a gravitating force, we shall hereafter call it gravity. He then proposes that all bodies gravitate towards; every Planet and that the Weights of bodies towards any the same Planet, at equal distances from the

center of the Planet, are proportional to the quantities of matter which they severally contain; and that there is a power of gravity tending to all bodies, proportional to the several quantities of matter which they contain; and that the force of gravity towards the several equal particles of any body, is reciprocally as the square of the distance of places from the particles; as appears from Cor. 3, Prop. LXXIV, Book I. In Proposition VI Newton provides his definition of *gravitational mass*, and in Proposition VII, together with its corollary 2, Newton restates his *universal law of gravitation.*

71 **Newton's calculation of Kepler's laws.**

74 **Newton's universal law of gravitation.**

78 **Henry Cavendish (October 10, 1731 – February 24, 1810).**

84 **Cavendish, H. (1798). Experiments to Determine the Density of Earth.** *Phil. Trans.*, 88, 469–526; doi:10.1098/rstl.1798.0022. While Newton was able to articulate his *Law of Universal Gravitation* and verify it experimentally, he could only calculate the relative gravitational force in comparison to another force. It was not until Henry Cavendish's verification of the gravitational constant that the Law of Universal Gravitation received its final form, $F = GMm/r^2 = 6.674 \times 10^{-11}\ Mm/r^2$ N (*SI units*), where F represents the force in Newtons, M and m represent the two masses in kilograms, and r represents the separation in meters. G represents the gravitational constant, which has a value of 6.674×10^{-11} N $(m/kg)^2$. Because of the magnitude of G, gravitational force is very small *unless large masses or short distances are involved.* Cavendish used apparatus, originally developed by Rev. John Michell, of the Royal Society, prior to his death, but which Cavendish improved. The apparatus is very simple; it consists of a wooden arm, 6 feet long, made so as to unite great strength with little weight. This arm is suspended in a horizontal position, by a slender wire 4.0 inches long, and to each extremity is hung a leaden ball, about 2 inches in diameter; and the whole is enclosed in a narrow wooden case, to defend it from the wind. As no more force is required to make this arm turn round on its center, than what is necessary to twist the suspending wire, it is plain, that if the wire is sufficiently slender, the most minute force, such as the attraction of a leaden weight a few inches in diameter, will be sufficient to draw the arm sensibly aside.

90 **Johann Carl Friedrich Gauss (April 30, 1777 – February 23, 1855).**

93 **Gauss's law for gravity.**

98 **Baron Siméon Denis Poisson (June 21, 1781 – April 25, 1840).**

101 **Poisson's equation.**

103 **Explanation for forward motion and direction of rotation of the planets around the Sun, and of the Moon around the Earth, and for the axial rotation of the Earth and the Sun, and the axial tilt of the Earth.**

106 **Einstein, A. (March, 1916). Die Grundlage der allgemeinen Relativitätstheorie. (The foundation of the general theory of relativity.)** *Ann. Phys.*, 49, 7, 769-822; http://dx.doi.org/10.1002/andp.19163540702; translation in A. Engel (translator), E. Schuckling (consultant). (1997). *The Collected Papers of Albert Einstein*, Volume 6: The Berlin Years: Writings, 1914-1917, Princeton University Press, Princeton, Doc. 30, 146-200; https://einsteinpapers.press.princeton.edu/vol6-trans/158; translation by T. G. Underwood; also, translation by S. N. Bose at https://en.wikisource.org/wiki/The_Foundation_of_ the_Generalised _Theory_of_Relativity; final consolidation by Einstein of his various papers on the subject - in particular, his three papers in November 1915. This was based on his conclusion in his *theory of general relativity* that space and time quantities could not be defined in such a way that spatial coordinate differences could be measured directly with the unit scale, or temporal ones with a normal clock. Einstein assumed that the general laws of nature should be expressed by equations that applied to all coordinate systems not just inertial systems, i.e. were covariant to arbitrary substitutions (generally covariant); and that the *theory of special relativity* was applicable for *infinitely small four-dimensional areas*. He assumed $ds^2 = \sum_{\mu\nu} g_{\mu\nu}\, dx_\mu\, dx_\nu$, where $g_{\mu\nu}$ is the "*fundamental tensor*", which described a curved surface, the *gravitational field*. He introduced the *extension* of the *fundamental tensor* $g_{\mu\nu}$, known as the *Riemann-Christoffel Tensor*, and equated the *equation of motion* of a freely moving body in a frame moving with uniform acceleration relative to the reference frame, i.e. along a *geodetic line* in space time, with the *equation of motion* of a material-point in a *gravitational field*. Einstein used the *field equations* of forces arising in an accelerated frame in the absence of matter, expressed in terms of the Hamiltonian, to obtain an equation corresponding to the *laws of conservation of momentum and energy*, in terms of the *energy components t_σ^{α} of the gravitation field*, adding an arbitrary factor -2κ, to obtain $\kappa t_\sigma^{\alpha} = \frac{1}{2}\, \delta_\sigma^{\alpha} g^{\mu\nu}\Gamma^{\lambda}_{\mu\beta}\, \Gamma^{\beta}_{\nu\lambda} - g^{\mu\nu}\Gamma^{\alpha}_{\mu\beta}\, \Gamma^{\beta}_{\nu\sigma}$,
where $\Gamma^{\tau}_{\mu\nu} = -\frac{1}{2}\, g^{\tau\alpha}\, (\partial g_{\mu\alpha}/\partial x_\nu + \partial g_{\nu\alpha}/\partial x_\mu - \partial g_{\mu\nu}/\partial x_\alpha)$. He then introduced matter into the *field equations* by adding an *energy-tensor* T_σ^{α} *associated with matter*, corresponding to the density ρ of Poisson's equation $\Delta\varphi = 4\pi\kappa\rho$, where φ was the gravitational potential and ρ was the density of matter, to obtain the *general field equations of gravitation* in the form
$\partial/\partial x_\alpha\, (g^{\sigma\beta}\Gamma^{\alpha}_{\mu\beta}) = -\kappa\{(t_\mu^{\sigma} + T_\mu^{\sigma}) - \frac{1}{2}\, \delta_\mu^{\sigma}\, (t + T)\}$, $(-g)^{1/2} = 1$,

or $\partial\Gamma^{\alpha}_{\mu\nu}/\partial x_{\alpha} + \Gamma^{\alpha}_{\mu\beta}\,\Gamma^{\beta}_{\nu\alpha} = -\,\kappa(T_{\mu\nu} - \tfrac{1}{2}\,g_{\mu\nu}T)$, with $(-g)^{1/2} = 1$, with the *sum of the energy components of matter and gravitation*, $t_{\mu}{}^{\sigma} + T_{\mu}{}^{\sigma}$ in place of the *energy components* $t_{\mu}{}^{\sigma}$, where $t = t^{\alpha}{}_{\alpha}$, and $T = T_{\mu}{}^{\mu}$ (Laue's scalar). Einstein introduced *Euler's equation of motion for a frictionless adiabatic liquid* in a *relativistic* form in which the *contravariant energy-tensor* of the liquid was
$T^{\alpha\beta} = -\,g^{\alpha\beta}p + \rho\; dx_{\alpha}/ds\; dx_{\beta}/ds$ in an attempt to provide a link between the *stress-energy tensor* defined in his *field equations* and matter. However, the force on matter in Euler's equation is much stronger, has nothing to do with the weak force of gravitational attraction between matter, *and is of opposite sign.* He then considered cases *when the velocity of the particle was very small compared with the speed of light* and the $g_{\mu\nu}$ differed from the values in an inertial frame under special relativity only by small magnitudes so that small quantities of the second and higher orders could be neglected (his "first aspect of the approximation") and dx_1/ds, dx_2/ds, dx_3/ds could be treated as small quantities, whereas dx_4/ds was equal to 1 (his "second point of view for approximation). This reduced his *equation of motion of a particle moving along the geodesic line* from $d^2x_{\tau}/ds^2 = \Gamma^{\tau}_{\mu\nu}\; dx_{\mu}/ds\; dx_{\nu}/ds$, where $\Gamma^{\tau}_{\mu\nu} = -\,\tfrac{1}{2}\,g^{\tau\alpha}\,(\partial g_{\mu\alpha}/\partial x_{\nu} + \partial g_{\nu\alpha}/\partial x_{\mu} - \partial g_{\mu\nu}/\partial x_{\alpha})$, to
$d^2x_{\tau}/dt^2 = -\,\tfrac{1}{2}\,\partial g_{44}/\partial x_{\tau}$ ($\tau = 1, 2, 3$), which Einstein considered represented the motion of a material point according to Newton's theory, in which $g_{44}/2$ played the part of the *gravitational potential.* Under a series of approximations to the *contravariant energy-tensor* of a frictionless adiabatic liquid $T^{\alpha\beta}$, all components vanished except $T_{44} = \rho = T$, from which Einstein obtained an equation for the *gravitational potential* in terms of the integral of the density of matter divide by the distance from the center of the matter $\varphi(r) = -\,\kappa/8\pi\,\int\rho d\tau/r$, of similar form to Newton's law of gravitation $\varphi(r) = -\,K/c^2\,\int\rho d\tau/r$. *In order to obtain a value for κ, Einstein set these two equations equal* giving $\kappa = 8\pi K/c^2 = 1.87 \times 10^{-29}$ (after correction for units), where K is the gravitation-constant 6.7×10^{-10}. He noted that according to his *theory of general relativity* $ds^2 = \sum g_{\mu\nu}dx_{\mu}dx_{\nu} = 0$, determining the velocity of light, so that light-rays are bent if the $g_{\mu\nu}$ were not constant. As in Einstein (November 18, 1915), his calculation of the bending of light, was obtained from his approximations for his equation of the *geodetic line* $\sum_{\alpha}\,\partial\Gamma^{\alpha}_{\mu\nu}/\partial x^{\alpha} + \sum_{\alpha\beta}\,\Gamma^{\alpha}_{\mu\beta}\,\Gamma^{\beta}_{\nu\alpha} = 0$, where $\Gamma^{\alpha}_{\mu\nu} = -\,\tfrac{1}{2}\sum_{\beta}\,g^{\alpha\beta}\,(\delta g_{\mu\beta}/\delta x_{\nu} + \delta g_{\nu\beta}/\delta x_{\mu} - \delta g_{\mu\nu}/\delta x_{\alpha})$, in which the link to the weak attractive force of gravitation was provided by *Newton's law of gravitation.* Einstein calculated the *deflection of light by the Sun* at a distance Δ,
$B = 2\alpha/\Delta = \kappa M/2\pi\Delta$, by substituting $\alpha = \kappa M/4\pi$, from his equation for the *gravitational potential* $\varphi(r) = -\,\tfrac{1}{2}\,\alpha/r = -\,\kappa/8\pi\,\int\rho d\tau/r = -\,\kappa M/8\pi r$, and setting
$\kappa = 8\pi K/c^2 = 1.87 \times 10^{-29}$. Consequently, as before, his computed value for the bending of light was the Newtonian value. He restated his formula for the addition to the precession of the perihelion of Mercury, but did not provide the derivation.

Why anyone gave credence to this is a mystery. By 1921 Einstein was already moving his research interests into superseding general relativity.

173 **Einstein, A. (February, 1917). Kosmologische Betrachtungen zur allgemeinen Relativitätstheorie. (Cosmological Considerations in the General Theory of Relativity.)** *Koniglich Preußische Akademie der Wissenschaften, Sitzungsberichte* (Berlin), 142–52; translation by W. Perrett & G. B. Jeffery in A. Engel (translator), E. Schuckling (consultant). (1997). *The Collected Papers of Albert Einstein*, Volume 6: The Berlin Years: Writings, 1914-1917, Princeton University Press, Princeton, Doc. 43, 421-32; https://einsteinpapers.press.princeton.edu/vol6-trans/433; describes Einstein's struggles with supplementing the *relativistic differential equations* by *limiting conditions* at *spatial infinity* in order to regard the universe as being of infinite spatial extent. In his treatment of the planetary problem, he chose these limiting conditions on the basis of the assumption that it is possible to select a system of reference so that at spatial infinity all the gravitational potentials $g_{\mu\nu}$ become constant, but it was by no means evident that the same limiting conditions could be applied to larger portions of the physical universe. Einstein attempts to resolve this using a method analogous to the extension of Poisson's equation used in the *non-relativistic case*, by adding to the left-hand side of field equation, the fundamental tensor $g_{\mu\nu}$, multiplied by a universal constant, $-\lambda$. As he noted, "*we admittedly had to introduce an extension of the field equations of gravitation which is not justified by our actual knowledge of gravitation*".

184 **Part II. What is Gravity.**

184 **The five fundamental interactions of physics.**

188 **Einstein, A. & Grommer, J. (1923). Proof of the Non-Existence of an Everywhere Regular, Centrally Symmetric Field According to the Field Theory of Kaluza.** *Scripta Universitatis atque Bibliothecae Hierosolymitanarum. Mathematica et Physica* 1 (1923), VII: 1–5 Kitvei ha-Universita ve-Beth-ha-Sfarim bi-Yerushalayim. Mathematica u'Fisica. A (5684), VII: 1–4 (Hebrew); translation in *The Collected Papers of Albert Einstein*, Volume 13: The Berlin Years: Writings & Correspondence January 1922 - March 1923 (English translation supplement); edited by Diana Kormos Buchwald, József Illy, Ze'ev Rosenkranz, & Tilman Sauer; Princeton University Press, Princeton, Doc. 12, 30-3; https://einsteinpapers.press.princeton.edu/ vol13-trans/60; Between 1923 until he died in 1955, Einstein published 31 papers on *classical unified field theory*, a unified theory of electromagnetism and gravity, without any success. This is the first.

PREFACE

This book picks up where my last book left off. Underwood, T. G. (2023). *General Relativity*: "Conclusion. … A detailed examination of Einstein's *theory of general relativity* reveals that it is not a *theory of gravity*; it is a *relativistic* theory about the *effects* of gravitation, or more strictly, of a uniformly accelerated reference frame. There is nothing in any version of this theory that represents or explains or provides any connection to the weak attractive gravitational force between matter. We are no further forward in understanding the origin of this fundamental force. … Einstein's theory of general relativity attempted to extend his theory of special relativity beyond space and time, to include matter and gravitational fields. Gravitation was introduced through the "equivalence principle", the equivalence of the *outcome* of the force of gravity and the acceleration of matter, first recognized in Newton's *Principia*. … In order to make calculations with his theory, Einstein had to import Newton's law of gravitation, which itself is an empirical law with no fundamental foundation. … A fundamental theoretical explanation for the weak attractive gravitational force between matter is also long overdue."

This volume attempts to address this omission. Without gravity there would be no Milky Way galaxy, no Solar System, no Sun, no planet Earth, no life. These all depend on the weak attractive force of gravity magnified in large dense masses. Yet we know less about gravity than any other force. The purpose of this book is to try to fill that void.

Part I reviews contributions to the understanding of the effects of gravity, starting with an extract from Johannes Kepler's seminal *Astronomia Nova ΑΙΤΙΟΛΟΓΗΤΟΣ seu physica coelestis, tradita commentariis de motibus stellae Martis ex observationibus G.V. Tychonis Brahe*. (New Astronomy, reasoned from Causes, or Celestial Physics, Treated by Means of Commentaries on the Motions of the Star Mars, from the Observations of the noble Tycho Brahe), published in (1609). This contains the results of the astronomer Johannes Kepler's ten-year-long investigation of the motion of Mars. One of the most significant books in the history of astronomy, the *Astronomia Nova* provided strong arguments for heliocentrism and contributed valuable insight into the movement of the planets. This included *the first mention of the planets' elliptical paths* and the change of their movement to the movement of free-floating bodies as opposed to objects on rotating spheres. It records the discovery of *the first two of the three principles known today as Kepler's laws of planetary motion; (1) That the planets move in elliptical orbits with the Sun at one focus*, and (2) *That the speed of the planet changes at each moment such that the time between two positions is always proportional to the area swept out on the orbit between these positions*.

This is followed by an extract from Kepler's *Harmonices mundi libri V*. (The Five Books of The Harmony of the World.), published in 1619, in which Kepler discusses harmony and congruence in geometrical forms and physical phenomena. The final section of this work relates his discovery of his "*third law of planetary motion*", a decade after the publication of the *Astronomia nova*. He found that (3) *The ratio of the cube of the length of the semi-major axis of each planet's orbit, to the square of time of its orbital period, is the same for all planets.*

Then it is Sir Isaac Newton's turn. In July, 1687, Newton published *Philosophiæ Naturalis Principia Mathematic,* (The Mathematical Principles of Natural Philosophy), a work in three books written in Latin. The *Principia* includes Newton's three laws of motion, laying the foundation for classical mechanics; Newton's law of universal gravitation; and a derivation of Johannes Kepler's laws of planetary motion (which Kepler had first obtained empirically).

In *Book I: The Motion of Bodies*, Newton addresses the motion of bodies attracted to each other by centripetal forces. He proposes that if two bodies, attracting each other with forces reciprocally proportional to the squares of their distance, revolve about their common center of gravity; that the principal axis of the ellipse which either of the bodies describes by this motion about the other, will be to the principal axis of the ellipse, which the same body may describe in the same periodical time about the other body. Of the motions of bodies tending to each other with centripetal forces, he proposes that two bodies attracting each other mutually, describe similar figures about their common center of gravity, and about each other mutually.

In *Book III: Of the System of the World,* Newton notes that the centripetal force which arises between planets is the same as the gravitational force attracting matter to the Earth and focusses on gravitational attraction. He observes that the circumjovial planets, by radii drawn to Jupiter's center, describe areas proportional to the times of description; and that their periodic times, the fixed stars being at rest, are in the sesquiplicate proportion of their distances from, its center, etc. That the fixed Stars being at rest, the periodic times of the five primary Planets, and (whether of the Sun about the Earth, or) of the Earth about the Sun, are in the sesquiplicate proportion of their mean distances from the Sun. He noted that this proportion, first observed by Kepler, is now accepted by all astronomers.

> [The *sesquiplicate* ratio of given terms is the ratio between the square roots of the cubes of those terms.]

He further observes that then the primary Planets, by the areas, which they describe by radii drawn to the Sun, are proportional to the times of description; and that the Moon by a radius drawn to the Earth's center, describes an area proportional to the time of description. From this he proposes that the forces by which the circumjovial planets are continually drawn off from rectilinear motions, and retained in their proper orbits, tend to Jupiter's center; and are reciprocally as the squares of the distances of the places of those planets from that center; that the forces by which the primary planets are continually drawn off from rectilinear motions, and retained in their proper orbits, tend to the Sun; and are reciprocally as the squares of the distances of the places of those planets from the Sun's center; and that the force by which the Moon is retained in its orbit tends to the Earth; and is reciprocally as the square of the distance of its place from the Earth's center. He then proposes that that the Moon gravitates towards the Earth, and by the force of gravity is continually drawn off from a rectilinear motion, and retained in its orbit; that the circumjovial planets gravitate towards Jupiter; the circumsaturnal towards Saturn; the circumsolar towards the Sun; and by the forces of their gravity are drawn off from rectilinear motions, and retained in curvilinear orbits; and concludes that the force which retains the celestial bodies in their orbits has been hitherto called centripetal force; but it being now made plain that it can be no other than a gravitating force, we shall hereafter call it gravity. He then proposes that all bodies gravitate towards; every Planet and that the Weights of bodies towards any the same Planet, at equal distances from the center of the Planet, are proportional to the quantities of matter which they severally contain; and that there is a power of gravity tending to all bodies, proportional to the several quantities of matter which they contain; and that the force of gravity towards the several equal particles of any body, is reciprocally as the square of the distance of places from the particles; as appears from Cor. 3, Prop. LXXIV, Book I.

In Proposition VI, Newton provides his definition of *gravitational mass*, and in Proposition VII, together with its corollary 2, Newton restates his *universal law of gravitation.*

Proposition VI. Theorem VI. *That all bodies gravitate towards; every Planet and that the Weights of bodies towards any the same Planet, at equal distances from the center of the Planet, are proportional to the quantities of matter which they severally contain.*

Proposition VII. Theorem VII. *That there is a power of gravity tending to all bodies, proportional to the several quantities of matter which they contain.*

...

Corollary 2. The force of gravity towards the several equal particles of any body, is reciprocally as the square of the distance of places from the particles; as appears from Cor. 3, Prop. LXXIV, Book I.

However, while Newton was able to articulate his *Law of Universal Gravitation* and verify it experimentally, he could only calculate the relative gravitational force in comparison to another force. It was not until Henry Cavendish's verification in 1798 of the Gravitational Constant, that the Law of Universal Gravitation received its final form,

$$F = GMm/r^2 = 6.674 \times 10^{-11}\ Mm/r^2\ N\ (\textit{SI units}),$$

where F represents the force in Newtons, M and m represent the two masses in kilograms, and r represents the separation in meters. G represents the gravitational constant, which has a value of 6.674×10^{-11} N $(m/kg)^2$.

Because of the magnitude of G, gravitational force is very small unless large masses or short distances are involved. Cavendish used apparatus, consisting of a wooden arm, 6 feet long, made so as to unite great strength with little weight. This arm was suspended in a horizontal position, by a slender wire 4.0 inches long, and to each extremity is hung a leaden ball, about 2 inches in diameter; and the whole is enclosed in a narrow wooden case, to defend it from the wind. As no more force is required to make this arm turn round on its center, than what is necessary to twist the suspending wire, it is plain, that if the wire is sufficiently slender, the most minute force, such as the attraction of a leaden weight a few inches in diameter, will be sufficient to draw the arm sensibly aside.

Part I also describes *Gauss's law for gravity*, named after Carl Friedrich Gauss, also known as Gauss's flux theorem for gravity, which is a law of physics that is equivalent to Newton's law of universal gravitation; and *Poisson's equation*, which provides the *potential field* caused by a given *mass density distribution* from which the *gravitational (force) field* can be calculated.

Gauss's law for gravity states that *the flux (surface integral) of the gravitational field over any closed surface is proportional to the enclosed mass.* It is often more convenient to work from than Newton's law.

The form of *Gauss's law for gravity* is mathematically similar to Gauss's law for electrostatics, one of Maxwell's equations. *Gauss's law for gravity* has the same mathematical relation to Newton's law that Gauss's law for electrostatics bears to Coulomb's law. This is because both Newton's law and Coulomb's law describe inverse-square interaction in a 3-dimensional space.

The *gravitational field* g (also called gravitational acceleration) is a vector field – a vector at each point of space (and time). It is defined so that the *gravitational force* experienced by a particle is equal to the mass of the particle multiplied by the *gravitational field* at that point.

The left-hand side of integral form of Gauss's law for gravity is called the *flux of the gravitational field*. According to the law it is always negative (or zero), and never positive. This can be contrasted with Gauss's law for electricity, where the flux can be either positive or negative. The difference is because charge can be either positive or negative, while mass can only be positive. *Gravitational flux* is a surface integral of the *gravitational field* over a closed surface, analogous to how magnetic flux is a surface integral of the magnetic field.

Poisson's equation is an elliptic partial differential equation of broad utility in theoretical physics, first published in 1813. The solution to Poisson's equation is the *potential field* caused by a given electric charge or mass density distribution; with the potential field known, one can then calculate the electrostatic or gravitational (force) field. If the mass density is zero, Poisson's equation reduces to Laplace's equation. The corresponding Green's function can be used to calculate the potential at distance r from a central point mass m (i.e., the fundamental solution). In three dimensions the *potential* is

$$\Phi(r) = -\,Gm/r$$

which is equivalent to Newton's law of universal gravitation.

Part I also includes the annotated copy of Einstein's March, 1916, paper, Einstein's final consolidation of his various papers on the subject, from Underwood, T. G. *General Relativity*, pp. 347-409. This shows how Einstein obtained an equation for the *gravitational potential* in terms of the integral of the density of matter divide by the distance from the center of the matter and substituted Newton's law of gravitation; and how his calculation of the bending of light, was obtained from his approximations for his equation of the *geodetic line*, in which the link to the weak attractive force of gravitation was also provided by *Newton's law of gravitation*. Consequently, his computed value for the bending of light was the Newtonian value.

A detailed examination of Einstein's *theory of general relativity* reveals that it is not a *theory of gravity*; it is a *relativistic* theory about the *effects* of gravitation, or more strictly, of a uniformly accelerated reference frame. There is nothing in any version of this theory that represents or explains or provides any connection to the weak attractive gravitational force between matter. In order to make calculations with his theory, Einstein had to import Newton's empirical law of gravitation. We are no further forward in understanding the origin of this fundamental force. Whilst Einstein's and others' objectives in removing a preferred reference frame and the existence of an ether from physics were admirable intentions, Einstein's subsequent fixation on the constancy of the speed of light, or some form of invariant space-time, in the face of reasonable alternatives, such as Ritz's emission theory on which quantum electrodynamics is founded, was not.

Finally, Part I includes an extract from Einstein, A. (February, 1917). *Kosmologische Betrachtungen zur allgemeinen Relativitätstheorie*. (Cosmological Considerations in the General Theory of Relativity.), which describes Einstein's struggles with supplementing the *relativistic differential equations* by *limiting conditions* at *spatial infinity* in order to regard the universe as being of infinite spatial extent. In his treatment of the planetary problem, he chose these limiting conditions on the basis of the assumption that it is possible to select a system of reference so that at spatial infinity all the gravitational potentials $g_{\mu\nu}$ become constant, but it was by no means evident that the same limiting conditions could be applied to larger portions of the physical universe. Einstein attempts to resolve this using a method analogous to the extension of Poisson's equation used in the *non-relativistic case*, by adding to the left-hand side of field equation, the fundamental tensor $g_{\mu\nu}$, multiplied by a universal constant, $-\lambda$. As he noted, "*we admittedly had to introduce an extension of the field equations of gravitation which is not justified by our actual knowledge of gravitation*".

Part II addresses "What is Gravity?". It begins by reviewing Einstein's unsuccessful attempts at producing a *classical unified field theory* between 1923 until he died in 1955, during which time Einstein published 31 papers on a unified theory of electromagnetism and gravity. An extract of the first of these, which he published jointly with Jacob Grommer is included.

Part II also includes Heinrich Weyl's attempt in 1929 to incorporate Dirac theory into the scheme of *general relativity* by introducing *gauge invariance* of *theory of coupled electromagnetic potentials* and Dirac *matter waves* [Weyl, H. (May, 1929). Elektron und Gravitation. (Electron and gravity.)] Weyl claimed that the barrier which hems progress of quantum theory is quantization of the field equations.

It also notes Einstein, A., Podolsky, B. & Rosen, N. May, 1935 paper, which suggests that the description of reality as given by a wave function in quantum mechanics is not complete.

But then it moves on to consider quantum entanglement, or some form of entanglement between matter, as a potential source of gravity and to examine the origin of gravity according to the Big Bang theory. In an attempt to answer the question: "Why does matter attract matter?", facts related to gravity and other forces of nature were are listed and reviewed. These include the following:

(1) The gravitational attractive force between two bodies obeys an inverse square law as if it were a radiated force in three-dimensional space;

(2) The gravitational attractive force between two bodies is proportional to the product of the masses, m_1 and m_2, of the bodies being attracted;

(3) The gravitational attractive force is additive in the sense that each of the constituents (one atom or one molecule) of one body are attracted by the each of the constituents of the other body in proportion to the product of each of their masses;

(4) The force of attraction between two objects with opposite charges also obeys an inverse square law;

(5) The gravitational force is about 10^{36} times weaker than the electric force. As both forces are subject to inverse square laws, this relationship will apply at all distances;

(6) The net inward gravitational force on 1 kg of matter at a distance s from the center of a planet or a star with radius r and density ρ is given by

$$F_{net} = 6.674 \text{ x } 10^{-11}\ 4/3\ \pi(2s - r)\ \rho\ N;$$

(7) There appears to be no equivalent gravitational repulsive force;

(8) The gravitational field of each body is present at every other body;

(9) According to Big Bang theory, the force of gravity separated from the other forces as the universe's temperature fell, immediately prior to the cosmic inflation during which the universe grew exponentially by a factor of at least 10^{78};

(10) According to Newton's universal law, the inward gravitational force F on a mass m at the at the surface of the universe,

$$F = GmM/r^2,$$

and the change in the inward gravitational force F on a mass m at the at the surface of the universe with the increase of the radius, due to the expansion of the universe, is inversely proportional to the cube of the radius of the universe.

$$dF/dr = - 2\ GmM/r^3;$$

(11) One way of looking at the expansion of the universe is as the expansion of the metric which determines the size of the universe as we observe it, in which all distances appear to have expanded and to continue to expand at the same rate,

separating matter without any apparent force. Gravitational attraction of matter may then simply be a reflection of the resistance of matter to this separation;

(12) On the other hand, the universe may have begun literally with a big bang which resulted in a uniform increase in the distance between matter as it evolved;

(13) The existence of a large amount of energy at the origin of the universe helps explain the subsequent expansion without the need to refer to it as dark energy;

(14) The fact that matter, comprising protons, neutrons and electrons, were formed whilst the distance between them was increasing may have something to do with the emergence of gravity;

(15) If the inward attraction of matter is a reflection of the resistance to the outward expansion of the universe, it may be possible to relate the gravitational constant G to the current rate of expansion of the universe;

(16) Gravity appears to be some sort of entanglement of the protons, neutrons and electrons comprising matter that has existed since soon after the Big Bang;

(17) Gravity appears to be similar to (non-relativistic) quantum entanglement between two particles that interact and then separate in such a way that the quantum state of each particle cannot be described independently of the state of the others?;

(18) The mass of the observable universe (including ordinary matter, the interstellar medium (ISM), and the intergalactic medium (IGM), but excluding dark matter and dark energy) is around 1.5×10^{53} kg;

(19) The actual density of atoms in the universe is equivalent to roughly 1 proton per 4 cubic meters = $1.6726 \times 10^{-27}/4 = 4.18 \times 10^{-28}$ kg m^{-3};

(20) This implies that the current volume of the observable universe is $(1.5 \times 10^{53})/(4.18 \times 10^{-28}) = 3.6 \times 10^{80}$ m^3, and its radius is 4.4×10^{26} m;

(21) The rate of expansion of the universe is estimated to be 73.3 kilometers per second per megaparsec (1 megaparsec = 3,260,000 light years = 3.0857×10^{19} km) or for every km from the Earth $73.3/3.0857 \times 10^{19} = 2.375 \times 10^{-18}$ km/sec;

(22) According to mass-energy equivalence, $E = mc^2$, the energy equivalent of one kilogram of mass is 8.99×10^{16} joules, so the energy equivalent of the mass of the observable universe is around $1.5 \times 10^{53} \times 8.99 \times 10^{16} = 1.35 \times 10^{70}$ joules;

(23) The zero-energy universe hypothesis proposes that the total amount of energy in the universe is exactly zero: its amount of positive energy in the form of matter is exactly canceled out by its negative energy in the form of gravity;

(24) Alternatively, a “closed” universe, where the density parameter $\Omega > 1$, and Ω is defined as the average matter density of the universe divided by a critical value of that density, in which positive energy dominates, will eventually collapse in a "Big Crunch"; while an "open" universe, where $\Omega < 1$, in which negative energy dominates, will either expand indefinitely or eventually disintegrate in a "Big Rip";

(25) According to the “closed” universe model, the universe might have started, after the initial Big Bang expansion, as a large sphere in space containing uniformly distributed matter, largely in the form of atoms and molecules, comprised of protons, neutrons and electrons, which continued to expand after the force causing the initial expansion ceased, based on the outward momentum of the matter;

(26) However, there appears to be a problem. Based on the actual density of atoms in the universe, the time taken for molecules of hydrogen or cosmic dust to accrete due to the force of gravitation is far too long;

(27) The existence of a very large amount of energy at the time of the origin of the current universe, makes the idea of a universe in which gravitational attractive forces eventually overcome the forces causing the expansion of the universe particularly attractive, in that it provides an explanation for this energy and for the Big Bang without invoking dark matter or dark energy;

(28) Under this theory, the current universe, which originated about 13.8 billion years ago, evolved for about 9.6 billion years before a primitive form of life originated under the particular conditions of a small rocky planet;

(29) Gravitational energy or gravitational potential energy U is the potential energy a massive object m has in relation to another massive object M due to gravity; $U = GmM/R$, where R is the distance between the centers. It is the potential energy associated with the gravitational field, which is released (converted into kinetic energy) when the objects fall towards each other. Gravitational potential energy increases when two objects are brought further apart.

In the common situation where a much smaller mass m is moving near the surface of a much larger object with mass M, the gravitational field is nearly constant and so the expression for gravitational energy can be simplified. The change in potential energy moving from the surface (a distance R from the center) to a height h above the surface is

$$\Delta U \approx GmM/R^2 \approx m(GM/R^2)\ h.$$

As the gravitational field is $g = GM/R^2$, this reduces to

$$\Delta U \approx mgh.$$

Taking U = 0 at the surface (instead of at infinity), the familiar expression for gravitational potential energy emerges:

$$U= mgh.$$

(30) This returns us to the question of whether it may be possible to relate the gravitational constant G to the current rate of expansion of the universe.

The gravitational field at the at the surface of the universe, is approximately equal to the gravitational constant.

The gravitational field $g = GM/R^2$. Substituting $M = 1.5 \times 10^{53}$ kg and $R = 4.4 \times 10^{26}$ m, gives

$$g = 0.7748\ G.$$

$g = G$ when $R^2 = M = 1.5 \times 10^{53}$; or $R = 3.873 \times 10^{26}$ m. This implies that the volume of the universe, $V = 4/3\ \pi R^3 = 2.4335 \times 10^{80}\ m^3$; which is in line with other estimates.

This is probably the closest that I can get.

I would like to acknowledge Wikipedia, in particular, which provided much of this material, as well as other referenced sources.

Trevor G. Underwood
18 SE 10th Ave
Fort Lauderdale, FL33301.

March 11, 2024.

PART I The effects of Gravity.

Johannes Kepler (December 27, 1571 – November 15, 1630).

Kepler was a German astronomer, mathematician, astrologer, natural philosopher and writer on music. He is a key figure in the 17th-century Scientific Revolution, best known for his laws of planetary motion, and his books *Astronomia nova*, *Harmonice Mundi*, and *Epitome Astronomiae Copernicanae*, influencing among others Isaac Newton, providing one of the foundations for his theory of universal gravitation. The variety and impact of his work made Kepler one of the founders and fathers of modern astronomy, the scientific method, natural and modern science.

Kepler was a mathematics teacher at a seminary school in Graz, where he became an associate of Prince Hans Ulrich von Eggenberg. Later he became an assistant to the astronomer Tycho Brahe in Prague, and eventually the imperial mathematician to Emperor Rudolf II and his two successors Matthias and Ferdinand II. He also taught mathematics in Linz, and was an adviser to General Wallenstein. Additionally, he did fundamental work in the field of optics, being named the father of modern optics, in particular for his *Astronomiae pars optica*. He also invented an improved version of the refracting telescope, the Keplerian telescope, which became the foundation of the modern refracting telescope, while also improving on the telescope design by Galileo Galilei, who mentioned Kepler's discoveries in his work.

Kepler lived in an era when there was no clear distinction between astronomy and astrology, but there was a strong division between astronomy (a branch of mathematics within the liberal arts) and physics (a branch of natural philosophy). Kepler also incorporated religious arguments and reasoning into his work, motivated by the religious conviction and belief that God had created the world according to an intelligible plan that is accessible through the natural light of reason. Kepler described his new astronomy as "celestial physics", as "an excursion into Aristotle's *Metaphysics*", and as "a supplement to Aristotle's *On the Heavens*", transforming the ancient tradition of physical cosmology by treating astronomy as part of a universal mathematical physics.

Kepler was born on 27 December 1571, in the Free Imperial City of Weil der Stadt (now part of the Stuttgart Region in the German state of Baden-Württemberg, 30 km west of Stuttgart's center). His grandfather, Sebald Kepler, had been Lord Mayor of the city. By the time Johannes was born, he had two brothers and one sister and the Kepler family fortune was in decline. His father, Heinrich Kepler, earned a precarious living as a mercenary, and he left the family when Johannes was five years old. He was believed to have died in the Eighty Years' War in the Netherlands. His mother, Katharina Guldenmann, an innkeeper's daughter, was a healer and herbalist. Born prematurely, Johannes claimed to have been weak and sickly as a child. Nevertheless, he often impressed travelers at his grandfather's inn with his phenomenal mathematical faculty.

As a child, Kepler witnessed the Great Comet of 1577, which attracted the attention of astronomers across Europe.

He was introduced to astronomy at an early age and developed a strong passion for it that would span his entire life. At age six, he observed the Great Comet of 1577, writing that he "was taken by [his] mother to a high place to look at it." In 1580, at age nine, he observed another astronomical event, a lunar eclipse, recording that he remembered being "called outdoors" to see it and that the Moon "appeared quite red". However, childhood smallpox left him with weak vision and crippled hands, limiting his ability in the observational aspects of astronomy.

In 1589, after moving through grammar school, Latin school, and seminary at Maulbronn, Kepler attended Tübinger Stift at the University of Tübingen. There, he studied philosophy under Vitus Müller and theology under Jacob Heerbrand (a student of Philipp Melanchthon at Wittenberg), who also taught Michael Maestlin while he was a student, until he became Chancellor at Tübingen in 1590. He proved himself to be a superb mathematician and earned a reputation as a skillful astrologer, casting horoscopes for fellow students. Under the instruction of Michael Maestlin, Tübingen's professor of mathematics from 1583 to 1631, he learned both the Ptolemaic system and the Copernican system of planetary motion. He became a Copernican at that time. In a student disputation, he defended heliocentrism from both a theoretical and theological perspective, maintaining that the Sun was the principal source of motive power in the universe. Despite his desire to become a minister, near the end of his studies, Kepler was recommended for a position as teacher of mathematics and astronomy at the Protestant school in Graz. He accepted the position in April 1594, at the age of 22.

Before concluding his studies at Tübingen, Kepler accepted an offer to teach mathematics as a replacement to Georg Stadius at the Protestant school in Graz (now in Styria, Austria). During this period (1594–1600), he issued many official calendars and prognostications that enhanced his reputation as an astrologer. Although Kepler had mixed feelings about astrology and disparaged many customary practices of astrologers, he believed deeply in a connection between the cosmos and the individual. He eventually published some of the ideas he had entertained while a student in the *Mysterium Cosmographicum* (1596), published a little over a year after his arrival at Graz.

In December 1595, Kepler was introduced to Barbara Müller, a 23-year-old widow (twice over) with a young daughter, Regina Lorenz, and he began courting her. Müller, an heiress to the estates of her late husbands, was also the daughter of a successful mill owner. Her father Jobst initially opposed a marriage. Even though Kepler had inherited his grandfather's nobility, Kepler's poverty made him an unacceptable match. Jobst relented after Kepler completed work on *Mysterium*, but the engagement nearly fell apart while Kepler was away tending to the details of publication. However, Protestant officials—who had helped set up the match—pressured the Müllers to honor their agreement. Barbara and Johannes were married on 27 April 1597. In the first years of their marriage, the Keplers had two children (Heinrich and Susanna), both of whom died in infancy. In 1602, they had a daughter (Susanna); in 1604, a son (Friedrich); and in 1607, another son (Ludwig).

Following the publication of *Mysterium* and with the blessing of the Graz school inspectors, Kepler began an ambitious program to extend and elaborate his work. He planned four additional books: one on the stationary aspects of the universe (the Sun and the fixed stars); one on the planets and their motions; one on the physical nature of planets and the formation of geographical features (focused especially on Earth); and one on the effects of the heavens on the Earth, to include atmospheric optics, meteorology, and astrology.

He also sought the opinions of many of the astronomers to whom he had sent *Mysterium*, among them Reimarus Ursus (Nicolaus Reimers Bär)—the imperial mathematician to Rudolf II and a bitter rival of Tycho Brahe. Ursus did not reply directly, but republished Kepler's flattering letter to pursue his priority dispute over (what is now called) the Tychonic system with Tycho. Despite this black mark, Tycho also began corresponding with Kepler, starting with a harsh but legitimate critique of Kepler's system; among a host of objections, Tycho took issue with the use of inaccurate numerical data taken from Copernicus. Through their letters, Tycho and Kepler discussed a broad range of astronomical problems, dwelling on lunar phenomena and Copernican theory (particularly its theological viability). But without the significantly more accurate data of Tycho's observatory, Kepler had no way to address many of these issues.

Instead, he turned his attention to chronology and "harmony," the numerological relationships among music, mathematics and the physical world, and their astrological consequences. By assuming the Earth to possess a soul (a property he would later invoke to explain how the Sun causes the motion of planets), he established a speculative system connecting astrological aspects and astronomical distances to weather and other earthly phenomena. By 1599, however, he again felt his work limited by the inaccuracy of available data—just as growing religious tension was also threatening his continued employment in Graz. In December of that year, Tycho invited Kepler to visit him in Prague;

on 1 January 1600 (before he even received the invitation), Kepler set off in the hopes that Tycho's patronage could solve his philosophical problems as well as his social and financial ones.

On 4 February 1600, Kepler met Tycho Brahe and his assistants Franz Tengnagel and Longomontanus at Benátky nad Jizerou (35 km from Prague), the site where Tycho's new observatory was being constructed. Over the next two months, he stayed as a guest, analyzing some of Tycho's observations of Mars; Tycho guarded his data closely, but was impressed by Kepler's theoretical ideas and soon allowed him more access. Kepler planned to test his theory from *Mysterium Cosmographicum* based on the Mars data, but he estimated that the work would take up to two years (since he was not allowed to simply copy the data for his own use). With the help of Johannes Jessenius, Kepler attempted to negotiate a more formal employment arrangement with Tycho, but negotiations broke down in an angry argument and Kepler left for Prague on 6 April. Kepler and Tycho soon reconciled and eventually reached an agreement on salary and living arrangements, and in June, Kepler returned home to Graz to collect his family.

Political and religious difficulties in Graz dashed his hopes of returning immediately to Brahe; in hopes of continuing his astronomical studies, Kepler sought an appointment as a mathematician to Archduke Ferdinand. To that end, Kepler composed an essay—dedicated to Ferdinand—in which he proposed a force-based theory of lunar motion: "In Terra inest virtus, quae Lunam ciet" ("There is a force in the earth which causes the moon to move"). Though the essay did not earn him a place in Ferdinand's court, it did detail a new method for measuring lunar eclipses, which he applied during the 10 July eclipse in Graz. These observations formed the basis of his explorations of the laws of optics that would culminate in *Astronomiae Pars Optica*.

On 2 August 1600, after refusing to convert to Catholicism, Kepler and his family were banished from Graz. Several months later, Kepler returned, now with the rest of his household, to Prague. Through most of 1601, he was supported directly by Tycho, who assigned him to analyzing planetary observations and writing a tract against Tycho's (by then deceased) rival, Ursus. In September, Tycho secured him a commission as a collaborator on the new project he had proposed to the emperor: the *Rudolphine Tables* that should replace the *Prutenic Tables* of Erasmus Reinhold. Two days after Tycho's unexpected death on 24 October 1601, Kepler was appointed his successor as the imperial mathematician with the responsibility to complete his unfinished work. The next 11 years as imperial mathematician would be the most productive of his life.

Kepler's primary obligation as imperial mathematician was to provide astrological advice to the emperor. Though Kepler took a dim view of the attempts of contemporary astrologers to precisely predict the future or divine specific events, he had been casting well-received detailed horoscopes for friends, family, and patrons since his time as a student in Tübingen. In addition to horoscopes for allies and foreign leaders, the emperor sought Kepler's advice in times of political trouble. Rudolf was actively interested in the work of many of his court scholars (including numerous alchemists) and kept up with Kepler's work in physical astronomy as well.

Officially, the only acceptable religious doctrines in Prague were Catholic and Utraquist, but Kepler's position in the imperial court allowed him to practice his Lutheran faith unhindered. The emperor nominally provided an ample income for his family, but the difficulties of the over-extended imperial treasury meant that actually getting hold of enough money to meet financial obligations was a continual struggle. Partly because of financial troubles, his life at home with Barbara was unpleasant, marred with bickering and bouts of sickness. Court life, however, brought Kepler into contact with other prominent scholars (Johannes Matthäus Wackher von Wackhenfels, Jost Bürgi, David Fabricius, Martin Bachazek, and Johannes Brengger, among others) and astronomical work proceeded rapidly.

The extended line of research that culminated in *Astronomia Nova* (A New Astronomy)—including the first two laws of planetary motion—began with the analysis, under Tycho's direction, of the orbit of Mars. In this work Kepler introduced the revolutionary concept of *planetary orbit*, a path of a planet in space resulting from the action of physical causes, distinct from previously held notion of planetary orb (a spherical shell to which planet is attached). As a result of this breakthrough astronomical phenomena came to be seen as being governed by physical laws. Kepler calculated and recalculated various approximations of Mars's orbit using an equant (the mathematical tool that Copernicus had eliminated with his system), eventually creating a model that generally agreed with Tycho's observations to within two arcminutes (the average measurement error). But he was not satisfied with the complex and still slightly inaccurate result; at certain points the model differed from the data by up to eight arcminutes. The wide array of traditional mathematical astronomy methods having failed him, Kepler set about trying to fit an ovoid orbit to the data.

In Kepler's religious view of the cosmos, the Sun (a symbol of God the Father) was the source of motive force in the Solar System. As a physical basis, Kepler drew by analogy on William Gilbert's theory of the magnetic soul of the Earth from *De Magnete* (1600) and on his own work on optics. Kepler supposed that the motive power (or motive species) radiated by the Sun weakens with distance, causing faster or slower motion as planets move closer or farther from it. [note 1] Perhaps this assumption entailed a mathematical relationship that would restore astronomical order. Based on measurements of the aphelion and perihelion of the Earth and Mars, he created a formula in *which a planet's rate of motion is inversely proportional to its distance from the Sun*. Verifying this relationship throughout the orbital cycle required very extensive calculation; to simplify this task, by late 1602 Kepler reformulated the proportion in terms of geometry: *planets sweep out equal areas in equal times*—his *second law of planetary motion.*

He then set about calculating the entire orbit of Mars, using the geometrical rate law and assuming an egg-shaped ovoid orbit. After approximately 40 failed attempts, *in late 1604 he at last hit upon the idea of an ellipse*, which he had previously assumed to be too simple a solution for earlier astronomers to have overlooked. Finding that an elliptical orbit fit the Mars data (the Vicarious Hypothesis), Kepler immediately concluded that *all planets move in ellipses, with the Sun at one focus*—his *first law of planetary motion.* Because he employed no calculating assistants, he did not extend the mathematical analysis beyond Mars. By the end of the year, he completed the manuscript for *Astronomia nova*, though it would not be published until 1609 due to legal disputes over the use of Tycho's observations, the property of his heirs.

As Kepler slowly continued analyzing Tycho's Mars observations—now available to him in their entirety—and began the slow process of tabulating the *Rudolphine Tables*, Kepler also picked up the investigation of the laws of optics from his lunar essay of 1600. Both lunar and solar eclipses presented unexplained phenomena, such as unexpected shadow sizes, the red color of a total lunar eclipse, and the reportedly unusual light surrounding a total solar eclipse. Related issues of atmospheric refraction applied to all astronomical observations. Through most of 1603, Kepler paused his other work to focus on optical theory; the resulting manuscript, presented to the emperor on 1 January 1604, was published as *Astronomiae Pars Optica* (The Optical Part of Astronomy). In it, Kepler described the inverse-square law governing the intensity of light, reflection by flat and curved mirrors, and principles of pinhole cameras, as well as the astronomical implications of optics such as parallax and the apparent sizes of heavenly bodies. He also extended his study of optics to the human eye, and is generally considered by neuroscientists to be the first to recognize that images are projected inverted and reversed by the eye's lens onto the retina. The solution to this dilemma was not of particular importance to Kepler as he did not see it as pertaining to optics, although he did suggest that the image was later corrected "in the hollows of the brain" due to the "activity of the Soul."

Today, *Astronomiae Pars Optica* is generally recognized as the foundation of modern optics (though the law of refraction is conspicuously absent). With respect to the beginnings of projective geometry, Kepler introduced the idea of continuous change of a mathematical entity in this work. He argued that if a focus of a conic section were allowed to move along the line joining the foci, the geometric form would morph or degenerate, one into another. In this way, an ellipse becomes a parabola when a focus moves toward infinity, and when two foci of an ellipse merge into one another, a circle is formed. As the foci of a hyperbola merge into one another, the hyperbola becomes a pair of straight lines. He also assumed that if a straight line is extended to infinity, it will meet itself at a single point at infinity, thus having the properties of a large circle.

In October 1604, a bright new evening star (SN 1604) appeared, but Kepler did not believe the rumors until he saw it himself. Kepler began systematically observing the supernova. Astrologically, the end of 1603 marked the beginning of a fiery trigon, the start of the about 800-year cycle of great conjunctions; astrologers associated the two previous such periods with the rise of Charlemagne (c. 800 years earlier) and the birth of Christ (c. 1600 years earlier), and thus expected events of great portent, especially regarding the emperor.

It was in this context, as the imperial mathematician and astrologer to the emperor, that Kepler described the new star two years later in his *De Stella Nova*. In it, Kepler addressed the star's astronomical properties while taking a skeptical approach to the many astrological interpretations then circulating. He noted its fading luminosity, speculated about its origin, and used the lack of observed parallax to argue that it was in the sphere of fixed stars, further undermining the doctrine of the immutability of the heavens (the idea accepted since Aristotle that the celestial spheres were perfect and unchanging). The birth of a new star implied the variability of the heavens. Kepler also attached an appendix where he discussed the recent chronology work of the Polish historian Laurentius Suslyga; he calculated that, if Suslyga was correct that accepted timelines were four years behind, then the Star of Bethlehem—analogous to the present new star—would have coincided with the first great conjunction of the earlier 800-year cycle.

Over the following years, Kepler attempted (unsuccessfully) to begin a collaboration with Italian astronomer Giovanni Antonio Magini, and dealt with chronology, especially the dating of events in the life of Jesus. Around 1611, Kepler circulated a manuscript of what would eventually be published (posthumously) as *Somnium* [The Dream]. Part of the purpose of *Somnium* was to describe what practicing astronomy would be like from the perspective of another planet, to show the feasibility of a non-geocentric system. The manuscript, which disappeared after changing hands several times, described a fantastic trip to the Moon; it was part allegory, part autobiography, and part treatise on interplanetary travel (and is sometimes described as the first work of science fiction). Years later, a distorted version of the story may have instigated the witchcraft trial against his mother, as the mother of the narrator consults a demon to learn the means of space travel. Following her eventual acquittal, Kepler composed 223 footnotes to the story—several times longer than the actual text—which explained the allegorical aspects as well as the considerable scientific content (particularly regarding lunar geography) hidden within the text.

In 1611, the growing political-religious tension in Prague came to a head. Emperor Rudolf—whose health was failing—was forced to abdicate as King of Bohemia by his brother Matthias. Both sides sought Kepler's astrological advice, an opportunity he used to deliver conciliatory political advice (with little reference to the stars, except in general statements to discourage drastic action). However, it was clear that Kepler's future prospects in the court of Matthias were dim.

Also in that year, Barbara Kepler contracted Hungarian spotted fever, then began having seizures. As Barbara was recovering, Kepler's three children all fell sick with smallpox; Friedrich, 6, died. Following his son's death, Kepler sent letters to potential patrons in Württemberg and Padua. At the University of Tübingen in Württemberg, concerns over Kepler's perceived Calvinist heresies in violation of the Augsburg Confession and the Formula of Concord prevented his return. The University of Padua—on the recommendation of the departing Galileo—sought Kepler to fill the mathematics professorship, but Kepler, preferring to keep his family in German territory, instead travelled to Austria to arrange a position as teacher and district mathematician in Linz. However, Barbara relapsed into illness and died shortly after Kepler's return.

Kepler postponed the move to Linz and remained in Prague until Rudolf's death in early 1612, though between political upheaval, religious tension, and family tragedy (along with the legal dispute over his wife's estate), Kepler could do no research. Instead, he pieced together a chronology manuscript, *Eclogae Chronicae*, from correspondence and earlier work. Upon succession as Holy Roman Emperor, Matthias re-affirmed Kepler's position (and salary) as imperial mathematician but allowed him to move to Linz.

In Linz, Kepler's primary responsibilities (beyond completing the *Rudolphine Tables*) were teaching at the district school and providing astrological and astronomical services. In his first years there, he enjoyed financial security and religious freedom relative to his life in Prague—though he was excluded from Eucharist by his Lutheran church over his theological scruples. It was also during his time in Linz that Kepler had to deal with the accusation and ultimate verdict of witchcraft against his mother Katharina in the Protestant town of Leonberg. That blow, happening only a few years after Kepler's excommunication, is not seen as a coincidence but as a symptom of the full-fledged assault waged by the Lutherans against Kepler.

His first publication in Linz was *De vero Anno* (1613), an expanded treatise on the year of Christ's birth. He also participated in deliberations on whether to introduce Pope Gregory's reformed calendar to Protestant German lands. On 30 October 1613, Kepler married the 24-year-old Susanna Reuttinger. Following the death of his first wife Barbara, Kepler had considered 11 different matches over two years (a decision process formalized later as the marriage problem). He eventually returned to Reuttinger (the fifth match) who, he wrote, "won me over with love, humble loyalty, economy of household, diligence, and the love she gave the stepchildren." The first three children of this marriage (Margareta Regina, Katharina, and Sebald) died in childhood. Three more survived into adulthood: Cordula (born 1621); Fridmar (born 1623); and Hildebert (born 1625). According to Kepler's biographers, this was a much happier marriage than his first.

Since completing the *Astronomia Nova*, Kepler had intended to compose an astronomy textbook that would cover all the fundamentals of heliocentric astronomy. Kepler spent the next several years working on what would become *Epitome Astronomiae Copernicanae* (Epitome of Copernican Astronomy). Despite its title, which merely hints at heliocentrism, the Epitome is less about Copernicus's work and more about Kepler's own astronomical system. The Epitome contained all *three laws of planetary motion* and attempted to explain heavenly motions through physical causes. Although it explicitly extended the first two laws of planetary motion (applied to Mars in *Astronomia nova*) to all the planets as well as the Moon and the Medicean satellites of Jupiter, it did not explain how elliptical orbits could be derived from observational data.

Originally intended as an introduction for the uninitiated, Kepler sought to model his *Epitome* after that of his master Michael Maestlin, who published a well-regarded book explaining the basics of geocentric astronomy to non-experts. Kepler completed the first of three volumes, consisting of Books I–III, by 1615 in the same question-answer format of Maestlin's and have it printed in 1617. However, the banning of Copernican books by the Catholic Church, as well as the start of the Thirty Years' War, meant that publication of the next two volumes would be delayed. In the interim, and to avoid being subject to the ban, Kepler switched the audience of the *Epitome* from beginners to that of expert astronomers and mathematicians, as the arguments became more and more sophisticated and required advanced mathematics to be understood. The second volume, consisting of Book IV, was published in 1620, followed by the third volume, consisting of Books V–VII, in 1621.

In 1619 Kepler published *Harmonices mundi*, which he had been working on since 1595. [Kepler, J. (1619). Harmonices mundi libri V (The Five Books of The Harmony of the World.] The fifth volume described the harmony of the motions of the planets. This included a long digression on astrology, which was immediately followed by *Kepler's third law of planetary motion*, which showed a constant proportionality between the cube of the semi-major axis of a planet's orbit and the square of the time of its orbital period.

In the years following the completion of *Astronomia Nova*, most of Kepler's research was focused on preparations for the *Rudolphine Tables* and a comprehensive set of ephemerides (specific predictions of planet and star positions) based on the table, though neither would be completed for many years. Kepler, at last, completed the Rudolphine Tables in 1623, which at the time was considered his major work. However, due to the publishing requirements of the emperor and negotiations with Tycho Brahe's heir, it would not be printed until 1627.

On 8 October 1630, Kepler set out for Regensburg, hoping to collect interest on work he had done previously. A few days after reaching Regensburg, Kepler became sick, and progressively became worse. On 15 November 1630, just over a month after his arrival, he died. He was buried in a Protestant churchyard that was completely destroyed during the Thirty Years' War.

Kepler, J. (1609). ***Astronomia Nova ΑΙΤΙΟΛΟΓΗΤΟΣ seu physica coelestis, tradita commentariis de motibus stellae Martis ex observationibus G.V. Tychonis Brahe*** **(New Astronomy, reasoned from Causes, or Celestial Physics, Treated by Means of Commentaries on the Motions of the Star Mars, from the Observations of the noble Tycho Brahe).**

Translation by W.H. Donahue (1992). Cambridge University Press. Also in Donahue, W. H. (2004). *Selections from Kepler's Astronomia Nova.* Santa Fe: Green Lion Press.

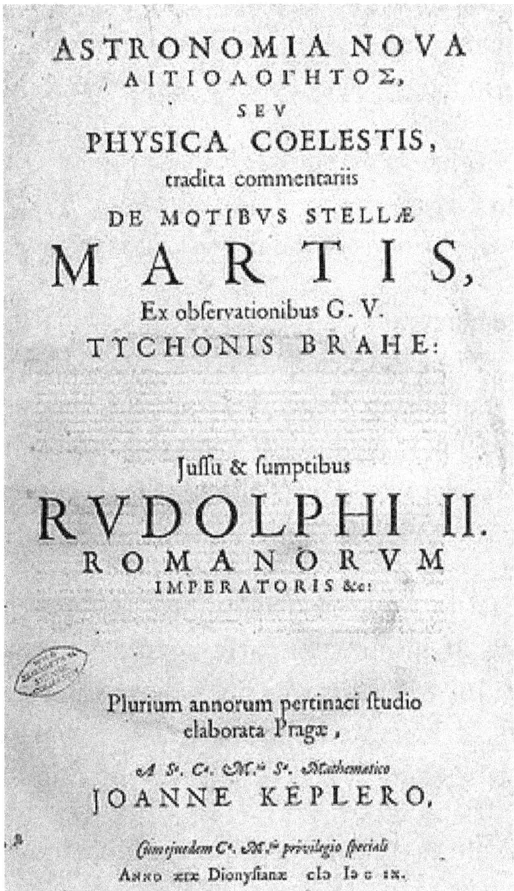
ASTRONOMIA NOVA
ΑΙΤΙΟΛΟΓΗΤΟΣ,
SEV
PHYSICA COELESTIS,
tradita commentariis
DE MOTIBVS STELLÆ
MARTIS,
Ex obſervationibus G. V.
TYCHONIS BRAHE:

Juſſu & ſumptibus
RVDOLPHI II.
ROMANORVM
IMPERATORIS &c:

Plurium annorum pertinaci ſtudio
elaborata Pragæ,
A S. C. M. S. Mathematico
JOANNE KEPLERO,

Anno æræ Dionyſianæ cIↄ Iↄ c ix.

Astronomia nova, published in 1609, contains the results of the astronomer Johannes Kepler's ten-year-long investigation of the motion of Mars. One of the most significant books in the history of astronomy, the *Astronomia nova* provided strong arguments for heliocentrism and contributed valuable insight into the movement of the planets. This included *the first mention of the planets' elliptical paths* and the change of their movement to the movement of free-floating bodies as opposed to objects on rotating spheres. The *Astronomia nova* records the discovery of *the first two of the three principles known today as Kepler's laws of planetary motion; (1) That the planets move in elliptical orbits with the Sun at one focus*, and (2) *That the speed of the planet changes at each moment such that the time between two positions is always proportional to the area swept out on the orbit between these positions.*

For over 650 pages (in the English translation), Kepler walks his readers, step by step, through his process of discovery. It is recognized as one of the most important works of the Scientific Revolution. Kepler sent Galileo the book while the latter was working on his *Dialogue Concerning the Two Chief World Systems* (published in 1632, two years after Kepler's death). Galileo had been trying to determine the path of an object falling from rest towards the center of the Earth, but used a semicircular orbit in his calculation.

Prior to Kepler, Nicolaus Copernicus proposed in 1543 that the Earth and other planets orbit the Sun. The Copernican model of the Solar System was regarded as a device to explain the observed positions of the planets rather than a physical description.

Kepler sought for and proposed physical causes for planetary motion. His work is primarily based on the research of his mentor, Tycho Brahe. The two, though close in their work, had a tumultuous relationship. Regardless, in 1601 on his deathbed, Brahe asked Kepler to make sure that he did not "die in vain," and to continue the development of his model of the Solar System. Kepler would instead write the *Astronomia nova*, in which he rejects the Tychonic system, as well as the Ptolemaic system and the Copernican system.

By 1602, Kepler set to work on determining the orbit pattern of Mars, keeping David Fabricius informed of his progress. He suggested the possibility of an oval orbit to Fabricius by early 1604, though was not believed. Later in the year, Kepler wrote back with his discovery of Mars's elliptical orbit. The manuscript for *Astronomia nova* was completed by September 1607, and was in print by August 1609.

The introduction outlines the four steps Kepler took during his research.

The first step is his claim that the Sun itself and not any imaginary point near the Sun (as in the Copernican system) is the point where all the planes of the eccentrics of the planets intersect, or the center of the orbits of the planets.

The second step consists of Kepler placing the Sun as the center and mover of the other planets. This step also contains Kepler's reply to objections against placing the Sun at the center of the universe, including objections based on scripture. In reply to scripture, he argues that it is not meant to claim physical dogma, and the content should be taken spiritually.

In the third step, he posits that the Sun is the source of the motion of all planets, using Brahe's proof based on comets that planets do not rotate on orbs.

The fourth step consists of describing the path of planets as not a circle, but an oval.

As the *Astronomia nova* proper starts, Kepler demonstrates that the Tychonic, Ptolemaic, and Copernican systems are indistinguishable on the basis of observations alone. The three models predict the same positions for the planets in the near term, although they diverge from historical observations, and fail in their ability to predict future planetary positions by a small, though absolutely measurable amount. Kepler here introduces his famous diagram of the movement of Mars in relation to Earth if Earth remained unmoving at the center of its orbit. The diagram shows that Mars's orbit would be completely imperfect and never follow along the same path.

Kepler discusses all his work at great length throughout the book. He addresses this length in the sixteenth chapter:

"If thou art bored with this wearisome method of calculation, take pity on me, who had to go through with at least seventy repetitions of it, at a very great loss of time".

Kepler, in a very important step, also questions the assumption that the planets move around the center of their orbit at a uniform rate. He finds that computing critical measurements based upon the Sun's actual position in the sky, instead of the Sun's "mean" position injects a significant degree of uncertainty into the models, opening the path for further investigations. The idea that the planets do not move at a uniform rate, but at a speed that varies as their distance from the Sun, was completely revolutionary and would become his second law (discovered before his first). Kepler, in his calculations leading to his second law, made multiple mathematical errors, which luckily cancelled each other out "as if by miracle."

Given this second law, he puts forth in Chapter 33 that the Sun is the engine that moves the planets. To describe the motion of the planets, he claims the Sun emits a physical species, analogous to the light it also emits, which pushes the planets along. *He also suggests a second force within every planet itself that pulls it towards the Sun to keep it from spiraling off into space*.

Kepler then attempts to find the *true shape of planetary orbits, which he determines is elliptical*. His initial attempt to define the orbit of Mars as a circle was off by only eight minutes of arc, but this was enough for him to dedicate six years to resolve the discrepancy. The data seemed to produce a symmetrical oviform curve inside of his predicted circle. He first tested an egg shape, then engineered a theory of an orbit which oscillates in diameter, and returned to the egg. Finally, *in early 1605, he geometrically tested an ellipse*, which he had previously assumed to be too simple a solution for earlier astronomers to have overlooked. Ironically, he had already derived this solution trigonometrically many months earlier. As he says:

"I laid [the original equation] aside, and fell back on ellipses, believing that this was quite a different hypothesis, whereas the two, as I shall prove in the next chapter, are one in [sic] the same... Ah, what a foolish bird I have been!"

The *Astronomia nova* records the discovery of *the first two of the three principles known today as Kepler's laws of planetary motion*, which are:

1. *That the planets move in elliptical orbits with the Sun at one focus.*

2. *That the speed of the planet changes at each moment such that the time between two positions is always proportional to the area swept out on the orbit between these positions.*

Kepler discovered the "second law" before the first. He presented his second law in two different forms: In Chapter 32 he states that the speed of the planet varies inversely based upon its distance from the Sun, and therefore he could measure changes in position of the planet by adding up all the distance measures, or looking at the area along an orbital arc. This is his so-called "*distance law*". In Chapter 59, he states that a radius from the Sun to a planet sweeps out equal areas in equal times. This is his so-called "*area law*".

However, Kepler's "area-time principle" did not facilitate easy calculation of planetary positions. Kepler could divide up the orbit into an arbitrary number of parts, compute the planet's position for each one of these, and then refer all questions to a table, but he could not determine the position of the planet at each and every individual moment because the speed of the planet was always changing. This paradox, referred to as the "Kepler problem," prompted the development of calculus.

Kepler, J. (1619). *Harmonices mundi libri V.* (The Five Books of The Harmony of the World.)

Linz, Austria; Johann Planck; translation by E. J. Aiton, A. M. Duncan, and J. V. Field. (1997). The Harmony of the World, Philadelphia, Pennsylvania: American Philosophical Society; see https://books.google.com/books?id=rEkLAAAAIAAJ&pg=PA 411#v=onepage&q&f=false.

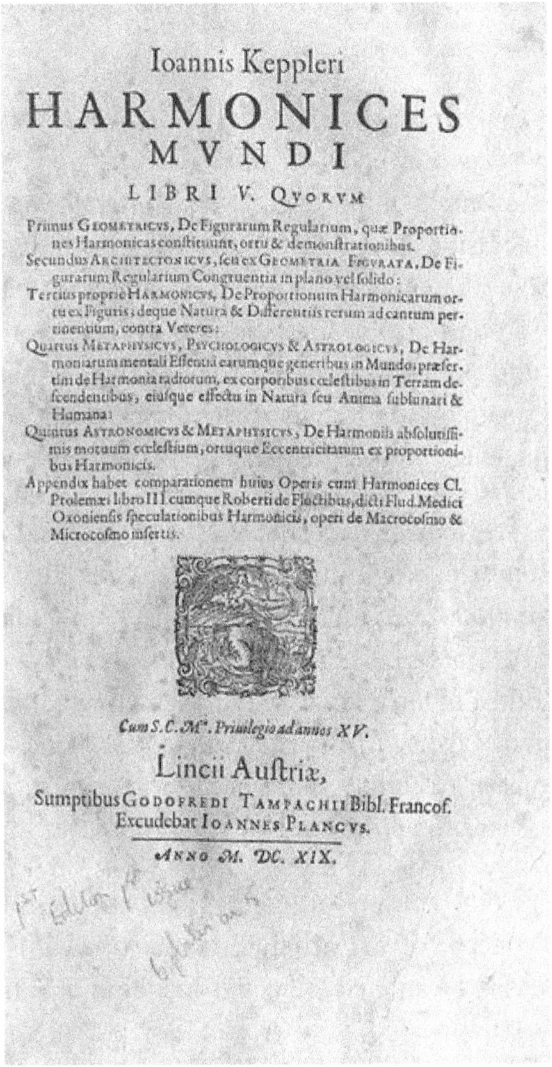
Ioannis Keppleri
HARMONICES
MVNDI
LIBRI V. QVORVM
Primus GEOMETRICVS, De Figurarum Regularium, quæ Proportiones Harmonicas constituunt, ortu & demonstrationibus.
Secundus ARCHITECTONICVS, seu ex GEOMETRIA FIGVRATA, De Figurarum Regularium Congruentia in plano vel solido:
Tertius propriè HARMONICVS, De Proportionum Harmonicarum ortu ex Figuris; deque Natura & Differentiis rerum ad cantum pertinentium, contra Veteres:
Quartus METAPHYSICVS, PSYCHOLOGICVS & ASTROLOGICVS, De Harmoniarum mentali Essentia earumque generibus in Mundo; præsertim de Harmonia radiorum, ex corporibus cœlestibus in Terram descendentibus, eiusque effectu in Natura seu Anima sublunari & Humana:
Quintus ASTRONOMICVS & METAPHYSICVS, De Harmoniis absolutissimis motuum cœlestium, ortuque Eccentricitatum ex proportionibus Harmonicis.
Appendix habet comparationem huius Operis cum Harmonices Cl. Ptolemæi libro III. cumque Roberti de Fluctibus, dicti Flud. Medici Oxoniensis speculationibus Harmonicis, operi de Macrocosmo & Microcosmo insertis.

Cum S. C. M^{tis}. Privilegio ad annos XV.
Lincii Austriæ,
Sumptibus GODOFREDI TAMPACHII Bibl. Francof.
Excudebat IOANNES PLANCVS.
ANNO M. DC. XIX.

In *Harmonice Mundi*, Kepler discusses harmony and congruence in geometrical forms and physical phenomena. The final section of the work relates his discovery of his "*third law of planetary motion*". A decade after the publication of the *Astronomia nova*, Kepler discovered his "third law", published in his 1619 *Harmonices Mundi* (Harmonies of the world). He found that (3)

The ratio of the cube of the length of the semi-major axis of each planet's orbit, to the square of time of its orbital period, is the same for all planets.

In his introductory discussion of a moving earth, Kepler addressed the question of how the Earth could hold its parts together if it moved away from the center of the universe which, according to Aristotelian physics, was the place toward which all heavy bodies naturally moved. *Kepler proposed an attractive force similar to magnetism*, which may have been known by Newton.

"*Gravity is a mutual corporeal disposition among kindred bodies to unite or join together; thus, the earth attracts a stone much more than the stone seeks the earth.* (The magnetic faculty is another example of this sort).... If two stones were set near one another in some place in the world outside the sphere of influence of a third kindred body, these stones, like two magnetic bodies, would come together in an intermediate place, each approaching the other by a space proportional to the bulk [moles] of the other.... For it follows that if the earth's power of attraction will be much more likely to extend to the moon and far beyond, and accordingly, that nothing that consists to any extent whatever of terrestrial material, carried up on high, ever escapes the grasp of this mighty power of attraction".

Kepler discusses the Moon's gravitational effect upon the tides as follows: "The sphere of the attractive virtue which is in the moon extends as far as the earth, and entices up the waters; but as the moon flies rapidly across the zenith, and the waters cannot follow so quickly, a flow of the ocean is occasioned in the torrid zone towards the westward. If the attractive virtue of the moon extends as far as the earth, it follows with greater reason that the attractive virtue of the earth extends as far as the moon and much farther; and, in short, nothing which consists of earthly substance anyhow constituted although thrown up to any height, can ever escape the powerful operation of this attractive virtue".

Kepler also clarifies the concept of lightness in terms of relative density, in opposition to the Aristotelian concept of the absolute nature or quality of lightness as follows. His argument could easily be applied today to something like the flight of a hot air balloon.

"Nothing which consists of corporeal matter is absolutely light, but that is comparatively lighter which is rarer, either by its own nature, or by accidental heat. And it is not to be thought that light bodies are escaping to the surface of the universe while they are carried upwards, or that they are not attracted by the earth. They are attracted, but in a less degree, and so are driven outwards by the heavy bodies; which being done, they stop, and are kept by the earth in their own place".

In reference to Kepler's discussion relating to gravitation, Walter William Bryant makes the following statement in his book Kepler (1920) "...the Introduction to Kepler's "Commentaries on the Motion of Mars," always regarded as his most valuable work, must have been known to Newton, so that no such incident as the fall of an apple was required to provide a necessary and sufficient explanation of the genesis of his Theory of Universal Gravitation. Kepler's glimpse at such a theory could have been no more than a glimpse, for he went no further with it.".

Kepler considered that this attraction was mutual and was proportional to the bulk of the bodies, but he considered it to have a limited range and he did not consider whether or how this force may have varied with distance. Furthermore, this attraction only acted between "kindred bodies"—bodies of a similar nature, a nature which he did not clearly define. Kepler's idea differed significantly from Newton's later concept of gravitation and it can be "better thought of as an episode in the struggle for heliocentrism than as a step toward Universal gravitation."

Kepler began working on *Harmonice Mundi* sometime near 1599, which was the year Kepler sent a letter to Michael Maestlin detailing the mathematical data and proofs that he intended to use for his upcoming text, which he originally planned to name *De harmonia mundi*. Kepler was aware that the content of *Harmonice Mundi* closely resembled that of the subject matter for Ptolemy's *Harmonica*, but was not concerned. The new astronomy Kepler would use (most notably the adoption of elliptic orbits in the Copernican system) allowed him to explore new theorems. Another important development that allowed Kepler to establish his celestial-harmonic relationships, was the abandonment of the Pythagorean tuning as the basis for musical consonance and the adoption of geometrically supported musical ratios; this would eventually be what allowed Kepler to relate musical consonance and the angular velocities of the planets.

The concept of musical harmonies intrinsically existing within the spacing of the planets existed in medieval philosophy prior to Kepler. *Musica universalis* was a traditional philosophical metaphor that was taught in the quadrivium, and was often called the "music of the spheres." Kepler was intrigued by this idea while he sought explanation for a rational arrangement of the heavenly bodies. When Kepler uses the term "harmony" it is not strictly referring to the musical definition, but rather a broader definition encompassing congruence in Nature and the workings of both the celestial and terrestrial bodies.

Kepler divides *Harmonice Mundi* into five long chapters: the first is on regular polygons; the second is on the congruence of figures; the third is on the origin of harmonic proportions in music; the fourth is on harmonic configurations in astrology; *the fifth is on the harmony of the motions of the planets*. This includes a long digression on astrology, which is immediately followed by *Kepler's third law of planetary motion*, which shows a constant proportionality between the cube of the semi-major axis of a planet's orbit and the square of the time of its orbital period.

Kepler's laws of planetary motion.

Kepler's laws of planetary motion, published by Kepler between 1609 and 1619, describe the orbits of planets around the Sun. The laws modified the heliocentric theory of Nicolaus Copernicus, replacing its circular orbits and epicycles with elliptical trajectories, and explaining how planetary velocities vary. The three laws state that:

(1) The orbit of a planet is an ellipse with the Sun at one of the two foci.

(2) A line segment joining a planet and the Sun sweeps out equal areas during equal intervals of time.

(3) The square of a planet's orbital period is proportional to the cube of the length of the semi-major axis of its orbit.

The elliptical orbits of planets were indicated by calculations of the orbit of Mars. From this, Kepler inferred that other bodies in the Solar System, including those farther away from the Sun, also have elliptical orbits. The second law helps to establish that when a planet is closer to the Sun, it travels faster. The third law expresses that the farther a planet is from the Sun, the slower its orbital speed, and vice versa.

Isaac Newton showed in 1687 that relationships like Kepler's would apply in the Solar System as a consequence of his own laws of motion and law of universal gravitation.

A more precise historical approach is found in *Astronomia nova* and *Epitome Astronomiae Copernicanae*.

***Kepler's first law*:**

The orbit of every planet is an ellipse with the Sun at one of the two foci.

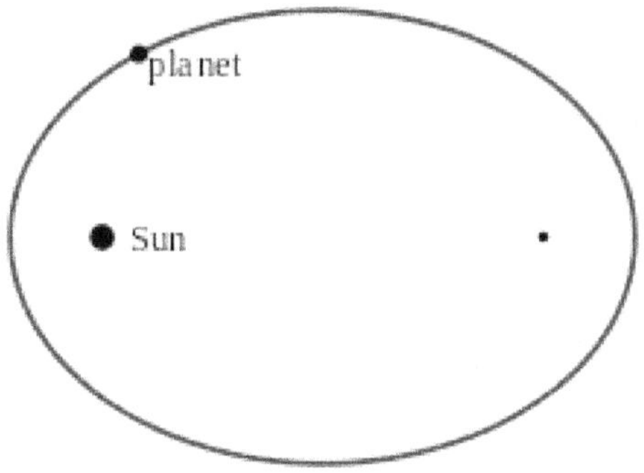

Figure 1: Kepler's first law placing the Sun at the focus of an elliptical orbit.

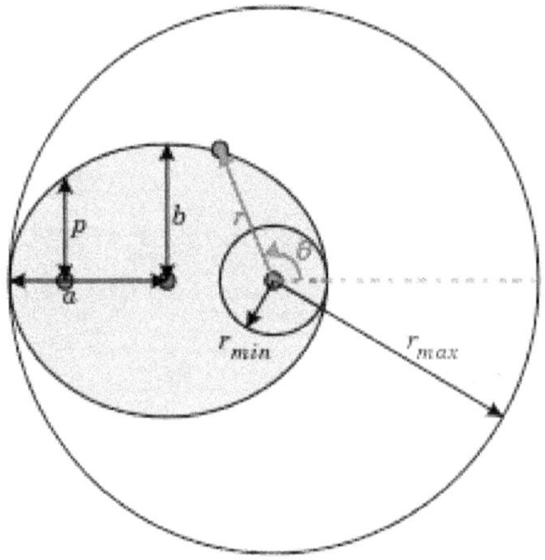

Figure 2: Heliocentric coordinate system (r, θ) for ellipse. Also shown are: semi-major axis *a*, semi-minor axis *b* and semi-latus rectum *p*; center of ellipse and its two foci marked by large dots. For θ = 0°, $r = r_{min}$ and for θ = 180°, $r = r_{max}$.

Mathematically, an *ellipse* can be represented by the formula:

$$r = p/(1 + \varepsilon \cos \theta)$$

where p is the *semi-latus rectum*, ε is the *eccentricity* of the ellipse, r is the distance from the Sun to the planet, and θ is the angle to the planet's current position from its closest approach, as seen from the Sun. So (r, θ) are *polar coordinates*.

For an *ellipse* $0 < \varepsilon < 1$; in the limiting case ε = 0, the orbit is a circle with the Sun at the center (i.e. where there is zero *eccentricity*).

At θ = 0°, *perihelion*, the distance is minimum

$$r_{min} = p/(1 + \varepsilon).$$

At θ = 90° and at θ = 270° the distance is equal to p.

At θ = 180°, *aphelion*, the distance is maximum (by definition, *aphelion* is – invariably – *perihelion* plus 180°)

$$r_{max} = p/(1 - \varepsilon).$$

The *semi-major axis a* is the arithmetic mean between r_{min} and r_{max}:

$$a = (r_{max} + r_{min})/2$$
$$a = p/(1 - \varepsilon^2).$$

The *semi-minor axis* b is the geometric mean between r_{min} and r_{max}:

$$b = \sqrt{(r_{max}\, r_{min})}$$
$$b = p/\sqrt{(1 - \varepsilon^2)}.$$

The *semi-latus rectum* p is the harmonic mean between r_{min} and r_{max}:

$$p = \{(r_{max}^{-1} + r_{min}^{-1})/2\}^{-1}$$
$$pa = r_{max}\, r_{min} = b^2.$$

The *eccentricity* ε is the coefficient of variation between r_{min} and r_{max}:

$$\varepsilon = (r_{max} - r_{min})/(r_{max} + r_{min}).$$

The area of the ellipse is

$$A = \pi ab.$$

The special case of a circle is $\varepsilon = 0$, resulting in $r = p = r_{min} = r_{max} = a = b$ and $A = \pi r^2$.

***Kepler's second law*:**

A line joining a planet and the Sun sweeps out equal areas during equal intervals of time.

The orbital radius and angular velocity of the planet in the elliptical orbit will vary. The planet travels faster when closer to the Sun, then slower when farther from the Sun.

In a small time dt the planet sweeps out a small triangle having base line r and height r dθ and area dA = ½ . r . r dθ, so the constant *areal velocity* is

$$dA/dt = r^2/2\; d\theta/dt.$$

The area enclosed by the elliptical orbit is πab. So the *period* T satisfies

$$T \,.\, r^2/2\; d\theta/dt = \pi ab$$

and the *mean motion* of the planet around the Sun

$$n = 2\pi/T$$

satisfies

$$r^2\, d\theta = abn\, dt.$$

And so,

$$dA/dt = abn/2 = \pi ab/T.$$

***Kepler's third law*:**

The ratio of the square of an object's orbital period with the cube of the semi-major axis of its orbit is the same for all objects orbiting the same primary. [a^3/T^2 = const.]

Upon finding this pattern Kepler wrote:
"I first believed I was dreaming... But it is absolutely certain and exact that the ratio which exists between the period times of any two planets is precisely the ratio of the 3/2th power of the mean distance."
—*translated from Harmonies of the World by Kepler (1619).*

This captures the relationship between the distance of planets from the Sun, and their orbital periods. Kepler enunciated this third law in 1619 in a laborious attempt to determine what he viewed as the "music of the spheres" according to precise laws, and express it in terms of musical notation. It was therefore known as the *harmonic law*. The following table shows the data used by Kepler to empirically derive his law:

Data used by Kepler (1618)

Planet	Mean distance to sun (AU)	Period (days)	R^3/T^2 (10^{-6} AU^3/day^2)
Mercury	0.389	87.77	7.64
Venus	0.724	224.70	7.52
Earth	1	365.25	7.50
Mars	1.524	686.95	7.50
Jupiter	5.20	4332.62	7.49
Saturn	9.510	10759.2	7.43

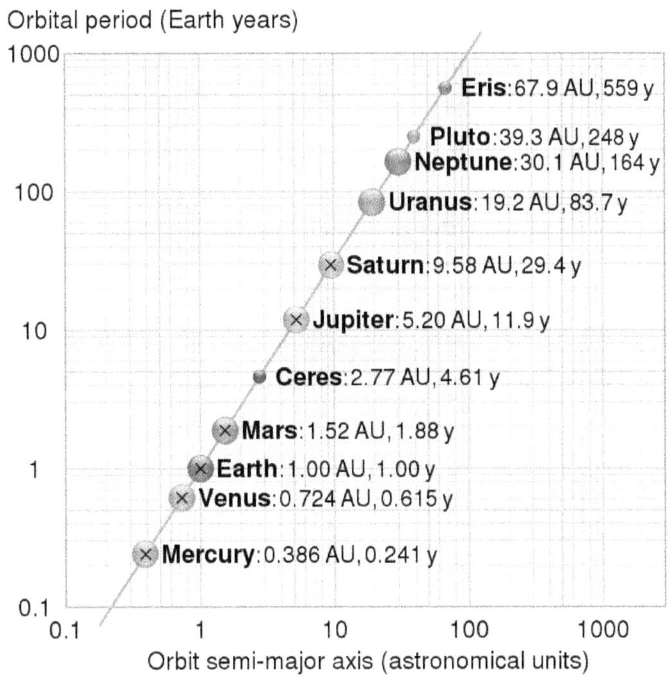

Log-log plot of period T vs semi-major axis *a* (average of aphelion and perihelion) of some Solar System orbits (crosses denoting Kepler's values) showing that a^3/T^2 is constant (green line).

Kepler used his two first laws to compute the position of a planet as a function of time. His method involved the solution of a transcendental equation called Kepler's equation.
The procedure for calculating the heliocentric polar coordinates (r, θ) of a planet as a function of the time t since perihelion, consisted in the following five steps:

1. Compute the *mean motion* $n = (2\pi \text{ rad})/P$, where P is the period.

2. Compute the *mean anomaly* $M = nt$, where t is the time since perihelion.

3. Compute the *eccentric anomaly* E by solving Kepler's equation:
   $$M = E - \varepsilon \sin E,$$
   where ε is the *eccentricity*.

4. Compute the *true anomaly* θ by solving the equation (but this is not elementary):
   $$(1 - \varepsilon) \tan^2 \theta/2 = (1 + \varepsilon) \tan^2 E/2$$

5. Compute the heliocentric distance r:
   $$r = a(1 - \varepsilon \cos E)$$
   where *a* is the semimajor axis.

The position polar coordinates (r, θ) can then be written as a Cartesian vector and the Cartesian velocity vector can then be calculated. The important special case of circular orbit, $\varepsilon = 0$, gives $\theta = E = M$. Because the uniform circular motion was considered to be normal, a deviation from this motion was considered an anomaly.

Below comes the detailed calculation of the acceleration of a planet moving according to Kepler's first and second laws.

From the heliocentric point of view consider the vector to the planet $r = r\,\underline{\mathbf{r}}$ where r is the distance to the planet and $\underline{\mathbf{r}}$ is a *unit vector* (*position vector*) pointing towards the planet,

$$d\underline{\mathbf{r}}/dt = d\theta/dt\ \underline{\boldsymbol{\theta}}, \qquad d\underline{\boldsymbol{\theta}}/dt = -\,d\theta/dt\ \underline{\mathbf{r}},$$

where θ is the *polar angle*, and $\underline{\boldsymbol{\theta}}$ is the *unit vector* (*angle vector*) whose direction is 90 degrees counterclockwise of $\underline{\mathbf{r}}$.

Differentiate the *position vector* twice to obtain the *velocity vector* and the *acceleration vector*:

$$d\mathbf{r}/dt = dr/dt\ \underline{\mathbf{r}} + r\ d\underline{\mathbf{r}}/dt = dr/dt\ \underline{\mathbf{r}} + r\ d\theta/dt\ \underline{\boldsymbol{\theta}},$$
$$d^2\mathbf{r}/dt^2 = \{d^2r/dt^2 - r\,(d\theta/dt)^2\}\ \underline{\mathbf{r}} + \{r\,d^2\theta/dt^2 + 2\,dr/dt\ d\theta/dt\}\ \underline{\boldsymbol{\theta}}.$$

So

$$d^2\mathbf{r}/dt^2 = a_r\ \underline{\mathbf{r}} + a_\theta\ \underline{\boldsymbol{\theta}},$$

where the *radial acceleration* is

$$a_r = \{d^2r/dt^2 - r\,(d\theta/dt)^2\}$$

and the *transversal acceleration* is

$$a_\theta = \{r\,d^2\theta/dt^2 + 2\,dr/dt\ d\theta/dt\}.$$

Inverse square law

Kepler's second law says that

$$r^2\,d\theta/dt = nab$$

is constant.

The *transversal acceleration* a_θ is zero:

$$d(r^2 d\theta/dt)/dt = r\,(2\,dr/dt\ d\theta/dt + d^2\theta/dt^2) = r a_\theta = 0.$$

So, the acceleration of a planet obeying *Kepler's second law* is directed towards the Sun.

Substituting $r^2\, d\theta/dt = nab$, the radial acceleration a_r is

$$a_r = d^2r/dt^2 - r\,(d\theta/dt)^2 = d^2r/dt^2 - n^2a^2b^2/r^3.$$

Kepler's first law states that the orbit is described by the equation:

$$p/r = 1 + \varepsilon \cos\theta.$$

Differentiating with respect to time

$$-\,p/r^2\,(dr/dt) = -\,\varepsilon \sin\theta\; d\theta/dt$$

and substituting $r^2\, d\theta/dt = nab$,

$$p\,(dr/dt) = nab\,\varepsilon \sin\theta.$$

Differentiating once more

$$p\, d^2r/dt^2 = n^2a^2b^2/r^2\, \varepsilon \cos\theta.$$

The radial acceleration a_r satisfies

$$p\, a_r = n^2a^2b^2/r^2\, \varepsilon \cos\theta - p\, n^2a^2b^2/r^3 = n^2a^2b^2/r^2\, (\varepsilon \cos\theta - p/r).$$

Substituting the equation of the ellipse gives

$$p\, a_r = -\, n^2a^2/r^2\, b^2.$$

The relation $b^2 = pa$ gives the simple final result

$$a_r = -\, n^2a^3/r^2.$$

This means that the acceleration vector $d^2\underline{\mathbf{r}}/dt^2$ of any planet obeying *Kepler's first and second law* satisfies the inverse square law

$$d^2\mathbf{r}/dt^2 = -\,\alpha/r^2\, \underline{\mathbf{r}}$$

where

$$\alpha = n^2a^3$$

is a constant, and $\underline{\mathbf{r}}$ is the unit vector pointing from the Sun towards the planet, and r is the distance between the planet and the Sun.

Since mean motion $n = 2\pi/T$ where T is the period, according to *Kepler's third law*, α has the same value for all the planets. So, the inverse square law for planetary accelerations applies throughout the entire Solar System.

The inverse square law is a *differential equation.* The solutions to this differential equation include the Keplerian motions, as shown, but they also include motions where the orbit is a hyperbola or parabola or a straight line.

Sir Isaac Newton (December 25, 1642 – March 20, 1727).

Newton was an English mathematician, physicist, astronomer, theologian, and author (described in his time as a "natural philosopher") who is widely recognized as one of the greatest mathematicians and most influential scientists of all time and as a key figure in the scientific revolution. His book *Philosophiæ Naturalis Principia Mathematica* (Mathematical Principles of Natural Philosophy), first published in 1687, established classical mechanics. Newton also made seminal contributions to optics, and shares credit with Gottfried Wilhelm Leibniz for developing the infinitesimal calculus.

Isaac Newton was born on Christmas Day, 25 December 1642 (according to the Julian calendar, in use in England at the time; NS 4 January 1643) at Woolsthorpe Manor in Woolsthorpe-by-Colsterworth, a hamlet near Grantham in Lincolnshire. His father, also named Isaac Newton, had died three months before. When Newton was three, his mother remarried and went to live with her new husband, leaving her son in the care of his maternal grandmother.

From the age of about twelve until he was seventeen, Newton was educated at The King's School, Grantham, which taught Latin and Greek and probably imparted a significant foundation of mathematics. He was removed from school and returned to Woolsthorpe-by-Colsterworth by October 1659. His mother, widowed for the second time, attempted to make him a farmer, an occupation he hated. Henry Stokes, master at The King's School, persuaded his mother to send him back to school. Motivated partly by a desire for revenge against a schoolyard bully, he became the top-ranked student.

In June 1661, he was admitted to Trinity College, Cambridge, on the recommendation of his uncle Rev William Ayscough, who had studied there. He started paying his way by performing valet's duties until he was awarded a scholarship in 1664, guaranteeing him four more years until he could get his MA. At that time, the college's teachings were based on those of Aristotle, whom Newton supplemented with modern philosophers such as Descartes, and astronomers such as Galileo and Thomas Street, through whom he learned of Kepler's work. He set down in his notebook a series of "Quaestiones" about mechanical philosophy as he found it. In 1665, he discovered the generalized binomial theorem and began to develop a mathematical theory that later became calculus. Soon after Newton had obtained his BA degree in August 1665, the university temporarily closed as a precaution against the Great Plague. Although he had been undistinguished as a Cambridge student, Newton's private studies at his home in Woolsthorpe over the subsequent two years saw the development of his theories on calculus, optics, and the law of gravitation.

In April 1667, he returned to Cambridge and in October was elected as a fellow of Trinity College. His studies had impressed the Lucasian professor Isaac Barrow, who was more anxious to develop his own religious and administrative potential (he became master of Trinity two years later). In 1669 Newton succeeded him and became the second Lucasian Professor of Mathematics at the University of Cambridge, only one year after receiving his MA, a position which he retained until 1701. He was elected a Fellow of the Royal Society (FRS) in 1672 at the age of 30.

It is known from his notebooks that Newton was grappling in the late 1660s with the idea that terrestrial gravity extends, in an inverse-square proportion, to the Moon; however, it took him two decades to develop the full-fledged theory. In 1679, Newton returned to his work on celestial mechanics by considering gravitation and its effect on the orbits of planets with reference to Kepler's laws of planetary motion. This followed stimulation by a brief exchange of letters in 1679–80 with Robert Hooke, who had been appointed to manage the Royal Society's correspondence, and who opened a correspondence intended to elicit contributions from Newton to Royal Society Transactions. [Document No. 235, letter from Hooke to Newton dated November 24, 1679. *Correspondence of Isaac Newton*, vol. 2, 1676–1687, ed. H.W. Turnbull, Cambridge University Press 1960] Newton's reawakening interest in astronomical matters received further stimulus by the appearance of a comet in the winter of 1680–1681. After the exchanges with Hooke, Newton worked out proof that the elliptical form of planetary orbits would result from a centripetal force inversely proportional to the square of the radius vector. Newton communicated his results to Edmond Halley and to the Royal Society in *De motu corporum in gyrum*, a tract written on about nine sheets which was copied into the Royal Society's Register Book in December 1684. [Whiteside, D.T., ed. (1974). *Mathematical Papers of Isaac Newton, 1684–1691.* Cambridge University Press. p. 30.] This tract contained the nucleus that Newton developed and expanded to form the *Principia*, which Newton published in 1687, when he was 35.

Newton built the first practical reflecting telescope and developed a sophisticated theory of color based on the observation that a prism separates white light into the colors of the visible spectrum. His work on light was collected in his highly influential book *Opticks*, published in 1704. He also formulated an empirical law of cooling, made the first theoretical calculation of the speed of sound, and introduced the notion of a Newtonian fluid. In addition to his work on calculus, as a mathematician Newton contributed to the study of power series, generalized the binomial theorem to non-integer exponents, developed a method for approximating the roots of a function, and classified most of the cubic plane curves.

Newton never married. In a note to Samuel Pepys, he complained that John Locke "endeavoured to embroil me with woemen". Newton served two brief terms as Member of Parliament for the University of Cambridge, in 1689–1690 and 1701–1702. He was knighted by Queen Anne in 1705 and spent the last three decades of his life in London, serving as Warden (1696–1699) and Master (1699–1727) of the Royal Mint, as well as president of the Royal Society (1703–1727). He died on March 20, 1727, and was buried in Westminster Abbey.

Newton, I. (July, 1687). *Philosophiæ Naturalis Principia Mathematica.* (The Mathematical Principles of Natural Philosophy.) Book I: The Motion of Bodies.

1st Edition, London; 2nd Edition, Cambridge, 1713; 3rd Edition, London, 1726. (In Latin); translation below of 3rd Edition by A. Motte, (1729). London.); https://en.wikisource.org/wiki/The_Mathematical_Principles_of_Natural_Philosophy_(1729); also https://ia601604.us.archive.org/1/items/newtonspmathema00newtrich/newtonspmathema00newtrich_bw.pdf.

Philosophiæ Naturalis Principia Mathematica (Mathematical Principles of Natural Philosophy) is a work in three books written in Latin, first published July 5, 1687, with encouragement and financial help from Edmond Halley. After annotating and correcting his personal copy of the first edition, Newton published two further editions, during 1713 with errors of the 1687 corrected, and an improved version in 1726. The *Principia* includes Newton's three laws of motion, laying the foundation for classical mechanics; Newton's law of universal gravitation; and a derivation of Johannes Kepler's laws of planetary motion (which Kepler had first obtained empirically). In Book I: The Motion of Bodies, Newton addresses the motion of bodies attracted to each by centripetal forces. Of the invention of *centripetal forces*, he propose that if two bodies S and P, attracting each other with forces reciprocally proportional to the squares of their distance, revolve about their common center of gravity; I say that the principal axis of the ellipse which either of the bodies as P describes by this motion about the other S, will be to the principal axis of the ellipse, which the same body P may describe in the same periodical time about the other body S quiescent, as the sum of the two bodies S+P to the first of two mean proportionals between that sum and the other body S. Of the motions of bodies tending to each other with centripetal forces, he proposes that two bodies attracting each other mutually, describe similar figures about their common center of gravity, and about each other mutually. Of the attractive forces of sphaerical bodies, he proposes that the same things supposed (if to the several points of a given sphere there tend equal centripetal forces decreasing in a duplicate ratio of the distances from the points), I say, that a corpuscle situate without the sphere is attracted with a force reciprocally proportional to the square of its distance from the center. He also proposes that if to the several points of a given sphere there tend equal centripetal forces decreasing in a duplicate ratio of the distances from the points; I say that another similar sphere will be attracted to it with a force reciprocally proportional to the square of the distance of the centers.

CONTENTS

...

Section II: Of the invention of centripetal forces

Proposition IX Theorem XXIII. *If two bodies S and P, attracting each other with forces reciprocally proportional to the squares of their distance, revolve about their common center of gravity; I say that the principal axis of the ellipse which either of the bodies as P describes by this motion about the other S, will be to the principal axis of the ellipse, which the same body P may describe in the same periodical time about the other body S quiescent, as the sum of the two bodies S+P to the first of two mean proportionals between that sum and the other body S.*

...

Section XI: Of the motions of bodies tending to each other with centripetal forces

Proposition LVII. Theorem XX. *Two bodies attracting each other mutually, describe similar figures about their common center of gravity, and about each other mutually.*
For the distances of the bodies from their common center of gravity are reciprocally as the bodies; and therefore, in a given ratio to each other; and thence by composition of ratio's, in a given ratio to the whole distance between the bodies. Now these distances revolve about their common term with an equable angular motion, because lying in the same right line they never change their inclination to each other mutually. But right lines that are in a given ratio to each other, and revolve about their terms with an equal angular motion, describe upon planes, which either rest with those terms, or move with any motion not angular, figures entirely similar round those terms. Therefore, the figures described by the revolution of these distances are similar. *Q. E. D.*"

> [Newton specifically considers forces of attraction between bodies which are equal to the reciprocal to the squares of the distance between them, which applies to the force between two electrically charged bodies, under Coulomb's law of electrical forces, as well as to gravitational forces. In Proposition LXXIV he argues that the if the force between particles is a reciprocal of the squares of the distance between them, then the force between a sphere and a particle external to the sphere is a reciprocal of the distance between the external particle and the center of the sphere.]

...
Section XII: Of the attractive forces of sphaerical bodies
Proposition LXXIV. Theorem XXXIV. *The same things supposed (if to the several points of a given sphere there tend equal centripetal forces decreasing in a duplicate ratio of the distances from the points), I say, that a corpuscle situate without the sphere is attracted with a force reciprocally proportional to the square of its distance from the center.*
For suppose the sphere to be divided into innumerable concentric spherical superficies, and the attractions of the corpuscle arising from the several superficies will be reciprocally proportional to the square of the distance of the corpuscle from the center of the sphere (by prop. 7 1.) And by composition, the sum of those attractions, that is, the attraction of the corpuscle towards the entire sphere, will be in the fame ratio. *Q. E. D.*

Cor. 1. Hence the attractions of homogeneous spheres at equal distances from the centers will be as the spheres themselves. For (by prop. 72) if the distances be proportional to the diameters of the spheres, the forces will be as the diameters. Let the greater distance be diminished in that ratio; and the distances now being equal, the attraction will be increased in the duplicate of that ratio; and therefore, will be to the other attraction in the triplicate of that ratio; that is, in the ratio of the spheres.

Cor. 2. At any distances whatever, the attractions are as the spheres applied to the squares of the distances.

Cor. 3. If a corpuscle placed without a homogeneous sphere is attracted by a force reciprocally proportional to the square of its distance from the center, and the sphere consists of attractive particles; the force of every particle will decrease in a duplicate ratio of the distance from each particle.

Proposition LXXV. Theorem XXXV. *If to the several points of a given sphere there tend equal centripetal forces decreasing in a duplicate ratio of the distances from the points; I say that another similar sphere will be attracted to it with a force reciprocally proportional to the square of the distance of the centers.*
For the attraction of every particle is reciprocally as the square of its distance from the center of the attracting sphere (by prop. 7.4.) and is therefore the same as if that whole attracting force issued from one single corpuscle placed in the center of this sphere. But this attraction is as great, as on the other hand the attraction of the same corpuscle would be, if that were itself attracted by the several particles of the attracted sphere with the same force with which they are attracted by it. But that attraction of the corpuscle would be (by prop. 7.4.) reciprocally proportional to the square of its distance from the center of the sphere; therefore, the attraction of the sphere, equal thereto, is also in the same ratio.
Q E.D.

Cor. 1. The attractions of spheres towards other homogeneous spheres, are as the attracting spheres applied to the squares of the distances of their centers from the centers of those which they attract.

Cor. 2. The case is the same when the attracted sphere does also attract. For the several points of the one attract the several points of the other with the same force with which they themselves are attracted by the others again; and therefore since in all attractions (by law 3.) the attracted and attracting point are both equally acted on, the force will be doubled by their mutual attractions, the proportions remaining.

> [In Book III of *Principia* (below), Newton notes *that the centripetal force which arises between planets is the same as the gravitational force attracting matter to the Earth* and focusses on gravitational attraction. Newton used the Latin word gravitas (weight) for the effect that would become known as *gravity*.
> "Book III. Proposition V. Theorem V. Scholium. *The force which retains the celestial bodies in their orbits has been hitherto called centripetal force; but it being now made plain that it can be no other than a gravitating force, we shall hereafter call it gravity.*
> For the cause of that centripetal force which retains the moon in its orbit will extend itself to all the planets."]

Newton, I. (July, 1687). *Philosophiæ Naturalis Principia Mathematica.* Book III: Of the System of the World.

> 1st Edition, London; 2nd Edition, Cambridge, 1713; 3rd Edition, London, 1726. (In Latin); translation below of 3rd Edition by A. Motte, (1729). London.); https://en.wikisource.org/wiki/The_Mathematical_Principles_of_Natural_Philosophy_(1729); also https://ia601604.us.archive.org/1/items/newtonspmathema00newtrich/newtonspmathema00newtrich_bw.pdf.

In Book III, Newton notes that the centripetal force which arises between planets is the same as the gravitational force attracting matter to the Earth and focusses on gravitational attraction.
He observes that the circumjovial planets, by radii drawn to Jupiter's center, describe areas proportional to the times of description; and that their periodic times, the fixed stars being at rest, are in the sesquiplicate proportion of their distances from, its center, etc. That the fixed Stars being at rest, the periodic times of the five primary Planets, and (whether of the Sun about the Earth, or) of the Earth about the Sun, are in the sesquiplicate proportion of their mean distances from the Sun. He noted that this proportion, first observed by Kepler, is now accepted by all astronomers.

> [The *sesquiplicate* ratio of given terms is the ratio between the square roots of the cubes of those terms.]

He further observes that then the primary Planets, by radii drawn to the Earth, describe areas no wise proportional to the times: But that the areas, which they describe by radii drawn to the Sun, are proportional to the times of description; and that the Moon by a radius drawn to the Earth's center, describes an area proportional to the time of description. From this he proposes that the forces by which the circumjovial planets are continually drawn off from rectilinear motions, and retained in their proper orbits, tend to Jupiter's center; and are reciprocally as the squares of the distances of the places of those planets from that center; that the forces by which the primary planets are continually drawn off from rectilinear motions, and retained in their proper orbits, tend to the sun; and are reciprocally as the squares of the distances of the places of those planets from the sun's center; and that the force by which the moon is retained in its orbit tends to the earth; and is reciprocally as the square of the distance of its place from the earths center. He then proposes that that the moon gravitates towards the earth, and by the force of gravity is continually drawn off from a rectilinear motion, and retained in its orbit; that the circumjovial planets gravitate towards Jupiter; the circumsaturnal towards Saturn; the circumsolar towards the sun; and by the forces of their gravity are drawn off from rectilinear motions, and retained in curvilinear orbits; and concludes that the force which retains the celestial bodies in their orbits has been hitherto called centripetal force; but it being now made plain that it can be no other than a gravitating force, we shall hereafter call it gravity. He then proposes that all bodies gravitate towards; every Planet and that the Weights of bodies towards any the same Planet, at equal distances from the center of the

Planet, are proportional to the quantities of matter which they severally contain; and that there is a power of gravity tending to all bodies, proportional to the several quantities of matter which they contain; and that the force of gravity towards the several equal particles of any body, is reciprocally as the square of the distance of places from the particles; as appears from Cor. 3, Prop. LXXIV, Book I.

[In Proposition VI, Newton provides his definition of *gravitational mass*, and in Proposition VII, together with its corollary 2, Newton restates his *universal law of gravitation*,

$$F = Gm_1m_2/r^2,$$

where F is the force, m_1 and m_2 are the masses of the objects interacting, r is the distance between the centers of the masses and G is the gravitational constant. Newton's law of gravitation states that every point mass in the universe attracts every other point mass with a force that is directly proportional to the product of their masses, and inversely proportional to the square of the distance between them.]

BOOK III: OF THE SYSTEM OF THE WORLD

...

Rules of Reasoning in Philosophy.

Rule I. *We are to admit no more causes of natural things than such as are both true and sufficient to explain their appearances.*

To this purpose the philosophers say that Nature does nothing in vain, and more is in vain when less will serve; for Nature is pleased with simplicity, and affects not the pomp of superfluous causes.

Rule II. *Therefore, to the same natural effects we must, as far as possible, assign the same causes.*

As to respiration in a man and in a beast; the descent of stones in Europe and in America; the light of our culinary fire and of the sun; the reflection of light in the earth, and in the planets.

Rule III. *The qualities of bodies, which admit neither intension nor remission of degrees, and which are found to belong to all bodies within the reach of our experiments, are to be esteemed the universal qualities of all bodies whatsoever.*

For since the qualities of bodies are only known to us by experiments, we are to hold for universal all such as universally agree with experiments; and such as are not liable to diminution can never be quite taken away. We are certainly not to relinquish the evidence of experiments for the sake of dreams and vain fictions of our own devising; nor are we to recede from the analogy of Nature, which uses to be simple, and always consonant to itself.
...
...

Rule IV. *In experimental philosophy we are to look upon propositions collected by general induction from, phenomena as accurately or very nearly true, notwithstanding any contrary hypotheses that may be imagined, till such time as other phenomena occur, by which they may either be made more accurate, or liable to exceptions.*

This rule we must follow, that the argument of induction may not be evaded by hypotheses

Phenomena, or Appearances.

Phenomenon I. *That the circumjovial planets, by radii drawn to Jupiter's center, describe areas proportional to the times of description; and that their periodic times, the fixed stars being at rest, are in the sesquiplicate proportion of their distances from, its center.* This we know from astronomical observations. For the orbits of these planets differ but insensibly from circles concentric to Jupiter; and their motions in those circles are found to be uniform. And all astronomers agree that *their periodic times are in the sesquiplicate proportion of the semi-diameters of their orbits*; and so it manifestly appears from the following table.

...

Phenomenon II. *That the circumsaturnal planets, by radii drawn, to Saturn's center, describe areas proportional to the times of description; and that their periodic times, the fixed stars being at rest, are in the sesqniplicata proportion of their distances from its center.*

For, as Cassini from his own observations has determined, their distances from Saturn's center and their periodic times are as follows.

...

Phenomenon III. *That the five primary planets, Mercury, Venus, Mars, Jupiter, and Saturn, with their several orbits, encompass the sun.*

That Mercury and Venus revolve about the sun, is evident from their moon-like appearances. When they shine out with a full face, they are, in respect of us, beyond or above the sun; when they appear half full, they are about the same height on one side or other of the sun; when horned, they are below or between us and the sun; and they are sometimes, when directly under, seen like spots traversing the sun's disk. That Mars surrounds the sun, is as plain from its full face when near its conjunction with the sun. and from the gibbous figure which it shews in its quadratures. And the same thing is demonstrable of Jupiter and Saturn, from their appearing full in all situations; for the shadows of their satellites that appear sometimes upon their disks make it plain that the light they shine with is not their own, but borrowed from the sun.

Phenomenon IV. *That the fixed Stars being at rest, the periodic times of the five primary Planets, and (whether of the Sun about the Earth, or) of the Earth about the Sun, are in the sesquiplicate proportion of their mean distances from the Sun.*

This proportion, first observ'd by Kepler, is now receiv'd by all astronomers. For the periodic times are the same, and the dimensions of the orbits are the same, whether the Sun revolves about the Earth, or the Earth about the Sun. And as to the measures of the periodic times, all astronomers are agreed about them. But for the dimensions of the orbits, Kepler and Bullialdus, above all others, have determin'd them from observations with the greatest accuracy: and the mean distances corresponding to the periodic times, differ but insensibly from those which they have assign'd, and for the most part fall in between them; as we may see from the following table. The periodic times, with respect to the fixed Stars, of the Planets and Earth revolving about the Sun, in days and decimal parts of a day. 10759,275. 4332,514. 686,9785. 365,2565. 224,6176. 87,9692. The mean distances of the Planets and of the Earth from the Sun. According to Kepler 951000. 519650. 152350. To Bullialdus 954198. 522520. 152350. To the periodic Times 954006. 520096. 152369. According to Kepler 100000. 72400. 38806. To Bullialdus 100000. 72398. 38585. To the periodic times 100000. 72333. 38710. As to Mercury and Venus, there can be no doubt about their distances from the Sun; for they are determin'd by the elongations of those Planets from the Sun. And for the distances of the superior Planets, all dispute is cut off by the eclipses of the satellites of Jupiter. For, by those eclipses, the position of the shadow, which Jupiter projects, is determin'd; whence we have the heliocentric longitude of Jupiter. And from its heliocentric and geocentric longitudes compar'd together, we determine its distance.

Phænomenon V. *Then the primary Planets, by radii drawn to the Earth, describe areas no wise proportional to the times: But that the areas, which they describe by radii drawn to the Sun, are proportional to the times of description.*

For to the Earth, they appear sometimes direct, sometimes stationary, nay and sometimes retrograde. But from the Sun they are always seen direct, and to proceed with a motion nearly uniform, that is to say, a little swifter in the perihelion and a little slower in the aphelion distances, so as to maintain an equality in the description of the areas. This is a noted proposition among astronomers, and particularly demonstrable in Jupiter, from the eclipses of his satellites; by the help of which eclipses, as we have said, the heliocentric longitudes of that Planet, and its distances from the Sun. are determined.

Phænomenon VI. *That the Moon by a radius drawn to the Earth's center, describes an area proportional to the time of description.*

This we gather from the apparent motion of the Moon, compar'd with its apparent diameter. It is true that the motion of the Moon is a little disturbed by the action of the Sun. But in laying down these phænomena, I neglect those small and inconsiderable errors.

Propositions.

Proposition I. Theorem I. *That the forces by which the circumjovial planets are continually drawn off from rectilinear motions, and retained in their proper orbits, tend to Jupiter's center; and are reciprocally as the squares of the distances of the places of those planets from that center.*

The former part of this Proposition appears from Phaen. I, and Prop. II or III, Book I: the latter from Phaen. I, and Cor. 6, Prop. IV, of the same Book. The same thing we are to understand of the planets which encompass Saturn, by Phaen. II.

Proposition II. Theorem II. *That the forces by which the primary planets are continually drawn off from rectilinear motions, and retained in their proper orbits, tend to the sun; and are reciprocally as the squares of the distances of the places of those planets from the sun's center.*

The former part of the Proposition is manifest from Phaen. V, and Prop. II, Book I; the latter from Phaen. IV, and Cor. 6, Prop. IV, of the same Book. But this part of the Proposition is, with great accuracy, demonstrable from the quiescence of the aphelion points; for a very small aberration from the reciprocal duplicate proportion would (by Cor. 1, Prop. XLV, Book I) produce a motion of the apsides sensible enough in every single revolution, and in many of them enormously great.

Proposition III. Theorem III. *That the force by which the moon is retained in its orbit tends to the earth; and is reciprocally as the square of the distance of its place from the earths center.*

The former part of the Proposition is evident from Phaen. VI, and Prop. II or III, Book I; the latter from the very slow motion of the moon's apogee; which in every single revolution amounting but to 3° 3' *in consequentia*, may be neglected. For (by Cor. 1. Prop. XLV, Book I) it appears, that, if the distance of the moon from the earth's center is to the semi-diameter of the earth as D to 1, the force, from which such a motion will result, is reciprocally as D^2 4/243, i.e., reciprocally as the power of D, whose exponent is 2 4/243; that is to say, in the proportion of the distance something greater than reciprocally duplicate, but which comes

59 ¾ times nearer to the duplicate than to the triplicate proportion. But in regard that this motion is owing to the action of the sun (as we shall afterwards shew), it is here to be neglected. The action of the sun, attracting the moon from the earth, is nearly as the moon's distance from the earth; and therefore (by what we have shewed in Cor. 2, Prop. XLV. Book I) is to the centripetal force of the moon as 2 to 357.45, or nearly so; that is, as 1 to 178 29/40. And if we neglect so inconsiderable a force of the sun, the remaining force, by which the moon is retained in its orb, will be reciprocally as D^2. This will yet more fully appear from comparing this force with the force of gravity, as is done in the next Proposition.

Corollary. If we augment the mean centripetal force by which the moon is retained in its orb, first in the proportion of 177 29/40 to 178 29/40, and then in the duplicate proportion of the semi-diameter of the earth to the mean distance of the centers of the moon and earth, we shall have the centripetal force of the moon at the surface of the earth; supposing this force, in descending to the earth's surface, continually to increase in the reciprocal duplicate proportion of the height.

Proposition IV. Theorem IV. *That the moon gravitates towards the earth, and by the force of gravity is continually drawn off from a rectilinear motion, and retained in its orbit.*

The mean distance of the moon from the earth in the syzygies in semidiameters of the earth, is, according to *Ptolemy* and most astronomers, 59: according to *Vendelin* and *Huygens*, 60; to *Copernicus*, 60 1/3 to *Street*, 60 2/3; and to *Tycho*, 56 1/2. But *Tycho*, and all that follow his tables of refraction, making the refractions of the sun and moon (altogether against the nature of light) to exceed the refractions of the fixed stars, and that by four or five minutes *near the horizon*, did thereby increase the moon's horizontal parallax by a like number of minutes, that is, by a twelfth or fifteenth part of the whole parallax. Correct this error, and the distance will become about 60 ½ semi-diameters of the earth, near to what others have assigned. Let us assume the mean distance of 60 diameters in the syzygies; and suppose one revolution of the moon, in respect of the fixed stars, to be completed in 27d. 7 h. 43', as astronomers have determined; and the circumference of the earth to amount to 123,249,600 *Paris* feet, as the French have found by mensuration. And now if we imagine the moon, deprived of all motion, to be let go, so as to descend towards the earth with the impulse of all that force by which (by Cor. Prop. III) it is retained in its orb, it will in the space of one minute of time, describe in its fall 15 1/12 *Paris* feet. This we gather by a calculus, founded either upon Prop. XXXVI, Book I, or (which comes to the same thing; upon Cor. 9, Prop. IV, of the same Book. For the versed sine of that arc, which the moon, in the space of one minute of time, would by its mean motion describe at the distance of 60 semi-diameters of the earth, is nearly 15 1/12 *Paris* feet, or more accurately 15 feet, 1 inch, and 1 line 4/9. Wherefore, since that force, in approaching to the earth, increases in the reciprocal duplicate proportion of the distance, and, upon that account, at the surface of the earth, is 60 x 60 times greater than at the moon, a body in our regions, falling with that force, ought in the space of one minute of time, to describe 60 x 60 x 15 1/12 *Paris* feet; and, in the space of one second of time, to describe 15 1/12 of those feet; or more accurately 15 feet, 1 inch, and 1 line 4/9. And with this very force we actually find that bodies here upon earth do really descend: for a pendulum oscillating seconds in the latitude of *Paris* will be 3 *Paris* feet, and 8 lines ½ in length, as Mr. *Huygens* has observed. And the space which a heavy body describes by falling in one second of time is to half the length of this pendulum in the duplicate ratio of the circumference of a circle to its diameter (as Mr. *Huygens* has also shewn), and is therefore 15 *Paris* feet, 1 inch, 1 line 7/9. And therefore, the force by which the moon is retained in its orbit becomes, at the very surface of the earth, equal to the force of gravity which we observe in heavy bodies there. And therefore (by Rule I and II) *the force by which the moon is retained in its orbit is that very same force which we commonly call gravity*; for, were gravity another force different from that, then bodies descending to the earth with the joint impulse of both forces would fall with a double velocity, and in the space of one second of time would describe 30 1/6 *Paris* feet; altogether against experience.

This calculus is founded on the hypothesis of the earth's standing still; for if both earth and moon move about the sun, and at the same time about their common center of gravity, the distance of the centers of the moon and earth from one another will be 60 ½ semi-diameters of the earth; as may be found by a computation from Prop. LX, Book I.

Scholium. The demonstration of this Proposition may be more diffusely explained after the following manner. Suppose several moons to revolve about the earth, as in the system of Jupiter or Saturn: the periodic times of these moons (by the argument of induction) would observe the same law which *Kepler* found to obtain among the planets; and therefore, their centripetal forces would be reciprocally as the squares of the distances from the center of the earth, by Prop. I, of this Book. Now if the lowest of these were very small, and were so near the earth as almost to touch the tops of the highest mountains, the centripetal force thereof, retaining it in its orb, would be very nearly equal to the weights of any terrestrial bodies that should be found upon the tops of those mountains, as may be known by the foregoing computation. Therefore, if the same little moon should be deserted by its centrifugal force that carries it through its orb, and so be disabled from going onward therein, it would descend to the earth; and that with the same velocity as heavy bodies do actually fall with upon the tops of those very mountains; because of the equality of the forces that oblige them both to descend. And if the force by which that lowest moon would descend were different from gravity, and if that moon were to gravitate towards the earth, as we find terrestrial bodies do upon the tops of mountains, it would then descend with twice the velocity, as being impelled by both these forces conspiring together. Therefore, since both these forces, that is, the gravity of heavy bodies, and the centripetal forces of the moons, respect the center of the earth, and are similar and equal between themselves, they will (by Rule I and II) have one and the same cause. And therefore, *the force which retains the moon in its orbit is that very force which we commonly call gravity*; because otherwise this little moon at the top of a mountain must either be without gravity, or fall twice as swiftly as heavy bodies are wont to do.

Proposition V. Theorem V. *That the circumjovial planets gravitate towards Jupiter; the circumsaturnal towards Saturn; the circumsolar towards the sun; and by the forces of their gravity are drawn off from rectilinear motions, and retained in curvilinear orbits.*

For the revolutions of the circumjovial planets about Jupiter, of the circumsaturnal about Saturn, and of Mercury and Venus, and the other circumsolar planets, about the sun, are appearances of the same sort with the revolution of the moon about the earth; and therefore, by Rule II, must be owing to the same sort of causes; especially since it has been demonstrated, that the forces upon which those revolutions depend tend to the centers of Jupiter, of Saturn, and of the sun; and that those forces, in receding from Jupiter, from Saturn, and from the sun, decrease in the same proportion, and according to the same law, as the force of gravity does in receding from the earth.

Corollary 1. There is, therefore, a power of gravity tending to all the planets; for, doubtless, Venus, Mercury, and the rest, are bodies of the same sort with Jupiter and Saturn. And since all attraction (by Law III) is mutual, Jupiter will therefore gravitate towards all his own satellites, Saturn towards his, the earth towards the moon, and the sun towards all the primary planets.

Corollary 2. The force of gravity which tends to any one planet is reciprocally as the square of the distance of places from that planet's center.

Corollary 3. All the planets do mutually gravitate towards one another, by Cor. 1 and 2. And hence it is that Jupiter and Saturn, when near their conjunction; by their mutual attractions sensibly disturb each other's motions. So, the sun disturbs the motions of the moon; and both sun and moon disturb our sea, as we shall hereafter explain.

Scholium. *The force which retains the celestial bodies in their orbits has been hitherto called centripetal force; but it being now made plain that it can be no other than a gravitating force, we shall hereafter call it gravity*. For the cause of that centripetal force which retains the moon in its orbit will extend itself to all the planets, by Rule I, II, and IV.

Proposition VI. Theorem VI. *That all bodies gravitate towards; every Planet and that the Weights of bodies towards any the same Planet, at equal distances from the center of the Planet, are proportional to the quantities of matter which they severally contain.*

It has been, now of a long time, observed by others, that all sorts of heavy bodies, (allowance being made for the inequality of retardation, which they suffer from a small power of resistance in the air) descend to the Earth from equal heights in equal times: and that equality of times we may distinguish to a great accuracy, by the help of pendulums. I tried the thing in gold, silver, lead, glass, sand, common salt, wood, water, and wheat. I provided two wooden boxes, round and equal. I filled the one with wood, and suspended an equal weight of gold (as exactly as I could) in the center of oscillation of the other. The Boxes hanging by equal threads of 11 feet, made a couple of pendulums perfectly equal in weight and figure, and equally receiving the resistance of the air. And placing the one by the other, I observed them to play together forwards and backwards, for a long time, with equal vibrations. And therefore, the quantity of matter in the gold (by Cor. 1. and 6. Prop. XXIV, Book II.) was to the quantity of matter in the wood, as the action of the motive force (or vis matrix) upon all the gold, to the action of the same upon all the wood; that is, as the weight of the one to the weight of the other. And the like happened in the other bodies. By these experiments, in bodies of the same weight, I could manifestly have discovered a difference of matter less than the thousandth part of the whole, had any such been. But without all doubt, the nature of gravity towards the Planets, is the same as towards the Earth. For, should we imagine our terrestrial bodies removed to the orb of the Moon, and there, together with the Moon, deprived of all motion, to be let go, so as to fall together towards the Earth: it is certain, from what we have demonstrated before, that, in equal times, they would describe equal spaces with the Moon, and of consequence are to the Moon, in quantity of matter, as their weights to its weight. Moreover, since the satellites of Jupiter perform their revolutions in times which observe the sesquiplicate proportion of their distances from Jupiter's center, their accelerative gravities towards Jupiter will be reciprocally as the squares of their distances from Jupiter's center; that is, equal, at equal distances. And therefore, these satellites, if supposed to fall *towards* Jupiter from equal heights, would describe equal spaces in equal times, in like manner as heavy bodies do on our Earth. And by the same argument, if the circumsolar Planets were supposed to be let sail at equal distances from the Sun, they would, in their descent towards the Sun, describe equal spaces in equal times. But forces, which equally accelerate unequal bodies, must be as those bodies; that is to say, the weights of the Planets *towards* the Sun must be as their quantities of matter. Further, that the weights of Jupiter and of his satellites towards the Sun are proportional to the several quantities of their matter, appears from the exceeding regular motions of the satellites, (by Cor. 3. Prop. LXV, Book I.) For if some of those bodies were more strongly attracted to the Sun in proportion to their quantity of matter, than others; the motions of the satellites would be disturbed by that inequality of attraction (by Cor. 2. Prop. LXV, Book I.) If, at equal distances from the Sun, any satellite in proportion to the quantity of its matter, did gravitate towards the Sun, with a force greater than Jupiter in proportion to his, according to any given proportion, suppose of d to e; then the distance between the centers of the Sun and of the satellite's orbit would be always greater than the distance between the centers of the Sun and of Jupiter, nearly in the subduplicate of that proportion; as by some computations I have found. And if the satellite did gravitate towards the Sun with a force, lesser in the proportion of e to d, the distance of the center of the satellite's orb from the Sun, would be less than the distance of the center of Jupiter from the Sun, in the subduplicate of the same proportion. Therefore if, at equal distances from the Sun, the accelerative gravity of any satellite towards the Sun were greater or less than the accelerative gravity of Jupiter towards the Sun, but by one 1/1000 part of the whole gravity; the distance of the center of the satellite's orbit from the Sun would be greater or less than the distance of Jupiter from the Sun, by one 1/2000 part of the whole distance; that is, by a fifth part of the distance of the utmost satellite from the center of Jupiter; an eccentricity of the orbit, which would be very sensible. But the orbits of the satellites are concentric to Jupiter, and therefore the accelerative gravities of Jupiter, and of all its satellites towards the Sun, are equal among themselves. And by the same argument, the weights of Saturn and of his satellites towards the Sun, at equal distances from the Sun, are as their several quantities of matter: and the weights of the Moon and of the Earth towards the Sun, are either none, or accurately proportional to the masses of matter which they contain. But some they are by Cor. 1. and 3, Prop. V.

But further, the weights of all the parts of every Planet towards any other Planet, are one to another as the matter in the several parts. For if some parts did gravitate more, others less, than for the quantity of their matter; then the whole Planet, according to the sort of parts with which it most abounds, would gravitate more or less, than in proportion to the quantity of matter in the whole. Nor is it of any moment, whether these parts are external or internal. For, if, for example, we would imagine the terrestrial bodies with us to be raised up to the orb of the Moon, to be there compared with its body: If the weights of such bodies were to the weights of the external parts of the Moon, as the quantities of matter in the one and in the other respectively; but to the weights of the internal parts, in a greater or less proportion, then likewise the weights of those bodies would be to the weight of the whole Moon, in a greater or less proportion; against what we have shewed above.

Corollary 1. Hence the weights of bodies do not depend upon their forms and textures. For if the weights could be altered with the forms, they would be greater or less, according to the variety of forms, in equal matter; altogether against experience.

Corollary 2. Universally, all bodies about the Earth, gravitate towards the Earth; and the weights of all, ar equal distances from the Earth's center, are as the quantities of matter which they severally contain. This is the quality of all bodies, within the reach of our experiments; and therefore, (by Rule III) to be affirmed of all bodies whatsoever. Is the *æther*, or any other body, were either altogether void of gravity, or were to gravitate less in proportion to its quantity of matter; then, because (according to *Aristotle*, *Des Cartes*, and others) there is no difference betwixt that and other bodies, but in mere form of matter, by a successive change from form to form, it might be changed at last into a body of the same condition with those which gravitate most in proportion to their quantity of matter; and, on the other hand, the heaviest bodies, acquiring the first form of that body, might by degrees, quite lose their gravity. And therefore, the weights would depend upon the forms of bodies, and with those forms might be changed, contrary to what was proved in the preceding Corollary.

Corollary 3. All spaces are not equally Full. For if all spaces were equally full, then the specific gravity of the fluid which fills the region of the air, on account of the extreme density of the matter, would fall nothing short of the specific gravity of quick-silver, or gold, or any other the most dense body; and therefore, neither gold, nor any other body, could descend in air. For bodies do not descend in fluids, unless they are specifically heavier than the fluids. And if the quantity of matter in a given space, can, by any rarefaction, be diminished, what should hinder a diminution to infinity?

Corollary 4. If all the solid particles of all bodies are of the same density, nor can be rarified without pores a void space or vacuum must be granted. By bodies of the same density, I mean those, whose *vires inertia* are in the proportion of their bulks.

Corollary 5. *The power of gravity is of a different nature from the power of magnetism.* For the magnetic attraction is not as the matter attracted. Some bodies are attracted more by the magnet, others less; most bodies not at all. The power of magnetism, in one and the same body, may be increased and diminished; and is sometimes far stronger, for the quantity of matter, than the power of gravity; and in receding from the magnet, decreases not in the duplicate, but almost in the triplicate proportion of the distance, as nearly as I could, judge from some rude observations.

Proposition VII. Theorem VII. *That there is a power of gravity tending to all bodies, proportional to the several quantities of matter which they contain.*

That all the Planets mutually gravitate one towards another, we have prov'd before; as well as that the force of gravity towards every one of them, consider'd apart, is reciprocally as the square of the distance of places from the center of the planet. And thence (by Prop. LXIX, Book I, and its Corollaries) it follows, that the gravity tending towards all the Planets, is proportional to the matter which they contain. Moreover, since all the parts of any planet A gravitate towards any other planet B; and the gravity of every part is to the gravity of the whole, as the matter of the part to the matter of the whole; and (by Law III) to every action corresponds an equal re-action: therefore the planet B will, on the other hand, gravitate towards all the parts of the planet A; and its gravity towards any one part will be to the gravity towards the whole, as the matter of the part to the matter of the whole. Q.E.D.

Corollary 1. Therefore, the force of gravity towards any whole planet, arises from, and is compounded of, the forces of gravity towards all its parts. Magnetic and electric attractions afford us examples of this. For all attraction towards the whole arises from the attractions towards the several parts. The thing may be easily understood in gravity, if we consider a greater planet, as form'd of a number of lesser planets, meeting together in one globe. For hence it would appear that the force of the whole must arise from the forces of the component parts. If it is objected, that, according to this law, all bodies with us must mutually gravitate one towards another, whereas no such gravitation anywhere appears: I answer, that since the gravitation towards these bodies is to the gravitation towards the whole Earth, as these bodies are to the whole Earth, the gravitation, towards them must be far less than to fall under the observation of our senses.

Corollary 2. The force of gravity towards the several equal particles of any body, is reciprocally as the square of the distance of places from the particles; as appears from Cor. 3, Prop. LXXIV, Book I.

> [In Proposition VI Newton provides his definition of gravitational mass, and in Proposition VII, together with its corollary 2, Newton restates his *universal law of gravitation*,
>
> $$F = Gm_1m_2/r^2,$$
>
> where F is the force, m_1 and m_2 are the masses of the objects interacting, r is the distance between the centers of the masses and G is the gravitational constant. Newton's law of gravitation states that every point mass in the universe attracts every other point mass with a force that is directly proportional to the product of their masses, and inversely proportional to the square of the distance between them.]

Proposition VIII. Theorem VIII. *In two spheres mutually gravitating each towards the other, if the matter in places on all sides round about and equidistant from the centers, is similar; the weight of either sphere towards the other, will be reciprocally as the square of the distance between their centers.*

After I had found that the force of gravity towards a whole planet did arise from, and was compounded of the forces of gravity towards all its parts; and towards every one part, was in the reciprocal proportion of the squares of the distances from the part: I was yet in doubt, whether that reciprocal duplicate proportion did accurately hold, or but nearly so, in the total force compounded of so many partial ones. For it might be that the proportion which accurately enough took place in greater distances, should be wide of the truth near the surface of the planet, where the distances of the particles are unequal, and their situation dissimilar. But by the help of Prop. LXXV and LXXVI, Book I, and their Corollaries, I was at last satisfy'd of the truth of the proposition, as it now lies before us.

Corollary 1. Hence, we may find and compare together the weights of bodies towards different planets. For the weights of bodies revolving in circles about planets, are (by Cor. 2, Prop. IV, Book I.) as the diameters of the circles directly, and the squares of their periodic times reciprocally; and their weights at the surfaces of the planets, or at any other distances from their centers, are (by this proposition) greater or less, in the reciprocal duplicate proportion of the distances. Thus from the periodic times of Venus, revolving about the Sun, in 224 d. 16 ¾ h. of the utmost circumjovial satellite revolving about Jupiter, in 16 d. 16 9/13 h; of the Hugenian satellite about Saturn in 15 d. 22 2/3 h; and of the Moon about the Earth in 27 d. 7h. 43'; compared with the mean distance of Venus from the Sun, and with the greatest heliocentric elongations of the outmost circumjovial satellite from Jupiter's center, 8' 16" of the Hugenian satellite from the center of Saturn, 3' 4", and of the Moon from the Earth, 10' 33"; by computation I found, that the weight of equal bodies, at equal distances from the centers of the Sun, of Jupiter, of Saturn, and of the Earth, towards the Sun, Jupiter, Saturn, and the Earth, were one to another, as 1, 1/1067, 1/3021, and 1/169282 respectively. Then because as the distances are increased or diminished, the weights are diminished or increased in a duplicate ratio; the weights of equal bodies towards the Sun, Jupiter, Saturn, and the Earth, at the distances 10000, 997, 791 and 109 from their centers, that is, at their very superficies, will be as 10000, 943, 529 and 435 respectively. How much the weights of bodies are at the superficies of the Moon, will be'shewn hereafter.

Corollary 2. Hence likewise we discover the quantity of matter in the several Planets. For their quantities of matter are as the forces of gravity at equal distances from their centers, that is, in the Sun, Jupiter, Saturn, and the Earth, as 1, 1/1067, 1/3021, and 1/169282 respectively. If the parallax of the Sun be taken greater or less than 10", 30"', the quantity of matter in the Earth must be augmented or diminished in the triplicate of that proportion.

Corollary 3. Hence also we find the densities of the Planets. For (by Prop. LXXII, Book I) the weights of equal and similar bodies towards similar spheres, are, at the surfaces of those spheres, as the diameters of the spheres. And therefore, the densities of dissimilar spheres are as those weights applied to the diameters of the spheres. But the true diameters of the Sun, Jupiter, Saturn, and the Earth, were one to another as 10000, 997, 791 and 109; and the weights towards the same, as 10000, 943, 529, and 435 respectively; and therefore, their densities are as 100, 94 ½, 67 and 400. The density of the Earth, which comes out by this computation, does not depend upon the parallax of the Sun, but is determined by the parallax of the Moon, and therefore is here truly defin'd. The Sun therefore is a little denser than Jupiter, and Jupiter than Saturn, and the Earth four times denser than the Sun; for the Sun, by its great heat, is kept in a sort of a rarefy'd state. The Moon is denser than the Earth, as shall appear afterwards.

Corollary 4. The smaller the Planets are, they are, *ceteris paribus*, of so much the greater density. For so the powers of gravity on their several surfaces, come nearer to equality. They are likewise, ceteris paribus, of the greater density, as they are nearer to the Sun. So, Jupiter is more dense than Saturn, and the Earth than Jupiter. For the Planets were to be placed at different distances from the Sun, that according to their degrees of density, they might enjoy a greater or less proportion of the Sun's heat. Our water, if it were remov'd as far as the orb of Saturn, would be converted into ice, and the orb of Mercury would quickly fly away in vapor. For the light of the Sun, to which its heat is proportional, is seven times denser in the orb of the Mercury than with us: and by the thermometer I have found, that a sevenfold heat of our summer-sun will make water boil. Nor are we to doubt, that the matter of Mercury is adapted to its heat, and is therefore more dense than the matter of our Earth; since, in a denser matter, the operations of nature require a stronger heat.

Proposition IX. Theorem IX. *That the force of gravity, consider'd downwards from the surface of the planets, decreases nearly in the proportion of the distances from their centers.*

If the matter of the planet were of a uniform density, this proposition would be accurately true, (by Prop. LXXIII, Book I). The error therefore can be no greater than what may arise from the inequality of the density.

…

General Scholium.

The hypotheses of Vortices is press'd with many difficulties. That every Planet by a radius drawn to the Sun may describe areas proportional to the times of description, the periodic times of the several parts of the Vortices should observe the duplicate proportion of their distances from the Sun. But that the periodic times of the Planets may obtain the sesquiplicate proportion of their distances from the Sun, the periodic times of the parts of the Vortex ought to be in sesquiplicate proportion of their distances. That the smaller Vortices may maintain their lesser revolutions about Saturn, Jupiter, and other Planets, and swim quietly and undisturb'd in the greater Vortex of the Sun, the periodic times of the parts of the Sun's Vortex should be equal. But the rotation of the Sun and Planets about their axes, which ought to correspond with the motions of their Vortices, recede far from all these proportions. The motions of the Comets are exceedingly regular, are govern'd by the same laws with the motions of the Planets, and can by no means be accounted for by the hypotheses of Vortices. For Comets are carry'd with very eccentric motions through all parts of the heavens indifferently, with a freedom that is incompatible with the notion of a Vortex. Bodies, projected in our air, suffer no resistance but from the air. Withdraw the air, as is done in Mr. Boyle's vacuum, and the resistance ceases. For in this void a bit of fine down and a piece of solid gold descend with equal velocity. And the parity of reason must take place in the celestial spaces above the Earth's atmosphere; in which spaces, where there is no air to resist their motions, all bodies will move with the greatest freedom; and the Planets and Comets will constantly pursue their revolutions in orbits given in kind and position, according to the laws above explain'd. But though these bodies may indeed persevere in their orbits by the mere laws of gravity, yet they could by no means have at first deriv'd the regular position of the orbits themselves from those laws. The six primary Planets are revolv'd about the Sun, in circles concentric with the Sun, and with motions directed towards the same parts and almost in the same plan. Ten Moons are revolv'd about the Earth, Jupiter and Saturn, in circles concentric with them, with the same direction of motion, and nearly in the planes of the orbits of those Planets. But it is not to be conceived that mere mechanical causes could give birth to so many regular motions: since the Comets range over all parts of the heavens, in very eccentric orbits. For by that kind of motion they pass easily through the orbits of the Planets, and with great rapidity; and in their aphelions, where they move the slowest, and are detain'd the longest, they recede to the greatest distances from each other, and thence suffer the least disturbance from their mutual attractions. …

…

Hitherto we have explain'd the phænomena of the heavens and of our sea, by the power of Gravity, *but have not yet assign'd the cause of this power*. This is certain, that it must proceed from a cause that penetrates to the very centers of the Sun and Planets, without suffering the least diminution of its force; that operates, not according to the quantity of surfaces of the particles upon which it acts, (as mechanical causes use to do,) but *according to the quantity of the solid matter which they contain*, and propagates its virtue on all sides, to immense distances, decreasing always in the duplicate proportion of the distances. Gravitation towards the Sun, *is made up out of the gravitations towards the several particles of which the body of the Sun is compos'd*; and in receding from the Sun, decreases accurately in the duplicate proportion of the distances, as far as the orb of Saturn, as evidently appears from the quiescence of the aphelions of the Planets; nay, and even to the remotest aphelions of the Comets, if those aphelions are also quiescent. *But hitherto I have not been able to discover the cause of those properties of gravity from phænomena, and I framed no hypotheses*. For whatever is not deduc'd from the phænomena, is to be called a hypothesis; and *hypotheses*, whether metaphysical or physical, whether of occult qualities or mechanical, *have no place in experimental philosophy*. In this philosophy particular propositions are inferr'd from the phænomena, and afterwards render'd general by induction. Thus, it was that the impenetrability, the mobility, and the impulsive force of bodies, and the laws of motion and of gravitation, were discovered. And to us it is enough, that gravity does really exist, and act according to the laws which we have explained, and abundantly serves to account for all the motions of the celestial bodies, and of our sea. And now we might add something concerning a certain most subtle Spirit, which pervades and lies hid in all gross bodies; by the force and action of which Spirit, the particles of bodies mutually attract one another at near distances, and cohere, if contiguous; and electric bodies operate to greater distances, as well repelling as attracting the neighboring corpuscles; and light is emitted, reflected, refracted, inflected, and heats bodies; and all sensation is excited, and the members of animal bodies move at the command of the will, namely, by the vibrations of this Spirit, mutually propagated along the solid filaments of the nerves, from the outward organs of sense to the brain, and from the brain into the muscles. But these are things that cannot be explain'd in few words, nor are we furnish'd with that sufficiency of experiments which is required to an accurate determination and demonstration of the laws by which this electric and elastic spirit operates.

Newton's calculation of Kepler's laws.

Newton used his mathematical description of gravity to derive Kepler's laws of planetary motion (which Kepler had first obtained empirically), to account for tides, the trajectories of comets, the precession of the equinoxes and other phenomena, eradicating doubt about the Solar System's heliocentricity. He demonstrated that the motion of objects on Earth and celestial bodies could be accounted for by the same principles. Newton's inference that the Earth is an oblate spheroid was later confirmed by the geodetic measurements of Maupertuis, La Condamine, and others, convincing most European scientists of the superiority of Newtonian mechanics over earlier systems.

In his *Principia*, Newton computed the acceleration of a planet moving according to Kepler's first and second laws.

1. The *direction* of the acceleration is towards the Sun.

2. The *magnitude* of the acceleration is inversely proportional to the square of the planet's distance from the Sun (the inverse square law).

Newton defined the force acting on a planet to be the product of its mass and the acceleration. So:

1. Every planet is attracted towards the Sun.

2. The force acting on a planet is directly proportional to the mass of the planet and is inversely proportional to the square of its distance from the Sun.

The Sun plays an unsymmetrical part, which is unjustified. So, Newton assumed, in his law of universal gravitation:

1. All bodies in the Solar System attract one another.

2. The force between two bodies is in direct proportion to the product of their masses and in inverse proportion to the square of the distance between them.

As the planets have small masses compared to that of the Sun, the orbits conform approximately to Kepler's laws. Newton's model improves upon Kepler's model, and fits actual observations more accurately.

By *Newton's second law*, the gravitational force that acts on the planet is:

$$F = m_{planet}\, d^2\mathbf{r}/dt^2 = -\, m_{planet}\, \alpha\, r^{-2}\, \underline{\mathbf{r}}$$

where m_{planet} is the mass of the planet and α has the same value for all planets in the Solar System. According to *Newton's third law*, the Sun is attracted to the planet by a force of the same magnitude. Since the force is proportional to the mass of the planet, under the symmetric consideration, it should also be proportional to the mass of the Sun, m_{Sun}. So

$$\alpha = G\, m_{Sun}$$

where G is the *gravitational constant.*

The acceleration of Solar System body number i is, according to Newton's laws:

$$d^2\mathbf{r}/dt^2 = G \sum_{j \neq I} m_j\, r_{ij}^{-2}\, \underline{\mathbf{r}}_{ij}$$

where m_j is the mass of body j, r_{ij} is the distance between body i and body j, $\underline{\mathbf{r}}_{ij}$ is the unit vector from body i towards body j, and the vector summation is over all bodies in the Solar System, besides i itself.

In the special case where there are only two bodies in the Solar System, Earth and Sun, the acceleration becomes

$$d^2\mathbf{r}/dt^2_{Earth} = G\, m_{Moon,Earth}\, r^{-2}\, \underline{\mathbf{r}}_{Moon,Earth}$$

which is the acceleration of the Kepler motion. So, this Earth moves around the Sun according to Kepler's laws.

Using *Newton's law of gravitation, Kepler's third law* (a^3/T^2 = const.) can be found in the case of a circular orbit *by setting the centripetal force equal to the gravitational force*:

$$mr\omega^2 = GmM/r^2.$$

Then, expressing the angular velocity ω in terms of the orbital period T and then rearranging, results in *Kepler's third law*:

$$mr\,(2\pi/T)^2 = GmM/r^2 \rightarrow T^2 = (4\pi^2/GM)\, r^3 \rightarrow T^2 \propto r^3.$$

A more detailed derivation can be done with general elliptical orbits, instead of circles, as well as orbiting the center of mass, instead of just the large mass. This results in replacing a circular radius, r, with the semi-major axis, *a*, of the elliptical relative motion of one mass relative to the other, as well as replacing the large mass M with M + m . However, with planet masses being so much smaller than the Sun, this correction is often ignored. The full corresponding formula is:

$$a^3/T^2 = G(M + m)/4\pi^2 \approx GM/4\pi^2 \approx 7.496 \times 10^{-6}\ AU^3 t^{-2} = \text{const.}$$

where M is the mass of the Sun (= 1.9885×10^{30} kg), m is the mass of the planet (in kg), G is the gravitational constant (= 6.674×10^{-11} $N \cdot m^2 . kg^{-2}$), T is the orbital period (in days), *a* is the elliptical semi-major axis (in meters), and AU is the astronomical unit (= 149,597,870,700 m), the average distance from earth to the sun.

Newton's universal law of gravitation.

While Newton was able to articulate his *Law of Universal Gravitation* and verify it experimentally, he could only calculate the relative gravitational force in comparison to another force. It was not until Henry Cavendish's verification of the Gravitational Constant that the Law of Universal Gravitation received its final form:

$$F = GMm/r^2 = 6.674 \times 10^{-11}\ Mm/r^2\ N\ (\textit{SI units})$$

where F represents the force in Newtons, M and m represent the two masses in kilograms, and r represents the separation in meters. G represents the Gravitational Constant, which has a value of $6.674 \times 10^{-11}\ N\ (m/kg)^2$. Because of the magnitude of G, gravitational force is very small *unless large masses or short distances are involved.*

[The *kilogram* is the base unit of mass in the International System of Units (SI), having the unit symbol kg. It is a widely used measure in science, engineering and commerce worldwide, and is often simply called a kilo colloquially. It means 'one thousand grams'.

The kilogram is defined in terms of the Planck constant, the second, and the meter, both of which are based on fundamental physical constants. This allows a properly equipped metrology laboratory to calibrate a mass measurement instrument such as a Kibble balance as the primary standard to determine an exact kilogram mass.

The kilogram was originally defined in 1795 during the French Revolution as the mass of one liter of water. The current definition of a kilogram agrees with this original definition to within 30 parts per million. In 1799, the platinum *Kilogramme des Archives* replaced it as the standard of mass. In 1889, a cylinder of platinum-iridium, the International Prototype of the Kilogram (IPK), became the standard of the unit of mass for the metric system and remained so for 130 years, before the current standard was adopted in 2019.

The kilogram is defined in terms of three fundamental physical constants:

- a specific atomic transition frequency $\Delta\nu_{Cs}$, which defines the duration of the second,
- the speed of light *c*, which when combined with the second, defines the length of the meter,
- and the Planck constant *h*, which when combined with the meter and second, defines the mass of the kilogram.

The formal definition according to the General Conference on Weights and Measures (CGPM) is: The *kilogram*, symbol kg, is the SI unit of mass. It is defined by taking the fixed numerical value of the Planck constant h to be 6.62607015×10^{-34} when expressed in the unit J·s, which is equal to $kg{\cdot}m^2{\cdot}s^{-1}$, where the meter and the second are defined in terms of c and $\Delta\nu_{Cs}$.

Defined in term of those units, the kg is formulated as:

$$1\ kg = (299792458)^2/(6.62607015\times10^{-34})(9192631770)h\Delta\nu_{Cs}/c^2$$
$$= 917097121160018/621541050725904751 0^{42}h\Delta\nu_{Cs}/c^2$$
$$\approx (1.475521399735270\times10^{40})h\Delta\nu_{Cs}/c^2 \ .$$

This definition is generally consistent with previous definitions: the mass remains within 30 ppm of the mass of one liter of water.

The *newton* (N) is the unit of force in the International System of Units (SI). It is defined as 1 kg . m . s^{-2}, the force which gives a mass of 1 kilogram an acceleration of 1 meter per second per second.]

[*Planck units*: *By definition*, the speed of light [or of gravity], c = 1, the gravitational constant, G = 1, $\hbar = h/2\pi = 1$, where h is Planck's constant, and the Boltzman constant, $k_B = 1$.

The *Planck time*, denoted by t_P, is the time required for light [or gravity] to travel a distance of 1 *Planck length* in a vacuum,

$t_P = \sqrt{(\hbar G/c^5)} = 1$(*plank unit*).

It is equal to approximately 5.3912×10^{-44} s. No current physical theory can describe timescales shorter than the Planck time, such as the earliest events after the Big Bang.

The *Planck length*, denoted ℓ_P, is a unit of length defined as:

$\ell_P = \sqrt{(\hbar G/c^3)} = 1$(*plank unit*).

It is equal to approximately 1.61626×10^{-35} m.

[Cf. diameter of neutron = 1.6×10^{-15} m]

$\ell_P/t_P = c^2 = 1$, c = 1.

The *speed of light*,

$$c \approx 2.99792 \times 10^{8}\ \text{m s}^{-1}$$
$$\approx 2.99792 \times 10^{8} \text{ x } 5.39 \times 10^{-44}/1.61626 \times 10^{-35}$$
$$= \textit{1 plank length/plank time } (= 1\ \ell_P\ t_P^{-1}).$$

The *Planck mass*, denoted m_P, is $m_P = \sqrt{(\hbar c/G)} = 1$(*plank unit*).

It is equal to approximately 2.17643×10^{-8} kg.

[Cf. mass of electron = 9.109×10^{-31} kg,
mass of proton = 1.6727×10^{-27} kg.]

The *Planck force*, denoted F_P, is

$$F_P = \hbar/\ell_P t_P = c^4/G = 1 \ (\textit{plank unit}).$$

It is equal to approximately $1.2103 \text{ x } 10^{44}$ N.

[Units of $F_P = GMm/r^2$
$= G\{\sqrt{(\hbar c/G)} \text{ x } \sqrt{(\hbar c/G)}\}/\{\sqrt{(\hbar G/c^3)} \text{ x } \sqrt{(\hbar G/c^3)}\} = c^4/G$.
Units of $c^4 = (m/s)^4$
Units of $F = N = kg\ .\ m\ .\ s^{-2}$
Units of $G = N\ (m/kg)^2 = kg\ .\ m\ .\ s^{-2}\ (m/kg)^2 = kg^{-1}\ .\ m^3\ .\ s^{-2}$
Units of $c^4/G = (m/s)^4/(kg^{-1}\ .\ m^3\ .\ s^{-2}) = kg\ .\ m\ .\ s^{-2}$.]

$$F_P = GMm/r^2 = 6.674 \text{ x } 10^{-11} \text{ x } (2.17643 \times 10^{-8}/1.61626 \times 10^{-35})^2$$
$$= 6.674 \text{ x } 10^{-11} \text{ x } (1.34658 \times 10^{27})^2$$
$$= 6.674 \text{ x } 10^{-11} \text{ x } 1.81328 \times 10^{54} = 1.2103 \times 10^{44}\ \text{N} = 1 \textit{ plank unit.}$$

$$F_P = G_P M_P m_P/r_P^2 = 1.2103 \times 10^{44}\ M_P m_P/r_P^2\ (\textit{plank units})$$

where $G_P = 1.2103 \times 10^{44}$ (*plank units*).

G_P is numerically so large because the *Planck mass* is much larger than the *Planck length.*]

[In terms of *atomic units*:

The *atomic time*, denoted by t_A, is the vibration time of the unperturbed ground-state hyperfine transition frequency of the caesium-133 atom. One second is equal to the time of 9192631770 vibrations, so one unit of atomic time is equal to approximately 1.0878×10^{-10} s.

The *atomic length*, denoted by ℓ_A, is a unit of length defined as the diameter of a neutron. It is equal to approximately 1.6×10^{-15} m.

The *atomic mass*, denoted m_A, is a unit of mass defined as the *mass of a neutron*

$$m_A = \sqrt{(\hbar c/G)} = 1(\textit{plank unit}).$$

It is equal to approximately 1.6749×10^{-27} kg.]

$$F = GMm/r^2 = 6.674 \text{ x } 10^{-11}\, Mm/r^2 \text{ N } (\textit{SI units}) = G_A M_A m_A/r_A^2$$
$$F = 6.674 \text{ x } 10^{-11} \text{ x } (1.6749 \times 10^{-27}/1.6 \times 10^{-15})^2\, M_A m_A/r_A^2 \text{ N}$$
$$F = 6.674 \text{ x } 10^{-11} \text{ x } 1.0958 \times 10^{-24}\, M_A m_A/r_A^2 \text{ N}$$
$$F = 7.3134 \times 10^{-35}\, M_A m_A/r_A^2 \text{ N},$$

or

$$F = G_A M_A m_A/r_A^2 \text{ N},$$

where $G_A = 7.3134 \times 10^{-35}$.

Now the *atomic length* $\ell_A = 1.6 \times 10^{-15}$ m, is numerically too large compared with the *atomic mass*.

Substituting the *Planck length*, $\ell_P = 1.61626 \times 10^{-35}$ m, the distance that light [or gravity] travels in a unit of *Planck time*, for the *atomic length* $\ell_A = 1.6 \times 10^{-15}$ m, gives

$$F = 6.674 \text{ x } 10^{-11} \text{ x } (1.6749 \times 10^{-27}/1.61626 \times 10^{-35})^2\, M_A m_A/r_A^2 \text{ N}$$
$$F = 6.674 \text{ x } 10^{-11} \text{ x } (1.03628 \times 10^{8})^2\, M_A m_A/r_A^2 \text{ N}$$
$$F = 6.674 \text{ x } 10^{-11} \text{ x } 1.07388 \times 10^{16}\, M_A m_A/r_A^2 \text{ N}$$
$$F = 7.1671 \text{ x } 10^{5}\, M_A m_A/r_A^2 \text{ N.]}$$

Henry Cavendish (October 10, 1731 – February 24, 1810)

Cavendish was an English natural philosopher and scientist who was an important experimental and theoretical chemist and physicist. He is noted for his discovery of hydrogen, which he termed "inflammable air". He described the density of inflammable air, which formed water on combustion, in a 1766 paper, *On Factitious Airs*. Antoine Lavoisier later reproduced Cavendish's experiment and gave the element its name.

A shy man, Cavendish was distinguished for great accuracy and precision in his researches into the composition of atmospheric air, the properties of different gases, the synthesis of water, *the law governing electrical attraction and repulsion*, a mechanical theory of heat, and *calculations of the density (and hence the mass) of the Earth*. His experiment to measure the density of the Earth (which, in turn, allows the gravitational constant to be calculated) has come to be known as the Cavendish experiment.

Cavendish was born on 10 October 1731 in Nice, where his family was living at the time. His mother was Lady Anne de Grey, fourth daughter of Henry Grey, 1st Duke of Kent, and his father was Lord Charles Cavendish, the third son of William Cavendish, 2nd Duke of Devonshire. The family traced its lineage across eight centuries to Norman times, and was closely connected to many aristocratic families of Great Britain. Henry's mother died in 1733, three months after the birth of her second son, Frederick, and shortly before Henry's second birthday, leaving Lord Charles Cavendish to bring up his two sons. Henry Cavendish was styled as "The Honourable Henry Cavendish".

From the age of 11 Henry attended Newcome's School, a private school near London. At the age of 18 (on 24 November 1748) he entered the University of Cambridge in St Peter's College, now known as Peterhouse, but left three years later on 23 February 1751 without taking a degree (at the time, a common practice). He then lived with his father in London, where he soon had his own laboratory.

Lord Charles Cavendish spent his life firstly in politics and then increasingly in science, especially in the Royal Society of London. In 1758, he took Henry to meetings of the Royal Society and also to dinners of the Royal Society Club. In 1760, Henry Cavendish was elected to both these groups, and he was assiduous in his attendance after that. He took virtually no part in politics, but followed his father into science, through his researches and his participation in scientific organizations. He was active in the Council of the Royal Society of London (to which he was elected in 1765).

His interest and expertise in the use of scientific instruments led him to head a committee to review the Royal Society's meteorological instruments and to help assess the instruments of the Royal Greenwich Observatory. His first paper, *Factitious Airs*, appeared in 1766. Other committees on which he served included the committee of papers, which chose the papers for publication in the *Philosophical Transactions of the Royal Society*, and the committees for the transit of Venus (1769), for the gravitational attraction of mountains (1774), and for the scientific instructions for Constantine Phipps's expedition (1773) in search of the North Pole and the Northwest Passage. In 1773, Henry joined his father as an elected trustee of the British Museum, to which he devoted a good deal of time and effort. Soon after the Royal Institution of Great Britain was established, Cavendish became a manager (1800) and took an active interest, especially in the laboratory, where he observed and helped in Humphry Davy's chemical experiments.

About the time of his father's death, Cavendish began to work closely with Charles Blagden, an association that helped Blagden enter fully into London's scientific society. In return, Blagden helped to keep the world at a distance from Cavendish. Cavendish published no books and few papers, but he achieved much. Several areas of research, including mechanics, optics, and magnetism, feature extensively in his manuscripts, but they scarcely feature in his published work. Cavendish is considered to be one of the so-called pneumatic chemists of the eighteenth and nineteenth centuries, along with, for example, Joseph Priestley, Joseph Black, and Daniel Rutherford. Cavendish found that a definite, peculiar, and highly inflammable gas, which he referred to as "Inflammable Air", was produced by the action of certain acids on certain metals. This gas was hydrogen, which Cavendish correctly guessed was proportioned two to one in water.

Although others, such as Robert Boyle, had prepared hydrogen gas earlier, Cavendish is usually given the credit for recognizing its elemental nature. In 1777, Cavendish discovered that air exhaled by mammals is converted to "fixed air" (carbon dioxide), not "phlogisticated air" as predicted by Joseph Priestley. Also, by dissolving alkalis in acids, Cavendish produced carbon dioxide, which he collected, along with other gases, in bottles inverted over water or mercury. He then measured their solubility in water and their specific gravity, and noted their combustibility. He concluded in his 1778 paper "General Considerations on Acids" that respirable air constitutes acidity. Cavendish was awarded the Royal Society's Copley Medal for this paper. Gas chemistry was of increasing importance in the latter half of the 18th century, and became crucial for Frenchman Antoine-Laurent Lavoisier's reform of chemistry, generally known as the chemical revolution.

In 1783, Cavendish published a paper on eudiometry (the measurement of the goodness of gases for breathing). He described a new eudiometer of his invention, with which he achieved the best results to date, using what in other hands had been the inexact method of measuring gases by weighing them. Then, after a repetition of a 1781 experiment performed by Priestley, Cavendish published a paper on the production of pure water by burning hydrogen in "dephlogisticated air" (air in the process of combustion, now known to be oxygen). Cavendish concluded that rather than being synthesized, the burning of hydrogen caused water to be condensed from the air. Some physicists interpreted hydrogen as pure phlogiston. Cavendish reported his findings to Priestley no later than March 1783, but did not publish them until the following year. The Scottish inventor James Watt published a paper on the composition of water in 1783; controversy about who made the discovery first ensued.

In 1785, Cavendish investigated the composition of common (i.e. atmospheric) air, obtaining impressively accurate results. He conducted experiments in which hydrogen and ordinary air were combined in known ratios and then exploded with a spark of electricity. Furthermore, he also described an experiment in which he was able to remove, in modern terminology, both the oxygen and nitrogen gases from a sample of atmospheric air until only a small bubble of unreacted gas was left in the original sample. Using his observations, Cavendish observed that, when he had determined the amounts of phlogisticated air (nitrogen) and dephlogisticated air (oxygen), there remained a volume of gas amounting to 1/120 of the original volume of nitrogen. By careful measurements he was led to conclude that "common air consists of one part of dephlogisticated air [oxygen], mixed with four of phlogisticated [nitrogen]".

In the 1890s (around 100 years later) two British physicists, William Ramsay and Lord Rayleigh, realized that their newly discovered inert gas, argon, was responsible for Cavendish's problematic residue; he had not made an error. What he had done was perform rigorous quantitative experiments, using standardized instruments and methods, aimed at reproducible results; taken the mean of the result of several experiments; and identified and allowed for sources of error. The balance that he used, made by a craftsman named Harrison, was the first of the precision balances of the 18th century, and as accurate as Lavoisier's (which has been estimated to measure one part in 400,000). Cavendish worked with his instrument makers, generally improving existing instruments rather than inventing wholly new ones.

Cavendish, as indicated above, used the language of the old phlogiston theory in chemistry. In 1787, he became one of the earliest outside France to convert to the new antiphlogistic theory of Lavoisier, though he remained skeptical about the nomenclature of the new theory. He also objected to Lavoisier's identification of heat as having a material or elementary basis. Working within the framework of Newtonian mechanism, Cavendish had tackled the problem of the nature of heat in the 1760s, explaining heat as the result of the motion of matter.

In 1783, he published a paper on the temperature at which mercury freezes and, in that paper, made use of the idea of latent heat, although he did not use the term because he believed that it implied acceptance of a material theory of heat. He made his objections explicit in his 1784 paper on air. He went on to develop a general theory of heat, and the manuscript of that theory has been persuasively dated to the late 1780s. His theory was at once mathematical and mechanical: it contained the principle of the conservation of heat (later understood as an instance of conservation of energy) and even included the concept (although not the label) of the mechanical equivalent of heat.

Following his father's death, Henry bought another house in town and also a house in Clapham Common (built by Thomas Cubitt), at that time to the south-west of London. The London house contained the bulk of his library, while he kept most of his instruments at Clapham Common, where he carried out most of his experiments. *The most famous of those experiments, published in 1798, was to determine the density of the Earth* and became known as the *Cavendish experiment*. The apparatus Cavendish used for weighing the Earth was a modification of the torsion balance built by Englishman and geologist John Michell, who died before he could begin the experiment. The apparatus was sent in crates to Cavendish, who completed the experiment in 1797–1798 and published the results. [Cavendish, Henry (1798). "Experiments to Determine the Density of Earth". Phil. Trans., 88, 469–526; doi:10.1098/rstl.1798.0022.]

The experimental apparatus consisted of a torsion balance with a pair of 2-inch

1.61-pound lead spheres suspended from the arm of a torsion balance and two much larger stationary lead balls (350 pounds). Cavendish intended to measure the force of gravitational attraction between the two. He noticed that Michell's apparatus would be sensitive to temperature differences and induced air currents, so he made modifications by isolating the apparatus in a separate room with external controls and telescopes for making observations.

Using this equipment, Cavendish calculated the attraction between the balls from the period of oscillation of the torsion balance, and then he used this value to calculate the *density of the Earth*. Cavendish found that the Earth's average density is 5.48 times greater than that of water. John Henry Poynting later noted that the data should have led to a value of 5.448, and indeed that is the average value of the twenty-nine determinations Cavendish included in his paper. The published number was due to a simple arithmetic error on his part. What was extraordinary about Cavendish's experiment was its elimination of every source of error and every factor that could disturb the experiment, and its precision in measuring an astonishingly small attraction, a mere 1/50,000,000 of the weight of the lead balls. *The result that Cavendish obtained for the density of the Earth is within 1 per cent of the currently accepted figure.*

Cavendish's work led others to accurate values for the *gravitational constant* (G) and *Earth's mass*. Based on his results, one can calculate a value for G of 6.754×10^{-11} N m^2/kg^2, which compares favorably with the modern value of 6.67428×10^{-11} N m^2/kg^2.

Books often describe Cavendish's work as a measurement of either G or the Earth's mass. Since these are related to the Earth's density by a trivial web of algebraic relations, none of these sources are wrong, but they do not match the exact word choice of Cavendish, and this mistake has been pointed out by several authors. Cavendish's stated goal was to measure the Earth's density. The first time that the constant got this name was in 1873, almost 100 years after the Cavendish experiment. *Cavendish's results also give the Earth's mass.*

Cavendish's electrical and chemical experiments, like those on heat, had begun while he lived with his father in a laboratory in their London house. Lord Charles Cavendish died in 1783, leaving almost all of his very substantial estate to Henry. Like his theory of heat, Cavendish's comprehensive theory of electricity was mathematical in form and was based on precise quantitative experiments. Working with his colleague, Timothy Lane, he created an artificial torpedo fish that could dispense electric shocks to show that the source of shock from these fish was electricity. He published an early version of his theory of electricity in 1771, based on an expansive electrical fluid that exerted pressure. He demonstrated that if the intensity of electric force were inversely proportional to distance, then the electric fluid more than that needed for electrical neutrality would lie on the outer surface of an electrified sphere; then he confirmed this experimentally. Cavendish continued to work on electricity after this initial paper, but he published no more on the subject.

Cavendish wrote papers on electrical topics for the Royal Society but the bulk of his electrical researches did not become known until long after other scientists had been credited with the same results, when they were collected and published a century later in 1879, by James Clerk Maxwell, the first Cavendish professor of experimental physics, to which Maxwell devoted the last five years of his life. [*The Scientific Papers of the Honourable Henry Cavendish, F.R.S.* Vol. 1: *The Electrical Researches.*] Among Cavendish's discoveries were the concept of *electric potential* (which he called the "degree of electrification"), an early unit of *capacitance* (that of a sphere one inch in diameter), the formula for the capacitance of a plate capacitor, the concept of the *dielectric constant* of a material, the relationship between *electric potential* and *current* (now called Ohm's law) (1781), laws for the division of current in parallel circuits (now attributed to Charles Wheatstone), and *the inverse square law of variation of electric force with distance*, now called *Coulomb's law*.

Cavendish died at Clapham on 24 February 1810 (as one of the wealthiest men in Britain) and was buried, along with many of his ancestors, in the church that is now Derby Cathedral. The road he used to live on in Derby has been named after him, as has a road near his house in Clapham, of which the north part is part of the South Circular Road. In honor of Henry Cavendish's achievements and due to an endowment granted by Henry's relative William Cavendish, 7th Duke of Devonshire, (Chancellor of the University from 1861 to 1891) the University of Cambridge's physics laboratory was named the Cavendish Laboratory by Maxwell.

Cavendish, H. (1798). Experiments to Determine the Density of Earth.

Phil. Trans., 88, 469–526; doi:10.1098/rstl.1798.0022.

Read June 21, 1798.

Cavendish used apparatus, originally developed by Rev. John Michell, of the Royal Society, prior to his death, but which Cavendish improved. The apparatus is very simple; it consists of a wooden arm, 6 feet long, made so as to unite great strength with little weight. This arm is suspended in a horizontal position, by a slender wire 4.0 inches long, and to each extremity is hung a leaden ball, about 2 inches in diameter; and the whole is enclosed in a narrow wooden case, to defend it from the wind. As no more force is required to make this arm turn round on its center, than what is necessary to twist the suspending wire, it is plain, that if the wire is sufficiently slender, the most minute force, such as the attraction of a leaden weight a few inches in diameter, will be sufficient to draw the arm sensibly aside.

Many years ago, the late Rev. John Michell, of this Society, contrived a method of determining the density of the earth, by rendering sensible the attraction of small quantities of matter; but, as he was engaged in other pursuits, he did not complete the apparatus till a short time before his death, and did not live to make any experiments with it. After his death, the apparatus came to the Rev. Francis John. Hyde Wollaston, Jacksonian Professor at Cambridge, who, not having conveniences for making experiments with it, in the manner he could wish, was so good as to give it to me.

The apparatus is very simple; it consists of a wooden arm, 6 feet long, made so as to unite great strength with little weight. This arm is suspended in a horizontal position, by a slender wire 4,0 inches long, and to each extremity is hung a leaden ball, about 2 inches in diameter; and the whole is enclosed in a narrow wooden case, to defend it from the wind.

As no more force is required to make this arm turn round on its center, than what is necessary to twist the suspending wire, it is plain, that if the wire is sufficiently slender, the most minute force, such as the attraction of a leaden weight a few inches in diameter, will be sufficient to draw the arm sensibly aside. The weights which Mr. Michell intended to use were 8 inches diameter. One of these was to be placed on one side the case, opposite to one of the balls, and as near it as could conveniently be done, and the other on the other side, opposite to the other ball, so that the attraction of both these weights would conspire in drawing the arm aside; and, when its position, as affected by these weights, was ascertained, the weights were to be removed to the other side of the case, so as to draw the arm the contrary way, and the position of the arm was to be again determined; and, consequently, half the difference of these positions would shew how much the arm was drawn aside by the attraction of the weights.

In order to determine from hence the density of the earth, it is necessary to ascertain what force is required to draw the arm aside through a given space. This Mr. Michell intended to do, by putting the arm in motion, and observing the time of its vibrations, from which it may easily be computed*.

* Mr. Coulomb has, in a variety of cases, used a contrivance of this kind for trying small attractions; but Mr. Michell informed me of his intention of making this experiment, and of the method he intended to use, before the publication of any of Mr. Coulomb's experiments.

Mr. Michell had prepared two wooden stands, on which the leaden weights were to be supported, and pushed forwards, till they came almost in contact with the case; but he seems to have intended to move them by hand.

As the force with which the balls are attracted by these weights is excessively minute, not more than 1/50.000,000 of their weight, it is plain, that a very minute disturbing force will be sufficient to destroy the success of the experiment; and, from the following experiments it will appear, that the disturbing force most difficult to guard against, is that arising from the variations of heat and cold; for, if one side of the case is warmer than the other, the air in contact with it will be rarefied, and, in consequence, will ascend, while that on the other side will descend and produce a current which will draw the arm sensibly aside.

As I was convinced of the necessity of guarding against this source of error, I resolved to place the apparatus in a room which should remain constantly shut, and to observe the motion of the arm from without, by means of a telescope; and to suspend the leaden weights in such manner, that I could move them without entering into the room. This difference in the manner of observing, rendered it necessary to make some alteration in Mr. Michell's apparatus; and, as there were some parts of it which I thought not so convenient as could be wished, I chose to make the greatest part of it afresh.

Fig. 1. (Tab. XXIII.) is a longitudinal vertical section through the instrument, and the building in which it is placed. ABCDDCBAEFFE, is the case; x and x are the two balls, which are suspended by the wires bx from the arm gbmb, which is itself suspended by the slender wire *gl*. This arm consists of a slender deal rod bmb, strengthened by a silver wire hgh; by which means it is made strong enough to support the balls, though very light.

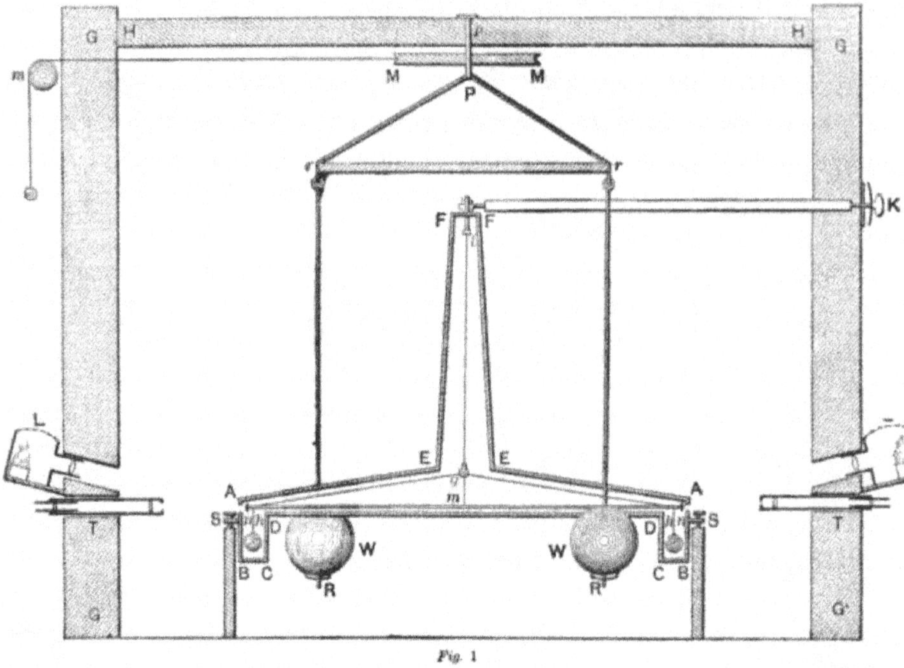

By Henry Cavendish - Cavendish,H.(1798), 'Experiments to determine the Density of the Earth' in McKenzie, A.S. ed. Scientific Memoirs Vol.9: The Laws of Gravitation, American Book Co. 1900, p.62 on Google Books., Public Domain, https://commons.wikimedia.org/w/index.php?curid=2621520.

The case is supported, and set horizontal, by four screws, resting on posts fixed firmly into the ground: two of them are represented in the figure, by S and S; the two others are not represented, to avoid confusion. GG and GG are the end walls of the building. W and W are the leaden weights; which are suspended by the copper rods RrPrR, and the wooden bar rr, from the center pin Pp .This pin passes through a hole in the beam HH, perpendicularly over the center of the instrument, and turns round in it, being prevented from falling by the plate p. MM is a pulley, fastened to this pin; and Mm, a cord wound round the pulley, and passing through the end wall; by which the observer may turn it round, and thereby move the weights from one situation to the other.

From this table it appears, that though the experiments agree pretty well together, yet the difference between them, both in the quantity of motion of the arm and in the time of vibration, is greater than can proceed merely from the error of observation. As to the difference in the motion of the arm, it may very well be accounted for, from the current of air produced by the difference of temperature; but, whether this can account for the difference in the time of vibration, is doubtful. If the current of air was regular, and of the same swiftness in all parts of the vibration of the ball, I think it could not; but, as there will most likely be much irregularity in the current, it may very likely be sufficient to account for the difference.

[The two large balls could be positioned either away from or to either side of the torsion balance rod. Their mutual attraction to the small balls caused the arm to rotate, twisting the suspension wire. The arm rotated until it reached an angle where the twisting force of the wire balanced the combined gravitational force of attraction between the large and small lead spheres. By measuring the angle of the rod and knowing the twisting force (torque) of the wire for a given angle, Cavendish was able to determine the force between the pairs of masses. Since the gravitational force of the Earth on the small ball could be measured directly by weighing it, the ratio of the two forces allowed the relative density of the Earth to be calculated, using Newton's law of gravitation.

Cavendish found that the Earth's density was 5.448 ± 0.033 times that of water (due to a simple arithmetic error, found in 1821 by Francis Baily, the erroneous value 5.480 ± 0.038 appears in his paper).] The current accepted value is 5.514 g/cm^3.

To find the wire's torsion coefficient, the torque exerted by the wire for a given angle of twist, Cavendish timed the natural oscillation period of the balance rod as it rotated slowly clockwise and counterclockwise against the twisting of the wire. For the first 3 experiments the period was about 15 minutes and for the next 14 experiments the period was half of that, about 7.5 minutes. The period changed because after the third experiment Cavendish put in a stiffer wire. The torsion coefficient could be calculated from this and the mass and dimensions of the balance. Actually, the rod was never at rest; Cavendish had to measure the deflection angle of the rod while it was oscillating.

Cavendish's equipment was remarkably sensitive for its time. The force involved in twisting the torsion balance was very small, 1.74×10^{-7} N, (the weight of only 0.0177 milligrams) or about 1⁄50,000,000 of the weight of the small balls. To prevent air currents and temperature changes from interfering with the measurements, Cavendish placed the entire apparatus in a mahogany box about 1.98 meters wide, 1.27 meters tall, and 14 cm thick, all in a closed shed on his estate. Through two holes in the walls of the shed, Cavendish used telescopes to observe the movement of the torsion balance's horizontal rod. The key observable was of course the deflection of the torsion balance rod, which Cavendish measured to be about 0.16" (or only 0.03" for the stiffer wire used mostly). Cavendish was able to measure this small deflection to an accuracy of better than 0.01 inches (0.25

mm) using vernier scales on the ends of the rod. The accuracy of Cavendish's result was not exceeded until C. V. Boys' experiment in 1895. In time, Michell's torsion balance became the dominant technique for measuring the gravitational constant (G) and most contemporary measurements still use variations of it.]

By a mean of the experiments made with the wire first used, the density of the earth comes out 5.48 times greater than that of water; and by a mean of those made with the latter wire, it comes out the same; and the extreme difference of the results of the 23 observations made with this wire, is only .75; so that the extreme results do not differ from the mean by more than .38, or 1/14 of the whole, and *therefore the density should seem to be determined hereby, to great exactness*. It, indeed, may be objected, that as the result appears to be influenced by the current of air, or some other cause, the laws of which we are not well acquainted with, this cause may perhaps act always, or commonly, in the same direction, and thereby make a considerable error in the result. But yet, as the experiments were tried in various weathers, and with considerable variety in the difference of temperature of the weights and air, and with the arm resting at different distances from the sides of the case, it seems very unlikely that this cause should act so uniformly in the same way, as to make the error of the mean result nearly equal to the difference between this and the extreme; and, therefore, it seems very unlikely that the density of the earth should differ from 5.48 by so much as 1/14 of the whole.

Another objection, perhaps, may be made to these experiments, namely, that it is uncertain whether, in these small distances, the force of gravity follows exactly the same law as in greater distances. There is no reason, however, to think that any irregularity of this kind takes place, until the bodies come within the action of what is called the attraction of cohesion, and which seems to extend only to very minute distances. With a view to see whether the result could be affected by this attraction, I made the 9th, 10th, 11th, and 15th experiments, in which the balls were made to rest as close to the sides of the case as they could; but there is no difference to be depended on, between the results under that circumstance, and when the balls are placed in any other part of the case.

According to the experiments made by Dr. Maskelyne, on the attraction of the hill Schehallien, the density of the earth is 4 ½ times that of water; which differs rather more from the preceding determination than I should have expected. But I forbear entering into any consideration of which determination is most to be depended on, till I have examined more carefully how much the preceding determination is affected by irregularities whose quantity I cannot measure.

[The formulation of Newtonian gravity in terms of a *gravitational constant* did not become standard until long after Cavendish's time. Indeed, one of the first references to G is in 1873, 75 years after Cavendish's work.

Cavendish expressed his result in terms of the density of the Earth. He referred to his experiment in correspondence as 'weighing the world'. Later authors reformulated his results in modern terms.

$$G = g\,R^2_{earth}/M_{earth} = 3g/(4\pi R_{earth}\rho_{earth}).$$

After converting to SI units, Cavendish's value for the Earth's density, 5.448 g cm^{-3}, gives

$$G = 6.74 \times 10^{-11}\ m^3\ kg^{-1}\ s^{-2},$$

which differs by only 1% from the 2014 CODATA value of 6.67408×10^{-11} m^3 kg^{-1} s^{-2}. Today, physicists often use units where the gravitational constant takes a different form. The Gaussian gravitational constant used in space dynamics is a defined constant and the Cavendish experiment can be considered as a measurement of this constant. In Cavendish's time, physicists used the same units for mass and weight, in effect taking g as a standard acceleration. Then, since R_{earth} was known, ρ_{earth} played the role of an inverse gravitational constant. The density of the Earth was hence a much sought-after quantity at the time, and there had been earlier attempts to measure it, such as the Schiehallion experiment in 1774.]

Johann Carl Friedrich Gauss (April 30, 1777 – February 23, 1855).

Guass was a German mathematician, geodesist, and physicist who made significant contributions to many fields in mathematics and science. Gauss ranks among history's most influential mathematicians. Gauss was a child prodigy in mathematics. While still a student at the University of Göttingen, he propounded several mathematical theorems. Gauss completed his masterpieces *Disquisitiones Arithmeticae and Theoria motus corporum coelestium* as a private scholar. Later he was director of the Göttingen Observatory and professor at the university for nearly half a century, from 1807 until his death in 1855.

Gauss published the second and third complete proofs of the fundamental theorem of algebra, made contributions to number theory and developed the theories of binary and ternary quadratic forms. He is credited with inventing the fast Fourier transform algorithm and was instrumental in the discovery of the dwarf planet Ceres. His work on the motion of planetoids disturbed by large planets led to the introduction of the Gaussian gravitational constant and the method of least squares, which he discovered before Adrien-Marie Legendre published on the method, and which is still used in all sciences to minimize measurement error. He also anticipated non-Euclidean geometry, and was the first to analyze it, even coining the term. He is considered one of its discoverers alongside Nikolai Lobachevsky and János Bolyai.

Johann Carl Friedrich Gauss was born on 30 April 1777 in Brunswick (Braunschweig), in the Duchy of Brunswick-Wolfenbüttel (now part of Lower Saxony, Germany), to a family of lower social status. His father Gebhard Dietrich Gauss (1744–1808) worked in several jobs, as butcher, bricklayer, gardener, and as treasurer of a death-benefit fund. Gauss characterized his father as an honorable and respected man, but rough and dominating at home. He was experienced in writing and calculating, but his wife Dorothea (1743–1839), Carl Friedrich's mother, was nearly illiterate. Carl Friedrich was christened and confirmed in a church near the school that he attended as a child. He had one elder brother from his father's first marriage.

Gauss was a child prodigy in the field of mathematics. When the elementary teachers noticed his intellectual abilities, they brought him to the attention of the Duke of Brunswick, who sent him to the local Collegium Carolinum, which he attended from 1792 to 1795 with Eberhard August Wilhelm von Zimmermann as one of his teachers. Thereafter the Duke granted him the resources for studies of mathematics, sciences, and classical languages at the Hanoverian University of Göttingen until 1798. It is not known why Gauss went to Göttingen and not to the University of Helmstedt near his native Brunswick, but it is assumed that the large library of Göttingen, where students were allowed to borrow books and take them home, was the decisive reason.

Though being a registered student at university, it is evident that he was a self-taught student in mathematics, since he independently rediscovered several theorems. He succeeded with a breakthrough in a geometrical problem that had occupied mathematicians since the days of the Ancient Greeks when he determined in 1796 which regular polygons can be constructed by compass and straightedge. This discovery was the subject of his first publication and ultimately led Gauss to choose mathematics instead of philology as a career. Gauss' mathematical diary shows that, in the same year, he was also productive in number theory. He made advanced discoveries in modular arithmetic, found the first proof of the quadratic reciprocity law, and dealt with the prime number theorem. Many ideas for his mathematical magnum opus *Disquisitiones arithmeticae*, published in 1801, date from this time.

Gauss graduated as a Doctor of Philosophy in 1799. He did not graduate from Göttingen, as is sometimes stated, but rather, at the Duke of Brunswick's special request, from the University of Helmstedt, the only state university of the duchy. The Duke then granted him his cost of living as a private scholar in Brunswick. Gauss showed his gratitude and loyalty for this bequest when he refused several calls from the Russian Academy of Sciences in St. Peterburg and from Landshut University. Later, the Duke promised him the foundation of an observatory in Brunswick in 1804. Architect Peter Joseph Krahe made preliminary designs, but one of Napoleon's wars cancelled those plans: the Duke was mortally wounded in the battle of Jena in 1806. The duchy was abolished in the following year, and Gauss's financial support stopped. He then followed a call to the University of Göttingen, an institution of the newly founded Kingdom of Westphalia under Jérôme Bonaparte, as full professor and director of the astronomical observatory.

The scientific activity of Gauss, besides pure mathematics, can be roughly divided into three periods: in the first two decades of the 19th century astronomy was the main focus, in the third decade geodesy, and in the fourth decade he occupied himself with physics, mainly magnetism.

Studying the calculation of asteroid orbits, Gauss established contact with the astronomical community of Bremen and Lilienthal, especially Wilhelm Olbers, Karl Ludwig Harding and Friedrich Wilhelm Bessel, an informal group of astronomers known as the Celestial police. One of their aims was the discovery of further planets, and they assembled data on asteroids and comets as a basis for Gauss's research. Gauss was thereby able to develop new, powerful methods for the determination of orbits, which he later published in his astronomical magnum opus *Theoria motus corporum coelestium* (1809).

Gauss arrived at Göttingen in November 1807. Gauss took on the directorate of the 60-year-old observatory, founded in 1748 by Prince-elector George II and built on a converted fortification tower, with usable, but partly out-of-date instruments. The construction of a new observatory had been approved by Prince-elector George III in principle since 1802, and the Westphalian government continued the planning, but the building was not finished until October 1816. It contained new up-to-date instruments, for instance two meridian circles from Repsold and Reichenbach, and a heliometer from Fraunhofer.

Gauss married Johanna Osthoff (1780–1809) on 9 October 1805. They had two sons and a daughter: Joseph (1806–1873), Wilhelmina (1808–1840) and Louis (1809–1810). Johanna died on 11 October 1809 one month after the birth of Louis, who himself died a few months later.

Gauss remarried within a year, on 4 August 1810, to Wilhelmine (Minna) Waldeck (1788–1831), a friend of his first wife. They had three more children: Eugen (later Eugene) (1811–1896), Wilhelm (later William) (1813–1879) and Therese Staufenau [de] (1816–1864). Minna Gauss died on 12 September 1831 after being seriously ill for more than a decade. Therese then took over the household and cared for Gauss for the rest of his life.

Gauss invented the heliotrope in 1821, a magnetometer in 1833 and, alongside Wilhelm Eduard Weber, invented the first electromagnetic telegraph in 1833.

Gauss remained mentally active into his old age, even while suffering from gout and general unhappiness. His last observation was the solar eclipse of July 28, 1851. On 23 February 1855, Gauss died of a heart attack in Göttingen; he is interred in the Albani Cemetery there.

Gauss's law for gravity.

Gauss's law for gravity, also known as Gauss's flux theorem for gravity, is a law of physics that is equivalent to Newton's law of universal gravitation. It is named after Carl Friedrich Gauss. It states that the flux (surface integral) of the gravitational field over any closed surface is proportional to the mass enclosed. *Gauss's law for gravity* is often more convenient to work from than Newton's law.

The form of *Gauss's law for gravity* is mathematically similar to *Gauss's law for electrostatics*, one of Maxwell's equations. *Gauss's law for gravity has the same mathematical relation to Newton's law that Gauss's law for electrostatics bears to Coulomb's law*. This is because both Newton's law and Coulomb's law describe inverse-square interaction in a 3-dimensional space.

Qualitative statement of the law

The *gravitational field* **g** (also called gravitational acceleration) is a vector field – a vector at each point of space (and time). It is defined so that the gravitational force experienced by a particle is equal to the mass of the particle multiplied by the gravitational field at that point.

Gravitational flux is a surface integral of the gravitational field over a closed surface, analogous to how magnetic flux is a surface integral of the magnetic field.

Gauss's law for gravity states: *The gravitational flux through any closed surface is proportional to the enclosed mass.*

The *integral form* of *Gauss's law for gravity* states:

$$\oiint_{\partial V} \mathbf{g} \cdot d\mathbf{A} = -4\pi GM$$

where

- $\oiint$ denotes a surface integral over a closed surface,
- ∂V is any closed surface (the *boundary* of an arbitrary volume V),
- $d\mathbf{A}$ is a vector, whose magnitude is the area of an infinitesimal piece of the surface ∂V, and whose direction is the outward-pointing surface normal (see surface integral for more details),
- **g** is the gravitational field,
- G is the universal gravitational constant, and
- M is the total mass enclosed within the surface ∂V.

The left-hand side of this equation is called the *flux of the gravitational field.* Note that according to the law it is always negative (or zero), and never positive. This can be contrasted with *Gauss's law for electricity*, where the flux can be either positive or negative. The difference is because *charge* can be either positive or negative, while *mass* can only be positive.

The *differential form* of *Gauss's law for gravity* states

$$\nabla \cdot \mathbf{g} = -4\pi G\rho$$

where $\nabla \cdot$ denotes divergence, G is the universal gravitational constant, and ρ is the *mass density* at each point.

Relation to the integral form

The two forms of *Gauss's law for gravity* are mathematically equivalent. The *divergence theorem* states:

$$\oiint_{\partial V} \mathbf{g} \cdot d\mathbf{A} = \int_V \nabla \cdot \mathbf{g}\, dV$$

where V is a closed region bounded by a simple closed oriented surface ∂V and dV is an infinitesimal piece of the volume V (see volume integral for more details). The *gravitational field* **g** must be a continuously differentiable vector field defined on a neighborhood of V.

Given also that $M = \int_V \rho\, dV$ we can apply the divergence theorem to the *integral form* of *Gauss's law for gravity*, from which we can obtain $\nabla \cdot \mathbf{g} = -4\pi G\rho$.

It is possible to derive the *integral form* from the *differential form* using the reverse of this method.

Deriving Gauss's law from Newton's law

Gauss's law for gravity can be derived from Newton's law of universal gravitation, which states that the gravitational field due to a point mass is:

$$\mathbf{g}(\mathbf{r}) = -GM\,\mathbf{e_r}/r^2$$

where

- $\mathbf{e_r}$ is the radial unit vector,
- r is the radius, $|\mathbf{r}|$.

- M is the mass of the particle, which is assumed to be a point mass located at the origin.

A proof using vector calculus is shown in the box below. It is mathematically identical to the proof of Gauss's law (in electrostatics) starting from Coulomb's law.

Outline of proof

$\mathbf{g}(\mathbf{r})$, the *gravitational field* at $\mathbf{r}$, can be calculated by adding up the contribution to $\mathbf{g}(\mathbf{r})$ due to every bit of mass in the universe (see *superposition principle*). To do this, we integrate over every point $\mathbf{s}$ in space, adding up the contribution to $\mathbf{g}(\mathbf{r})$ associated with the mass (if any) at $\mathbf{s}$, where this contribution is calculated by Newton's law. The result is:

$$\mathbf{g}(\mathbf{r}) = -G \int \rho(\mathbf{s})\, (\mathbf{r} - \mathbf{s}) / |\mathbf{r} - \mathbf{s}|^3\, d^3\mathbf{s}$$

($d^3\mathbf{s}$ stands for $ds_x ds_y ds_z$, each of which is integrated from $-\infty$ to $+\infty$.) If we take the divergence of both sides of this equation with respect to $\mathbf{r}$, and use the known theorem

$$\nabla \cdot (\mathbf{r}/|\mathbf{r}|^3) = 4\pi\, \delta(\mathbf{r})$$

where $\delta(\mathbf{r})$ is the Dirac delta function, the result is

$$\nabla \cdot \mathbf{g}(\mathbf{r}) = -4\pi G \int \rho(s)\, \delta(\mathbf{r} - \mathbf{s})\, d^3\mathbf{s}.$$

Using the "sifting property" of the Dirac delta function, we arrive at

$$\nabla \cdot \mathbf{g}(\mathbf{r}) = -4\pi G \rho(\mathbf{r})$$

which is the *differential form* of *Gauss's law for gravity*, as desired.

Deriving Newton's law from Gauss's law and irrotationality.

Newton's Law of Gravitation can be derived from *Gauss's law for gravity* and *irrotationality* under certain other assumptions. *It is impossible mathematically to prove Newton's law from Gauss's law alone, because Gauss's law specifies the divergence of* **g** *but does not contain any information regarding the curl of* **g**. In addition to Gauss's law, the assumption is used that **g** is *irrotational* (has zero curl), as gravity is a conservative force:

$$\nabla \times \mathbf{g} = 0$$

where $\nabla \times$ is the *curl of a vector*.

[The *curl of a vector* is a vector operator that describes the infinitesimal circulation of a vector field in three-dimensional Euclidean space. The curl at a point in the field is represented by a vector whose length and direction denote the magnitude and axis of the maximum circulation.]

Even these are not enough: *Boundary conditions on* **g** *are also necessary to prove Newton's law*, such as the assumption that the field is zero infinitely far from a mass.

The proof of Newton's law from these assumptions is as follows. Start with the integral form of Gauss's law:

$$\iint_{\delta V} g \, . \, dA = -4\pi GM.$$

Apply this law to the situation where the volume V is a sphere of radius r = |r| centered on a point-mass M located at the origin. It is reasonable to expect the gravitational field from a point mass to be spherically symmetric. By making this assumption, **g** takes the following form:

$$\mathbf{g}(r) = g(\mathbf{r})\mathbf{e}_r$$

where e_r is the radial unit vector. (i.e., the direction of **g** is parallel to the direction of **r**, and the magnitude of **g** depends only on the magnitude, not direction, of **r**). Using the fact that ∂V is a spherical surface with constant r and area $4\pi r^2$,

$$g(r) \iint_{\delta V} \mathbf{e}_r \, . \, d\mathbf{A} = -4\pi GM,$$
$$g(r) = -GM/r^2,$$
or $$\mathbf{g}(r) = -GM\,\mathbf{e}_r/r^2$$ which is Newton's law.

Poisson's equation and gravitational potential

Since the *gravitational field* has zero curl (equivalently, gravity is a conservative force) as mentioned above, it can be written as the *gradient of a scalar potential*, called the *gravitational potential*:

$$\mathbf{g} = -\nabla \varphi.$$

Then the *differential form* of *Gauss's law for gravity* becomes *Poisson's equation*:

$$\nabla^2 \varphi = 4\pi G\rho.$$

This provides an alternate means of calculating the *gravitational potential* and *gravitational field.* Although computing **g** via Poisson's equation is mathematically equivalent to computing **g** directly from Gauss's law, one or the other approach may be an easier computation in a given situation.

Baron Siméon Denis Poisson (June 21, 1781 – April 25, 1840).

Poisson was a French mathematician and physicist who worked on statistics, complex analysis, partial differential equations, the calculus of variations, analytical mechanics, electricity and magnetism, thermodynamics, elasticity, and fluid mechanics. Moreover, he predicted the Poisson spot in his attempt to disprove the wave theory of Augustin-Jean Fresnel, which was later confirmed.

Poisson was born in Pithiviers, Loiret district in France, the son of Siméon Poisson, an officer in the French army.

In 1798, he entered the École Polytechnique in Paris as first in his year, and immediately began to attract the notice of the professors of the school, who left him free to make his own decisions as to what he would study. In his final year of study, less than two years after his entry, he published two memoirs, one on Étienne Bézout's method of elimination, the other on the number of integrals of a finite difference equation and this was so impressive that he was allowed to graduate in 1800 without taking the final examination. The latter of the memoirs was examined by Sylvestre-François Lacroix and Adrien-Marie Legendre, who recommended that it should be published in the Recueil des savants étrangers, an unprecedented honor for a youth of eighteen. This success at once procured entry for Poisson into scientific circles. Joseph Louis Lagrange, whose lectures on the theory of functions he attended at the École Polytechnique, recognized his talent early on, and became his friend. Meanwhile, Pierre-Simon Laplace, in whose footsteps Poisson followed, regarded him almost as his son. The rest of his career, until his death in Sceaux near Paris, was occupied by the composition and publication of his many works and in fulfilling the duties of the numerous educational positions to which he was successively appointed.

Immediately after finishing his studies at the École Polytechnique, he was appointed répétiteur (teaching assistant) there, a position which he had occupied as an amateur while still a pupil in the school; for his schoolmates had made a custom of visiting him in his room after an unusually difficult lecture to hear him repeat and explain it. He was made deputy professor (professeur suppléant) in 1802, and, in 1806 full professor succeeding Jean Baptiste Joseph Fourier, whom Napoleon had sent to Grenoble. In 1808 he became astronomer to the Bureau des Longitudes; and when the Faculté des sciences de Paris was instituted in 1809 he was appointed a professor of rational mechanics (professeur de mécanique rationelle). He went on to become a member of the Institute in 1812, examiner at the military school (École Militaire) at Saint-Cyr in 1815, graduation examiner at the École Polytechnique in 1816, councillor of the university in 1820, and geometer to the Bureau des Longitudes succeeding Pierre-Simon Laplace in 1827.

In 1817, he married Nancy de Bardi and with her, he had four children. His father, whose early experiences had led him to hate aristocrats, bred him in the stern creed of the First Republic. Throughout the Revolution, the Empire, and the following restoration, Poisson was not interested in politics, concentrating instead on mathematics. He was appointed to the dignity of baron in 1825, but he neither took out the diploma nor used the title. In March 1818, he was elected a Fellow of the Royal Society, in 1822 a Foreign Honorary Member of the American Academy of Arts and Sciences, and in 1823 a foreign member of the Royal Swedish Academy of Sciences. The revolution of July 1830 threatened him with the loss of all his honors; but this disgrace to the government of Louis-Philippe was adroitly averted by François Jean Dominique Arago, who, while his "revocation" was being plotted by the council of ministers, procured him an invitation to dine at the Palais-Royal, where he was openly and effusively received by the citizen king, who "remembered" him. After this, of course, his degradation was impossible, and seven years later he was made a peer of France, not for political reasons, but as a representative of French science.

As a teacher of mathematics Poisson is said to have been extraordinarily successful, as might have been expected from his early promise as a répétiteur at the École Polytechnique. As a scientific worker, his productivity has rarely if ever been equaled. Notwithstanding his many official duties, he found time to publish more than three hundred works, several of them extensive treatises, and many of them memoirs dealing with the most abstruse branches of pure mathematics, applied mathematics, mathematical physics, and rational mechanics. (Arago attributed to him the quote, "Life is good for only two things: doing mathematics and teaching it.")

A list of Poisson's works, drawn up by himself, is given at the end of Arago's biography. All that is possible is a brief mention of the more important ones. It was in the application of mathematics to physics that his greatest services to science were performed. Perhaps the most original, and certainly the most permanent in their influence, were his memoirs on the theory of electricity and magnetism, which virtually created a new branch of mathematical physics.

Next (or in the opinion of some, first) in importance stand the memoirs on celestial mechanics, in which he proved himself a worthy successor to Pierre-Simon Laplace. The most important of these are his memoirs *Sur les inégalités séculaires des moyens mouvements des planètes*, *Sur la variation des constantes arbitraires dans les questions de mécanique*, both published in the Journal of the École Polytechnique (1809); *Sur la libration de la lune*, in Connaissance des temps (1821), etc.; and *Sur le mouvement de la terre autour de son centre de gravité*, in Mémoires de l'Académie (1827), etc. In the first of these memoirs, Poisson discusses the famous question of the stability of the planetary orbits, which had already been settled by Lagrange to the first degree of approximation for the disturbing forces. Poisson showed that the result could be extended to a second approximation, and thus made an important advance in planetary theory. The memoir is remarkable inasmuch as it roused Lagrange, after an interval of inactivity, to compose in his old age one of the greatest of his memoirs, entitled *Sur la théorie des variations des éléments des planètes, et en particulier des variations des grands axes de leurs orbites*. So highly did he think of Poisson's memoir that he made a copy of it with his own hand, which was found among his papers after his death. Poisson made important contributions to the theory of attraction.

As a tribute to Poisson's scientific work, which stretched to more than 300 publications, he was awarded a French peerage in 1837. His is one of the 72 names inscribed on the Eiffel Tower.

Poisson's equation.

Poisson's equation is an elliptic partial differential equation of broad utility in theoretical physics. For example, *the solution to Poisson's equation is the potential field caused by a given electric charge or mass density distribution*; with the potential field known, one can then calculate the *electrostatic* or *gravitational* (force) *field*. It is a generalization of *Laplace's* second order partial differential *equation* for potential

$$\nabla^2 \varphi = 4\pi G\rho$$

where ∇^2 is the *Laplace operator*.

[In three-dimensional Cartesian coordinates, $\nabla^2 = \partial^2/\partial x^2 + \partial^2/\partial y^2 + \partial^2/\partial z^2$.]

It was first published in the *Bulletin de la société philomatique* (1813). If $\rho = 0$, we retrieve *Laplace's equation*

$$\nabla^2 \varphi = 0.$$

If $\rho(x, y, z)$ is a continuous function and if for $r \to \infty$ (or if a point 'moves' to infinity) a function φ goes to 0 fast enough, a solution of Poisson's equation is the Newtonian potential of a function $\rho(x, y, z)$

$$\varphi = -1/4\pi \iiint \rho(x, y, z)/r \, dV$$

where is a distance between a volume element dV and a point P. The integration runs over the whole space.

Poisson's two most important memoirs on the subject are *Sur l'attraction des sphéroides* (*Connaiss. ft. temps*, 1829), and *Sur l'attraction d'un ellipsoide homogène* (*Mim. ft. l'acad.*, 1835). Poisson discovered that *Laplace's equation* is valid only outside of a solid. A rigorous proof for masses with variable density was first given by Carl Friedrich Gauss in 1839. Poisson's equation is applicable in not just gravitation, but also electricity and magnetism.

In the case of a *gravitational field* **g** due to an attracting massive object of density ρ, *Gauss's law for gravity in differential form* can be used to obtain the corresponding *Poisson equation for gravity*:

$$\nabla \cdot \mathbf{g} = -4\pi G\rho.$$

Since the *gravitational field* is conservative (and irrotational), it can be expressed in terms of a scalar potential φ:

$$\mathbf{g} = -\nabla \varphi.$$

Substituting this into Gauss's law,

$$\nabla \cdot (-\nabla \varphi) = -4\pi G\rho$$

yields *Poisson's equation for gravity*,

$$\nabla^2 \varphi = 4\pi G\rho.$$

If the mass density is zero, Poisson's equation reduces to Laplace's equation. The corresponding Green's function can be used to calculate the potential at distance *r* from a central point mass *m* (i.e., the fundamental solution). In three dimensions the potential is

$$\Phi(r) = -Gm/r$$

which is equivalent to *Newton's law of universal gravitation.*

Explanation for forward motion and direction of rotation of the planets around the Sun, and of the Moon around the Earth, and for the axial rotation of the Earth and the Sun, and the axial tilt of the Earth.

The closer a planet is to the sun, the faster it needs to move to maintain its orbit. The *average orbital speeds* of the planets in km/s are:

Mercury:	47.87
Venus:	35.02
Earth:	29.78
Mars:	24.13
Jupiter:	13.07

The speed of the Sun around the Galaxy is about 230 km/s (143 mi/s).

> [The average orbital speed *of the Earth around the Sun* in terms of the speed of light (or gravity) = 29.78/299,792 km/s = 9.93×10^{-5} c.
>
> The average orbital speed *of the Sun around the Galaxy* in terms of the speed of light (or gravity) = 200-240/299,792 km/s = $6.67\text{-}8.00 \times 10^{-4}$ c.
>
> *Consider the impact of the dramatic movement of the Earth on experiments conducted on the Earth, and of observations of the stars.*]

The most widely accepted explanation of how the solar system formed is called the *nebular hypothesis*. According to this hypothesis, the Sun and the planets of our solar system formed about 4.6 billion years ago from the collapse of a giant cloud of gas and dust, called a nebula.

The nebula was drawn together by gravity, which released gravitational potential energy. As small particles of dust and gas smashed together to create larger ones, they released kinetic energy. As the nebula collapsed, the gravity at the center increased and the cloud started to spin because of its angular momentum. As it collapsed further, the spinning got faster, much as an ice skater spins faster when he pulls his arms to his sides during a spin.

Much of the cloud's mass migrated to its center but the rest of the material flattened out in an enormous disk. The disk contained hydrogen and helium, along with heavier elements and even simple organic molecules. As gravity pulled matter into the center of the disk, the density and pressure at the center became intense. When the pressure in the center of the disk was high enough, nuclear fusion began. A star was born—the Sun. The burning star stopped the disk from collapsing further.

Meanwhile, the outer parts of the disk were cooling off. Matter condensed from the cloud and small pieces of dust started clumping together. These clumps collided and combined with other clumps. Larger clumps, called planetesimals, attracted smaller clumps with their gravity. Gravity at the center of the disk attracted heavier particles, such as rock and metal and lighter particles remained further out in the disk. Eventually, the planetesimals formed protoplanets, which grew to become the planets and moons that we find in our solar system today.

Because of the gravitational sorting of material, the inner planets — Mercury, Venus, Earth, and Mars — formed from dense rock and metal. The outer planets — Jupiter, Saturn, Uranus and Neptune — condensed farther from the Sun from lighter materials such as hydrogen, helium, water, ammonia, and methane. Out by Jupiter and beyond, where it's very cold, these materials form solid particles.

The *nebular hypothesis* was designed to explain some of the basic features of the solar system:

1. The orbits of the planets lie in nearly the same plane with the Sun at the center;
2. The planets revolve in the same direction;
3. The planets mostly rotate in the same direction;
4. The axes of rotation of the planets are mostly nearly perpendicular to the orbital plane;
5. The oldest moon rocks are 4.5 billion years.

Earth's axial tilt.

Axial tilt, also known as *obliquity*, is the angle between an object's rotational axis and its orbital axis, which is the line perpendicular to its orbital plane; equivalently, it is the angle between its equatorial plane and orbital plane. It differs from orbital inclination.

At an *obliquity* of 0 degrees, the two axes point in the same direction; that is, the rotational axis is perpendicular to the orbital plane.

The rotational axis of Earth, for example, is the imaginary line that passes through both the North Pole and South Pole, whereas the Earth's orbital axis is the line perpendicular to the imaginary plane through which the Earth moves as it revolves around the Sun; the Earth's *obliquity or axial tilt* is the angle between these two lines.

Earth's *orbital plane* is known as the *ecliptic plane*, and Earth's *tilt* is known to astronomers as the *obliquity of the ecliptic*, being *the angle between the ecliptic and the celestial equator on the celestial sphere*. It is denoted by the Greek letter ε.

Earth currently has an axial tilt of about 23.44°. This value remains about the same relative to a stationary orbital plane throughout the cycles of axial precession. But the *ecliptic* (i.e., Earth's orbit) moves due to planetary perturbations, and the obliquity of the ecliptic is not a fixed quantity. At present, it is decreasing at a rate of about 46.8″ per century.

Earth's axis remains tilted in the same direction with reference to the background stars throughout a year (regardless of where it is in its orbit) due to the gyroscope effect. This means that one pole (and the associated hemisphere of Earth) will be directed away from the Sun at one side of the orbit, and half an orbit later (half a year later) this pole will be directed towards the Sun. This is the cause of Earth's seasons. Summer occurs in the Northern hemisphere when the north pole is directed toward the Sun. Variations in Earth's axial tilt can influence the seasons and is likely a factor in long-term climatic change.

From 1984, the Jet Propulsion Laboratory's DE series of computer-generated ephemerides took over as the fundamental ephemeris of the Astronomical Almanac. *Obliquity* based on DE200, which analyzed observations from 1911 to 1979, was calculated:

$$\varepsilon = 23°26'21.448'' - 46.8150''\,T - 0.00059''\,T^2 + 0.001813''\,T^3$$

where hereafter T is Julian centuries from J2000.0.

Einstein, A. (March, 1916). Die Grundlage der allgemeinen Relativitätstheorie. (The foundation of the general theory of relativity.)

Ann. Phys., 49, 7, 769-822; http://dx.doi.org/10.1002/andp.19163540702; translation in A. Engel (translator), E. Schuckling (consultant). (1997). *The Collected Papers of Albert Einstein*, Volume 6: The Berlin Years: Writings, 1914-1917, Princeton University Press, Princeton, Doc. 30, 146-200; https://einsteinpapers.press.princeton.edu/vol6-trans/158; translation below by T. G. Underwood; also, translation by S. N. Bose at https://en.wikisource.org/wiki/The_Foundation_of_the_Generalised_Theory_of_Relativity.

Final consolidation by Einstein of his various papers on the subject - in particular, his three papers in November 1915. This was based on his conclusion in his *theory of general relativity* that space and time quantities could not be defined in such a way that spatial coordinate differences could be measured directly with the unit scale, or temporal ones with a normal clock. Einstein assumed that the general laws of nature should be expressed by equations that applied to all coordinate systems not just inertial systems, i.e. were covariant to arbitrary substitutions (generally covariant); and that the *theory of special relativity* was applicable for *infinitely small four-dimensional areas*. He assumed $ds^2 = \sum_{\mu\nu} g_{\mu\nu}\, dx_\mu\, dx_\nu$, where $g_{\mu\nu}$ is the "*fundamental tensor*", which described a curved surface, the *gravitational field*. He introduced the *extension* of the *fundamental tensor* $g_{\mu\nu}$, known as the *Riemann-Christoffel Tensor*, and equated the *equation of motion* of a freely moving body in a frame moving with uniform acceleration relative to the reference frame, i.e. along a *geodetic line* in space time, with the *equation of motion* of a material-point in a *gravitational field*. Einstein used the *field equations* of forces arising in an accelerated frame in the absence of matter, expressed in terms of the Hamiltonian, to obtain an equation corresponding to the *laws of conservation of momentum and energy*, in terms of the *energy components t_σ^α of the gravitation field*, adding an arbitrary factor -2κ, to obtain $\kappa t_\sigma^\alpha = \frac{1}{2}\, \delta_\sigma^\alpha g^{\mu\nu}\Gamma^\lambda_{\mu\beta}\, \Gamma^\beta_{\nu\lambda} - g^{\mu\nu}\Gamma^\alpha_{\mu\beta}\, \Gamma^\beta_{\nu\sigma}$,
where $\Gamma^\tau_{\mu\nu} = -\ \frac{1}{2}\, g^{\tau\alpha}\, (\partial g_{\mu\alpha}/\partial x_\nu + \partial g_{\nu\alpha}/\partial x_\mu - \partial g_{\mu\nu}/\partial x_\alpha)$. He then introduced matter into the *field equations* by adding an *energy-tensor* T_σ^α *associated with matter*, corresponding to the density ρ of Poisson's equation $\Delta\varphi = 4\pi\kappa\rho$, where φ was the gravitational potential and ρ was the density of matter, to obtain the *general field equations of gravitation* in the form
$\partial/\partial x_\alpha\, (g^{\sigma\beta}\Gamma^\alpha_{\mu\beta}) = -\,\kappa\{(t_\mu^\sigma + T_\mu^\sigma) -\ \frac{1}{2}\, \delta_\mu^\sigma\, (t + T)\}$, $(-g)^{1/2} = 1$,
or $\partial\Gamma^\alpha_{\mu\nu}/\partial x_\alpha + \Gamma^\alpha_{\mu\beta}\, \Gamma^\beta_{\nu\alpha} = -\,\kappa(T_{\mu\nu} -\ \frac{1}{2}\, g_{\mu\nu}T)$ with $(-g)^{1/2} = 1$, with the *sum of the energy components of matter and gravitation*, $t_\mu^\sigma + T_\mu^\sigma$ in place of the *energy components* t_μ^σ, where $t = t^\alpha_\alpha$, and
$T = T_\mu^\mu$ (Laue's scalar). Einstein introduced *Euler's equation of motion for a frictionless adiabatic liquid* in a *relativistic* form in which the *contravariant energy-tensor* of the liquid was
$T^{\alpha\beta} = -\ g^{\alpha\beta}p + \rho\ dx_\alpha/ds\ dx_\beta/ds$ in an attempt to provide a link between the *stress-energy tensor* defined in his *field equations* and matter. However, the force on matter in Euler's equation is much stronger, has nothing to do with the weak force of gravitational attraction between matter, *and is of opposite sign*. He then considered cases *when the velocity of the particle was very small*

compared with the speed of light and the $g_{\mu\nu}$ differed from the values in an inertial frame under special relativity only by small magnitudes so that small quantities of the second and higher orders could be neglected (his "first aspect of the approximation") and dx_1/ds, dx_2/ds, dx_3/ds could be treated as small quantities, whereas dx_4/ds was equal to 1 (his "second point of view for approximation). This reduced his *equation of motion of a particle moving along the geodesic line* from $d^2x_\tau/ds^2 = \Gamma^\tau_{\mu\nu}\, dx_\mu/ds\, dx_\nu/ds$, where $\Gamma^\tau_{\mu\nu} = -\tfrac{1}{2} g^{\tau\alpha} (\partial g_{\mu\alpha}/\partial x_\nu + \partial g_{\nu\alpha}/\partial x_\mu - \partial g_{\mu\nu}/\partial x_\alpha)$, to $d^2x_\tau/dt^2 = -\tfrac{1}{2}\, \partial g_{44}/\partial x_\tau$ ($\tau = 1, 2, 3$), which Einstein considered represented the motion of a material point according to Newton's theory, in which $g_{44}/2$ played the part of the *gravitational potential*. Under a series of approximations to the *contravariant energy-tensor* of a frictionless adiabatic liquid $T^{\alpha\beta}$, all components vanished except $T_{44} = \rho = T$, from which Einstein obtained an equation for the *gravitational potential* in terms of the integral of the density of matter divide by the distance from the center of the matter $\varphi(r) = -\kappa/8\pi \int\rho d\tau/r$, of similar form to Newton's law of gravitation $\varphi(r) = -K/c^2 \int\rho d\tau/r$. *In order to obtain a value for κ, Einstein set these two equations equal* giving $\kappa = 8\pi K/c^2 = 1.87 \times 10^{-29}$ (after correction for units), where $K = 6.7 \times 10^{-10}$ is the gravitation-constant. He noted that according to his *theory of general relativity* $ds^2 = \sum g_{\mu\nu}dx_\mu dx_\nu = 0$, determining the velocity of light, so that light-rays are bent if the $g_{\mu\nu}$ were not constant. As in Einstein (November 18, 1915), his calculation of the bending of light, was obtained from his approximations for his equation of the *geodetic line* $\sum_\alpha \partial\Gamma^\alpha_{\mu\nu}/\partial x^\alpha + \sum_{\alpha\beta} \Gamma^\alpha_{\mu\beta}\, \Gamma^\beta_{\nu\alpha} = 0$, where $\Gamma^\alpha_{\mu\nu} = -\tfrac{1}{2} \sum_\beta g^{\alpha\beta} (\delta g_{\mu\beta}/\delta x_\nu + \delta g_{\nu\beta}/\delta x_\mu - \delta g_{\mu\nu}/\delta x_\alpha)$, in which the link to the weak attractive force of gravitation was provided by *Newton's law of gravitation*. Einstein calculated the *deflection of light by the Sun* at a distance Δ, $B = 2\alpha/\Delta = \kappa M/2\pi\Delta$, by substituting $\alpha = \kappa M/4\pi$, from his equation for the *gravitational potential* $\varphi(r) = -\tfrac{1}{2}\, \alpha/r = -\kappa/8\pi \int\rho d\tau/r = -\kappa M/8\pi r$, and setting $\kappa = 8\pi K/c^2 = 1.87 \times 10^{-29}$. Consequently, as before, his computed value for the bending of light was the Newtonian value. He restated his formula for the addition to the precession of the perihelion of Mercury, but did not provide the derivation. Why anyone gave credence to this is a mystery. By 1921 Einstein was already moving his research interests into superseding general relativity.

> [Janssen, M. (2004.) Einstein's first systematic exposition of General Relativity. *PhilSci Archive*; https://philsci-archive.pitt.edu/2123/1/annalen.pdf: "In March 1916, Einstein sent his new review article, with a title almost identical to that of the one it replaced, to Wilhelm Wien, editor of the *Annalen*. This is why … , unlike the papers mentioned so far, can be found in the volume before you[10].
>
> > [10] The article is still readily available in English translation in the anthology The Principle of Relativity (Lorentz et al. 1952). Unfortunately, this reprint omits the one-page introduction to the paper in which Einstein makes a number of interesting points. He emphasizes the importance of Minkowski's geometric formulation of special relativity, which he had originally dismissed as "superfluous erudition" ("überflüssige Gelehrsamkeit;"), and the differential geometry of Riemann and

others for the development of general relativity. He also acknowledges the help of Grossmann in the mathematical formulation of the theory.

Many elements of Einstein's responses to Ehrenfest's queries ended up in this article. Even though there is no mention of the hole argument, for instance, Einstein does present the so-called "point-coincidence argument", which he had premiered in letters to Ehrenfest and Michele Besso explaining where the hole argument went wrong. The introduction of the field equations and the discussion of energy-momentum conservation in the crucial Part C of the paper—*which is very different from the corresponding Part D of Einstein, A. (1914)*. Die formale Grundlage der allgemeinen Relativitätstheorie. (The formal foundations of the general theory of relativity) —closely follows another letter to Ehrenfest, in which Einstein gave a self-contained statement of the energy-momentum considerations leading to the final version of the field equations. Initially, his readers had been forced to piece this argument together from his papers of November 1914 and 1915. As Einstein announced at the beginning of his letter to Ehrenfest: "I shall not rely on the papers at all but show you all the calculations." He closed the letter asking his friend: "Could you do me a favor and send these sheets back to me as I do not have this material so neatly in one place anywhere else." Einstein may very well have had this letter in front of him as he was writing the relevant sections of [this paper].
...

In [this paper] *the field equations and energy-momentum conservation are not developed in generally-covariant form but only in special coordinates*. Einstein had found the Einstein field equation in terms of these coordinates in November 1915. As explained above, this part of [this paper] is basically a sanitized version of the argument that had led Einstein to these equations in the first place. ...

... The 1916 review article preserves the physical considerations, especially concerning energy-momentum conservation, that originally led him to the Einstein field equations, arguably the crowning achievement of his scientific career."]

[Bacelar Valente, M. (2018). Einstein's redshift derivations: its history from 1907 to 1921. *Circumscribere: International Journal for the History of Science*, 22, 1-16: "*Einstein next redshift derivation* was made in his review paper on the general theory of relativity from 1916. After arriving at the Newtonian approximation in his theory of general relativity, Einstein considers a unit measuring rod for which $ds^2 = -1$; for a particular choice of orientation, we have $-1 = g_{11}dx_1^2$. According to Einstein, "the unit measuring-rod appears a little shortened in relation to the system of coordinates by the presence of the gravitational field, if the rod is laid along a radius"[39].

[39] Einstein, A. (1916). The Foundation of the General Theory of Relativity. *Ann. Phys.*, 49, 7, 769-822; http://dx.doi.org/10.1002/andp.19163540702, p. 197.

For the case of a unit clock "arranged to be at rest in a static gravitational field", ds = 1. (Ibid., 197.) Therefore $1 = g_{44}dx_4^2$; and so $dx_4 = 1 - (g_{44} - 1)/2$. Accordingly:

"*The clock goes more slowly if set up in the neighborhood of ponderable masses*. From this it follows that the spectral lines of light reaching us from the surface of large stars must appear displaced towards the red end of the spectrum"[41].

[41] *Ibid*., p. 198.

Again, like in the Entwurf theory, the standard or unit measuring clock is taken to be at rest in the gravitational field and being affected by it so that its physically meaningful coordinate time is changed in relation to its natural time duration ds. Again, we relate to the same clock (rod) ds and dt (dx); and *it is the change in dt that gives rise to the redshift*. The measuring clock is not considered as in "free fall" having a proper time ds but is treated as a clock from what later will be called the reference mollusk giving a coordinate time, to which is given a physical meaning. In fact, this derivation of the clock's rate is the equivalent in *general relativity* to that of the Entwurf theory.']

The theory presented below is the most far-reaching generalization of the theory commonly referred to today as the "*theory of relativity*"; In order to distinguish it from the former, I will call the latter "*special theory of relativity*" and assume that it is known. The generalization of the theory of relativity was greatly facilitated by the shape given to the *special theory of relativity* by Minkowski, who first clearly recognized the formal equivalence of spatial coordinates and the time coordinate and used this in the construction of the theory. The mathematical tools necessary for the *general theory of relativity* were ready in the "*absolute differential calculus*", which is based on the research of Gauss, Riemann and Christoffel on non-Euclidean manifolds and was brought into a system by Ricci and Levi-Civita and already applied to problems of theoretical physics. *In Section B of the present treatise, I have developed all the mathematical aids necessary for us*, which are not to be assumed to be known to the physicist, in the simplest and most transparent manner possible, so that a study of mathematical literature is not necessary for the understanding of the present treatise. Finally, at this point, I would like to thank my friend, the mathematician Grossmann, who, with his help, not only saved me the trouble of studying the relevant mathematical literature, but also supported me in my search for the field equations of gravitation.

[Einstein's starting point in the development of his field equation for *gravitation* is the Lorentz transformation on which his *theory of special relativity* was founded. From Einstein's *second postulate of relativity* (invariance of c) it follows that:

$$c^2(t_2 - t_1)^2 - (x_2 - x_1)^2 - (y_2 - y_1)^2 - (z_2 - z_1)^2 = 0$$

in all inertial frames for events connected by light signals., where the quantity on the left is called the spacetime interval between events (t_1, x_1, y_1, z_1) and (t_2, x_2, y_2, z_2). The simplest example of a Lorentzian manifold is *flat spacetime*, which can be given as $\mathbf{R}^4$ with coordinates (t, x, y, z) and the metric

$$ds^2 = - c^2dt^2 + dx^2 + dy^2 + dz^2 = \eta_{\mu\nu}\, dx^\mu dx^\nu$$

The flat space metric (or Minkowski metric) is often denoted by the symbol η and is the metric used in special relativity. In the above coordinates, the matrix representation of η is

$$\eta = \begin{matrix} -c^2 & 0 & 0 & 0 \\ 0 & 1 & 0 & 0 \\ 0 & 0 & 1 & 0 \\ 0 & 0 & 0 & 1. \end{matrix}$$

Einstein attempts to extend this metric from an inertial frame to a uniformly accelerated frame. His objective was to incorporate *gravity* into his theory whilst preserving special relativity in the case of no gravitational field.]

A. Principal considerations on the postulate of relativity.

§ 1. Observations on the special theory of relativity.

The *special theory of relativity* is based on the following postulate, which is also satisfied by Galileo-Newtonian mechanics:

If a coordinate system K is chosen in such a way that the laws of physics apply in their simplest form in relation to it, the same laws also apply to any other coordinate system K', which is in uniform translational motion relative to K. We call this postulate the "special *principle* of relativity". The word "special" is intended to imply that the *principle* is limited to the case where K' performs *a uniform translational motion* against K, but that the equivalence of K' and K does not extend to the case of non-uniform motion of K' against K. Thus, the *special theory of relativity* does not deviate from classical mechanics by the *postulate of relativity, but solely by the postulate of the constancy of the vacuum speed of light*, from which, in conjunction with the *special principle of relativity*, the relativity of simultaneity as well as the Lorentz transformation and the associated laws on the behavior of moving rigid bodies and clocks follow in a well-known way.

The modification that the theory of space and time has undergone by the *special theory of relativity* is indeed a profound one; but one important point remained untouched. According to the *special theory of relativity*, the theorems of geometry are to be interpreted directly as the laws about the possible relative positions of (resting) solid bodies, more generally the theorems of kinematics as theorems *describing the behavior of measuring bodies and clocks*. Two highlighted material points of a stationary (rigid) body always correspond to a distance of a very specific length, independent of the location and orientation of the body as well as of time; Two highlighted hand positions of a clock at rest relative to the (justified) reference frame always corresponds to a period of time of a certain length, regardless of place and time. *It will soon become apparent that the general theory of relativity cannot adhere to this simple physical interpretation of space and time.*

§ 2. The need for an extension of the relativity postulate.

Classical mechanics, and no less the *special theory of relativity*, has an epistemological defect, which, perhaps, for the first time, was clearly emphasized by E. Mach. We explain it with the following example. Two liquid bodies of the same size and type float freely in space at such a great distance from each other (and from all other masses) that only those gravitational forces which the parts of one of these bodies exert on each other need be taken into account. The distance between the bodies is unchangeable. Relative movements of the parts of one of the bodies against each other should not occur. But each mass is supposed to rotate from an observer at rest relative to the other mass around the line connecting the masses at a constant angular velocity (this is a detectable relative motion of both masses). Now we think of the surfaces of both bodies (S1 and S2) measured with the help of (relatively resting) scales; it follows that the surface of S1 is a sphere and that of S2 is an ellipsoid of rotation. We now ask: For what reason do the bodies S1 and S2 behave differently? An answer to this question can only be recognized as epistemologically satisfactory[1] if the thing given as the reason, is an *observable empirical fact*; because the law of causality only has the meaning of a statement about the world of experience if in the end only *observable facts* appear as causes and effects.

[1] Such an epistemologically satisfactory answer can, of course, still be physically inaccurate if it contradicts other experiences.

Newtonian mechanics does not give a satisfactory answer to this question. It says the following. The laws of mechanics probably apply to a space R_1, against which the body S_1 is at rest, but not to a space R_2, against which S_2 is at rest. However, the privileged Galilean space R_1, which is introduced here, is merely a *factitious* cause, not an observable thing. It is clear, then, that Newtonian mechanics does not really satisfy the requirement of causality in the case under consideration, but only apparently, by making the merely fictitious cause R_1 responsible for the observable different behavior of the bodies S_1 and S_2.

A satisfactory answer to the question raised above can only be as follows: The physical system consisting of S_1 and S_2 does not on its own show any conceivable cause to which the different behavior of S_1 and S_2 could be attributed. The cause must therefore lie outside this system. It is believed that the general laws of motion, which in particular determine the shapes of S_1 and S_2, must be such that the mechanical behavior of S_1 and S_2 must be essentially determined by distant masses, which we had not included in the system under consideration. These distant masses (and their relative movements against the bodies under consideration) are then to be regarded as carriers of in principle observable causes for the different behavior of our bodies under consideration; they take on the role of the fictitious cause R_1. Of all conceivable spaces R_1, R_2, etc., which are arbitrarily moving relative to each other, none may be regarded as preferred *a priori* if the epistemological objection presented is not to be revived. *The laws of physics must be such that they apply in relation to arbitrarily moving frames of reference. In this way, we arrive at an extension of the postulate of relativity.*

In addition to this serious epistemological argument, however, *there is also a well-known physical fact in favor of an extension of the theory of relativity*. Let K be a Galilean frame of reference, i.e. one relative to which (at least in the four-dimensional area under consideration) a mass sufficiently distant from others moves in a straight line and uniformly. Let K' be a second coordinate system, which is relative to K in uniformly accelerated translational motion. Relative to K', a mass sufficiently separated from others then carried out an accelerated motion in such a way that its acceleration and direction of acceleration is independent of its material composition and physical state.

Can an observer at rest relative to K' conclude from this that he is on a "really" accelerated frame of reference? The answer to this question is in the negative; for the above-mentioned behavior of freely moving masses relative to K' can just as well be interpreted in the following way. The reference frame K' is unaccelerated; however, there is a *gravitational field* in the time-spatial area under consideration, which generates the accelerated motion of the bodies relative to K'.

This view is made possible by the fact that experience has taught us the existence of a force field (namely, the *gravitational field*), which has the curious property of giving all bodies the same acceleration[1].

[1] That the gravitational field has this property with great accuracy has been experimentally proven by Eötvös.

The mechanical behavior of bodies relative to K' is the same as it is to experience with systems which we are accustomed to regard as "resting" or "justified" systems; therefore, from a physical point of view, it is reasonable to assume that systems K and K' can both be regarded as "stationary" with the same right, or that they are on an equal footing as reference systems for the physical description of the processes. From these considerations it can be seen that the implementation of the *general theory of relativity* must at the same time lead to a *theory of gravitation*; because you can "create" a *gravitational fi*eld by simply changing the coordinate system. *It can also be seen immediately that the principle of the constancy of the vacuum-speed of light must be modified. For it is easy to see that the trajectory of a ray of light with respect to K' must generally be a curvilinear one if the light propagates in a straight line with respect to K and at a definite constant speed.*

[These are inadequate arguments for rejecting Newtonian mechanics.]

§ 3. The space-time continuum. Requirement of general covariance for the equations expressing the general laws of nature.

In classical mechanics, as well as in *special relativity*, the coordinates of space and time have a direct physical meaning. A point event has the X_1 coordinate x_1, means: The one according to the rules of Euclidean geometry determined by means of rigid rods projection of the point event on the X_1 axis is obtained by plotting a certain member, the unit scale, x_1 times from the starting point of the coordinate body on the (positive) X_1 axis. A point has the X_4 coordinate $x_4 = t$, which means: A unit clock arranged at rest relative to the coordinate system, spatially (practically) coinciding with the point event, which is directed according to certain rules, has covered $x_4 = t$ periods when the point event occurs[1].

[1] We assume that "simultaneity" can be ascertained for spatially immediately adjacent events, or - more precisely for spatiotemporal immediate proximity (coincidence) - without giving a definition for this fundamental concept.

This conception of space and time has always been in the minds of physicists, albeit mostly unconsciously, as is clearly recognizable from the role, which play these terms in measuring physics; the reader had to take this view as the basis for the second consideration of the last paragraph in order to be able to connect a meaning with these remarks. *But we now want to show that it must be dropped and replaced by a more general one in order to be able to carry out the postulate of general relativity if the special theory of relativity applies to the borderline case of the absence of a gravitational field.*

We introduce a Galilean reference system K (x, y, z, t) in a space that is free of gravitational fields, and also *a coordinate system K' (x', y', z' t') that rotates uniformly relative to K.* Let the origins of both systems, as well as their Z-axes, permanently coincide. We want to show that for a space-time measurement in the system K', the above determination for the physical significance of lengths and times cannot be maintained. For reasons of symmetry, it is clear that a circle around the starting point in the X-Y plane of K can at the same time be understood as a circle in the X'-Y' plane of K'. We now think of measuring the circumference and diameter of this circle with a unit scale (infinitely small relative to the radius) and forming the quotient of both measurement results. If one were to compare this experiment with one relative to the Galilean system, K at rest, the number π would be obtained as a quotient.

The result of the determination executed with a scale at rest relative to K' would be a number greater than x. This is easily recognizable *if one judges the whole measurement process from the "stationary" system K and takes into account that the peripherally applied scale suffers a Lorentz shortening, but the radial scale does not. Therefore, Euclidean geometry does not apply with respect to K'*; the concept of coordinates defined above, which presupposes the validity of Euclidean geometry, thus fails with reference to the system K'. Nor is it possible to introduce in K' a time corresponding to physical needs, which is indicated by clocks of the same nature that are at rest relative to K'. To understand this, imagine that one of two clocks of the same nature is arranged in the coordinate origin and on the periphery of the circle and viewed from the "resting" system K. According to a well-known result of the *special theory of relativity*, K's starting point is that the clock arranged on the periphery of the circle is judged more slowly than the clock arranged in the starting point, because the former clock is moving, but the latter is not. An observer at the common coordinate origin, who would also be able to observe the clock located on the periphery by means of light, would therefore see the clock arranged on the periphery go slower than the clock arranged next to him. Since he will not decide to make the speed of light depend explicitly on time in the way in question, he will interpret his observation as meaning that the clock at the periphery "really" goes slower than the one arranged in the origin. He will therefore not be able to avoid defining time in such a way that the speed of a clock depends on its location. *We thus come to the conclusion: In the general theory of relativity, space and time quantities cannot be defined in such a way that spatial coordinate differences can be measured directly with the unit scale, or temporal ones with a normal clock.*

The previous method of placing coordinates in the temporal continuum in a certain way thus fails, and *there seems to be no other way to adapt coordinate systems to the four-dimensional world in such a way that a particularly simple formulation of the laws of nature could be expected when they were used.* Therefore, there is no choice but to regard all conceivable[1]

> [1] We do not want to speak here of certain restrictions which correspond to the requirement of unambiguous assignment and that of continuity.

coordinate systems as in principle equal for the description of nature. This boils down to the requirement:

The general laws of nature are to be expressed by equations that apply to all coordinate systems, i.e. are covariant to arbitrary substitutions (generally covariant).

> [A *covariant* transformation is a rule that specifies how certain entities, such as vectors or tensors, change under a change of basis. The transformation that describes the new basis vectors *as a linear combination* of the old basis vectors is defined as a *covariant transformation*. Conventionally, *indices identifying the basis vectors are placed as lower indices* and so are all entities that transform in the same way.
>
> The inverse of a covariant transformation is a *contravariant transformation.* Whenever a vector should be *invariant* under a change of basis, that is to say it should represent the same geometrical or physical object having the same magnitude and direction as before, its components must transform according to the *contravariant* rule. Conventionally, *indices identifying the components of a vector are placed as upper indices and so are all indices of entities that transform in the same way.* The sum over pairwise matching indices of a product with the same lower and upper indices are invariant under a transformation.]

It is clear that a physical theory which satisfies this postulate does justice to the *general postulate of relativity*. In any case, all substitutions also contain those that correspond to all relative movements of the (three-dimensional) coordinate systems. *That this requirement of general covariance, which deprives space and time of the last remnant of physical objectivity, is a natural requirement, is evident from the following consideration.* All our temporal observations always boil down to the determination of temporal coincidences. If, for example, the event consisted only in the movement of material points, then in the end nothing would be observable but the encounters of two or more of these points. The results of our measurements are also nothing more than the observation of such encounters of material points of our scales with other material points or coincidences between clock hands, dial points and envisaged point events taking place at the same place and at the same time.

The introduction of a reference system serves nothing more than to facilitate the description of the totality of such coincidences. Four temporal variables x_1, x_2, x_3, x_4 are assigned to the world, in such a way that each point event has a value system of the variable $x_1 .... x_4$. To two coincident point-events there corresponds to one system of values of the variables $x_1 .... x_4$; i.e. the coincidence is characterized by the identity of the coordinates. If in place of the variable $x_1 .... x_4$, we introduce arbitrary functions of the same, x_1', x_2', x_3', x_4' as a new coordinate system, so that the value systems are uniquely assigned to each other, the equality of all four coordinates is also the expression for the spatiotemporal coincidence of two point-events in the new system. Since all our physical experiences can ultimately be traced back to such coincidences, there is initially no reason to prefer certain coordinate systems over others, i.e. we arrive at the requirement of *general covariance*.

§ 4. The relation of the four co-ordinates to measurements in space and time.

Analytical term for the gravitational field.

It is not important to me in this paper to present the *general theory of relativity* as a logical system that is as simple as possible with a minimum of axioms. Rather, it is my main aim to develop this theory in such a way that the reader feels the psychological naturalness of the path taken and that the underlying conditions appear to be as secure as possible through experience. With this in mind, the premise has now been introduced:

For infinitely small four-dimensional areas, the theory of relativity in the narrower sense is applicable if the coordinates are appropriately chosen.

The acceleration state of the infinitely small ("local") coordinate system is to be chosen in such a way that a gravitational field does not occur; this is possible for an infinitesimal area. Let X_1, X_2, X_3 be the spatial coordinates; X_4 is the corresponding time coordinate, measured at an appropriate scale[1].

[1] The unit of time shall be chosen in such a way that the vacuum-light velocity - measured in the "local" coordinate system - becomes equal to 1.

These coordinates, if a rigid rod is thought of as a unit scale, have a direct physical meaning in the sense of the *special theory of relativity* given the orientation of the coordinate system. The expression

$$ds^2 = -dX_1^2 - dX_2^2 - dX_3^2 + dX_4^2 \qquad (1)$$

has then, according to the *special theory of relativity*, a value which may be obtained by space-time measurement, and which is independent of the orientation of the local coordinate system. We call ds the magnitude of the line element belonging to the infinitely adjacent points of four-dimensional space. If the ds belonging to the line element dX_1 .... dX_4 is positive, we follow Minkowski in calling it time-like; if it is negative, we call it space-like.

The "line element" under consideration or the two infinitely adjacent point events also include certain differentials dx_1 .... dx_4 of the four-dimensional coordinates of the selected reference frame. If this, as well as a "local" system of the above kind, is given for the position under consideration, then the dX_v can be represented here by definite linear homogeneous expressions of the dx_σ:

$$dX_v = \sum_\sigma \alpha_{v\tau}\, dx_\sigma \qquad (2)$$

If we substitute the expression in (1) we obtain

$$[ds^2 = \sum_v (\sum_\sigma \alpha_{v\tau}\, dx_\sigma)^2$$

or]

$$ds^2 = \sum_{\sigma\tau} g_{\sigma\tau}\, dx_\sigma\, dx_\tau \qquad (3)$$

where $g_{\sigma\tau}$ will be functions of x_σ, but will no longer depend upon the orientation and motion of the "local" co-ordinates; for ds^2 is a definite magnitude belonging to two point-events infinitely near in space and time and can be got by measurements with rods and clocks. The $g_{\sigma\tau}$ are hereto so chosen that $g_{\sigma\tau} = g_{\tau\sigma}$; the summation must be extended over all the values of σ and τ so that the sum consists of 4 x 4 terms, 12 of which are equal in pairs.

The case of the ordinary *theory of relativity* is evident from what has been considered here, if, by virtue of the special behavior of the $g_{\sigma\tau}$ in a finite domain, it is possible to choose the frame of reference in a finite domain in such a way that the $g_{\sigma\tau}$ assume the constant values

$$\begin{matrix} -1 & 0 & 0 & 0 \\ 0 & -1 & 0 & 0 \\ 0 & 0 & -1 & 0 \\ 0 & 0 & 0 & +1 \end{matrix} \qquad (4)$$

We shall see later that the choice of such coordinates is generally not possible for finite regions.

It is clear from the considerations of §§ 2 and 3 that, from the physical point of view, the quantities $g_{\sigma\tau}$ are to be regarded as those quantities which describe the *gravitational field* in relation to the chosen reference frame. Let us first assume that the *special theory of relativity* is valid for a certain four-dimensional area under consideration if the coordinates are chosen appropriately. The $g_{\sigma\tau}$ then have the values specified in (4). A free material point then moves in a straight line with respect to this system. Then if we introduce new space-time coordinates x_1 .... x_4, by an arbitrary substitution the $g_{\sigma\tau}$ in this new system will no longer be constants, but functions of space-time. At the same time, the motion of the free mass point in the new coordinates will be a curvilinear, non-uniform one, whereby this law of motion will be independent of the nature of the moving mass point. *We will therefore interpret this movement as one under the influence of a gravitational field.* We thus see the occurrence of a *gravitational field* linked to a spatiotemporal variability of the $g_{\sigma\tau}$. Even in the general case, when we are no longer able to bring about the validity of the *special theory of relativity* in a finite region with a suitable choice of coordinates, *we shall adhere to the view that the $g_{\sigma\tau}$ describe the gravitational field.*

Thus, according to the general theory of relativity, gravity plays an exceptional role compared to the other, especially the electromagnetic forces, in that the 10 functions representing the gravitational field even determine the metric properties of the four-dimensional space measured at the same time.

Einstein then extends this principle from the *flat space-time* of special relativity to *curved space-time* by expressing the square of the line element ds in terms of the "*fundamental tensor*", $ds^2 = \sum_{\mu\nu} g_{\mu\nu}\, dx_\mu\, dx_\nu$, which describes a curved surface.

B. Mathematical tools for the establishment of generally covariant equations.

Having seen above that the *general postulate of relativity* leads to the demand that the systems of equations of physics shall be covariant in the face of arbitrary substitutions of the coordinates x_1 .... x_4, we have to consider how such generally covariant equations can be obtained. We now turn to this purely mathematical task; it will be shown that the invariant ds given in equation (3) plays a fundamental role in solving it, which we have called the "line element" in accordance with Gaussian surface theory.

The basic idea of this *general theory of covariants* is as follows. Let certain things ("tensors") with respect to each coordinate system be defined by a number of space functions, which are called the "components" of the tensor. There are then certain rules according to which these components are calculated for a new coordinate system if they are known for the original system, and if the transformation linking the two systems is known. The things hereafter referred to as tensors are also characterized by the fact that the transformation equations for their components are linear and homogenic. Accordingly, all components in the new system disappear if they all disappear in the original system. *Thus, if a law of nature is formulated by zeroing all the components of a tensor, it is generally covariant*; by examining the laws of formation of tensors, we obtain the means to establish generally covariant laws.

§ 5. Contravariant and covariant four-vectors.

Contravariant four-vector. The line element is defined by the four "components" dx_ν, whose law of transformation is expressed by the equation

$$dx'_\sigma = \sum_\nu \partial x'_\sigma/\partial x_\nu \, dx_\nu \qquad (5)$$

The dx'_σ are expressed linearly and homogeneously through the dx_ν; we can therefore consider these coordinate differentials as the components of a "tensor", which we specifically call a *contravariant four-vector*. Anything that is defined with respect to the coordinate system by four quantities A^ν, and which is transformed according to the same law

$$A'^\sigma = \sum_\nu \partial x'_\sigma/\partial x_\nu \, A^\nu. \qquad (5a)$$

we also call a *contravariant four-vector*. From (5a) it follows at once that the sums $A^\sigma \pm B^\sigma$ are also components of a four-vector, if A^σ and B^σ are such. Corresponding relations hold for all "tensors" subsequently to be introduced. (Rule for the addition and subtraction of tensors.)

Covariant Four-vectors. We call four quantities A_ν the components of a *covariant four-vector*, if for any arbitrary choice of the *contravariant four-vector* B^ν

$$\sum_\nu A_\nu B^\nu = \text{Invariant} \qquad (6)$$

The law of transformation of a covariant four-vector follows from this definition. For if we replace B^ν on the right-hand side of the equation

$$\sum_\sigma A'_\sigma B'^\sigma = \sum_\nu A_\nu B^\nu$$

by the expression resulting from the inversion of (5a),

$$\sum_\sigma \partial x_\nu/\partial x'_\sigma \, B'^\sigma,$$

we obtain

$$\sum_\sigma B'^\sigma \sum_\nu \partial x_\nu/\partial x'_\sigma A_\nu = \sum_\sigma B'^\sigma A'_\sigma.$$

Since this equation is true for arbitrary values of the B'^σ, it follows that the *law of transformation* is

$$A'_\sigma = \sum_\nu \partial x_\nu/\partial x'_\sigma A_\nu. \qquad (7)$$

Note on a Simplified Way of Writing the Expressions.

A glance at the equations of this paragraph shows that there is always a summation with respect to the indices which occur twice under a sign of summation (e.g. the index ν in (5)), and only with respect to indices which occur twice. It is therefore possible, without loss of clearness, to omit the sign of summation. In its place we introduce the convention: *If an index occurs twice in one term of an expression, it is always to be summed unless the contrary is expressly stated.*

The difference between covariant and contravariant four-vectors lies in the law of transformation ((7) or (5a) respectively).

$$[A'_\sigma = \sum_\nu \partial x_\nu/\partial x'_\sigma A_\nu. \qquad (7)$$
$$A'^\sigma = \sum_\nu \partial x'_\sigma/\partial x_\nu A^\nu. \qquad (5a)]$$

Both forms are tensors in the sense of the general remark above. Therein lies their importance. Following Ricci and Levi-Civita, *we denote the contravariant character by placing the index above, the covariant by placing it below.*

§ 6 Tensors of the second and higher rank.

Contravariant tensors. If we form all the sixteen products $A^{\mu\nu}$ of the components A^μ and B^ν of two *contravariant four-vectors*

$$A^{\mu\nu} = A^\mu B^\nu \qquad (8)$$

then by (8) and (5a) $A^{\mu\nu}$ satisfies the law of transformation

$$A'^{\sigma\tau} = \partial x'_\sigma/\partial x_\mu \, \partial x'_\tau/\partial x_\nu \, A^{\mu\nu} \qquad (9)$$

We call a thing which is described relatively to any system of reference by sixteen quantities, satisfying the law of transformation (9), a *contravariant tensor of the second rank.* Not every such tensor allows itself to be formed in accordance with (8) from two four-vectors, but it is easily shown that any given sixteen $A^{\mu\nu}$ can be represented as the sums of the $A^\mu B^\nu$ of four appropriately selected pairs of *four-vectors*. Hence, we can prove nearly all the laws which apply to the *tensor of the second rank* defined by (9) in the simplest manner by demonstrating them for the *special tensors* of the type (8).

Contravariant Tensors of Any Rank. It is clear that, on the lines of (8) and (9), contravariant tensors of the third and higher ranks may also be defined with 4^3 components, and so on. In the same way it follows from (8) and (9) that the contravariant four-vector may be taken in this sense as a contravariant tensor of the first rank.

Covariant Tensors. On the other hand, if we take the sixteen products $A_{\mu\nu}$ of two *covariant four-vectors* A_μ and B_ν,

$$A_{\mu\nu} = A_\mu B_\nu, \tag{10}$$

the law of transformation for these is

$$A'_{\sigma\tau} = \partial x_\mu/\partial x'_\sigma \, \partial x_\nu/\partial x'_\tau \, A_{\mu\nu} \tag{11}$$

This law of transformation defines the *covariant tensor of the second rank.* All our previous remarks on contravariant tensors apply equally to covariant tensors.

Note. It is convenient to treat the scalar (or invariant) both as a contravariant and a covariant tensor of zero rank.

Mixed Tensors. We may also define a tensor *of the second rank* of the type

$$A_\mu{}^\nu = A_\mu B^\nu \tag{12}$$

which is covariant with respect to the index μ, and contravariant with respect to the index ν. Its law of transformation is

$$A'_\sigma{}^\tau = \partial x'_\tau/\partial x_\nu \, \partial x'_\mu/\partial x_\sigma \, A_\mu{}^\nu \tag{13}$$

Naturally there are mixed tensors with any number of indices of covariant character, and any number of indices of contravariant character. Covariant and contravariant tensors may be looked upon as special cases of mixed tensors.

Symmetrical Tensors. A *contravariant*, or a *covariant* tensor, *of the second or higher rank* is said to be symmetrical if two components, which are obtained the one from the other by the interchange of two indices, are equal. The tensor $A^{\mu\nu}$, or the tensor $A_{\mu\nu}$, is thus symmetrical if for any combination of the indices μ, ν,

$$A^{\mu\nu} = A^{\nu\mu}, \tag{14}$$

or respectively,

$$A_{\mu\nu} = A_{\nu\mu} \tag{14a}$$

...

Antisymmetrical Tensors. A *contravariant* or a *covariant* tensor *of the second, third, or fourth rank* is said to be *antisymmetrical* if two components, which are obtained the one from the other by the interchange of two indices, are equal and of opposite sign. The tensor $A^{\mu\nu}$, or the tensor $A_{\mu\nu}$ is therefore *antisymmetrical*, if always

$$A^{\mu\nu} = -A^{\nu\mu} \tag{15}$$

or respectively,

$$A_{\mu\nu} = -A_{\nu\mu} \tag{15a}$$

...

§ 7. *Multiplication of tensors.*

Outer multiplication of tensors.

...

§ 8. *Some aspects of the Fundamental Tensor $g_{\mu\nu}$.*

The covariant fundamental tensor. In the invariant expression of the square of the linear element

$$ds^2 = \sum_{\mu\nu} g_{\mu\nu}\, dx_\mu\, dx_\nu$$

dx_μ plays the role of any arbitrarily chosen *contravariant* vector, since further $g_{\mu\nu} = g_{\nu\mu}$, it follows from the considerations of the last paragraph that $g_{\mu\nu}$ is a symmetrical covariant tensor of the second rank. We call it the "*fundamental tensor*". In what follows we deduce some properties of this tensor *which, it is true, apply to any tensor of the second rank.* But *as the fundamental tensor plays a special part in our theory, which has its physical basis in the peculiar effects of gravitation,* it so happens that the relations to be developed are of importance to us only in the case of the *fundamental tensor.*

> [The *first fundamental form* is a quadratic form in the differentials of the coordinates on the surface that determines the intrinsic geometry of any surface in a neighborhood of a given point. Let the surface be defined by the equation $\mathbf{r} = \mathbf{r}(u, v)$, where u and v are coordinates on the surface, while
>
> $$d\mathbf{r} = \mathbf{r}_u du + \mathbf{r}_v dv$$

is the differential of the position vector **r** along the chosen direction du:dv of displacement from a point M to an infinitesimally close point M′.

The square of the principal linear part of the increment of the length of the arc MM′ can be expressed in terms of the square of the differential d**r**:

$$I = ds^2 = d\mathbf{r}^2 = \mathbf{r}^2{}_u du^2 + 2\mathbf{r}_u\mathbf{r}_v du dv + r^2{}_v dv^2,$$

and is called the *first fundamental form of the surface*. The coefficients in the first fundamental form are usually denoted by

$$E = \mathbf{r}^2{}_u,\quad F = (\mathbf{r}_u, \mathbf{r}_v),\quad G = \mathbf{r}^2{}_v,$$

or, in tensor symbols,

$$d\mathbf{r}^2 = g_{11}du^2 + 2g_{12}du dv + g_{22}dv^2.$$

The tensor g_{ij} is called the *first fundamental, or metric, tensor* of the surface.]

The contravariant fundamental tensor. If, in the *determinant* formed by the elements $g_{\mu\nu}$, we take the co-factor of each of the $g_{\mu\nu}$ and divide it by the *determinant* $g = | g_{\mu\nu} |$, we obtain certain quantities $g^{\mu\nu}$ ($= g^{\nu\mu}$), which as we shall demonstrate, form a *contravariant tensor*.

[The *determinant* is a scalar value that is a function of the entries of a square matrix. The determinant of a matrix A is commonly denoted det(A), det A, or |A|. Its value characterizes some properties of the matrix and the linear map represented by the matrix. The *determinant* is the unique function defined on the n × n matrices that has the four following properties.

(1) *The determinant of the identity matrix is 1*; det (I) = 1, where I is an *identity matrix*. When the *determinant is equal to one*, *the linear mapping defined by the matrix is equi-areal and orientation-preserving*; (2) the exchange of two rows multiplies the determinant by −1; (3) multiplying a row by a number multiplies the determinant by this number; (4) the determinant is *multilinear* adding to a row a multiple of another row does not change the determinant; (5) the determinant is *alternating*: whenever two columns of a matrix are identical, its determinant is 0.]

According to a known property of determinants

$$g_{\mu\nu}\, g^{\nu\mu} = \delta_\mu{}^\nu \qquad (16)$$

where the symbol δ_μ^ν means 1 or 0, depending on $\mu = \nu$ or $\mu \neq \nu$, [and $g^{\nu\mu}$ is a *contravariant tensor*].

…

It also follows from (16) that δ_μ is also a tensor, which we can call the *mixed fundamental tensor*.

…

The Determinant of the Fundamental Tensor. By the rule for the multiplication of determinants

$$| g_{\mu\alpha} g^{\alpha\nu} | = | g_{\mu\alpha} | \, | g^{\alpha\nu} |.$$

On the other hand

$$| g_{\mu\alpha} g^{\alpha\nu} | = | \delta_\mu^\nu | = 1.$$

It therefore follows that

$$| g_{\mu\nu} | \, | g^{\mu\nu} | = 1. \qquad (17)$$

The Volume Scalar. We seek first the law of transformation of the determinant $g = | g_{\mu\nu} |$. In accordance with (11)

$$[A'_{\sigma\tau} = \partial x_\mu/\partial x'_\sigma \, \partial x_\nu/\partial x'_\tau \, A_{\mu\nu} \qquad (11)]$$

$$g' = | \partial x_\mu/\partial x'_\sigma \, \partial x_\nu/\partial x'_\tau \, g_{\mu\nu} |$$

Hence, by a double application of the rule for the multiplication of determinants, it follows that

$$g' = | \partial x_\mu/\partial x'_\sigma | \, . \, | \partial x_\nu/\partial x'_\tau | \, . \, | g_{\mu\nu} | = | \partial x_\mu/\partial x'_\sigma |^2 \, g,$$

or

$$\sqrt{g'} = | \partial x_\mu/\partial x'_\sigma | \sqrt{g}$$

On the other hand, the law of transformation of the element of volume

$$d\tau = \int dx_1 \, dx_2 \, dx_3 \, dx_4$$

is, in accordance with the theorem of Jacobi,

$$d\tau' = | \partial x'_\sigma/\partial x_\mu | \, d\tau.$$

By multiplication of the last two equations, we obtain

$$\sqrt{g'}\, d\tau' = \sqrt{g}\, d\tau. \qquad (18)$$

Instead of $\sqrt{(g)}$, we introduce in what follows the quantity $\sqrt{(-g)}$, which is always real on account of the hyperbolic character of the space-time continuum. *The invariant $\sqrt{(-g)}\, d\tau$ is equal to the magnitude of the four-dimensional element of volume in the "local" system of reference, as measured with rigid rods and clocks in the sense of the special theory of relativity.*

Note on the Character of the Space-time Continuum. Our assumption that the *special theory of relativity* can always be applied to an infinitely small region, implies that ds^2 can always be expressed in accordance with (1)

$$[ds^2 = -\, dX_1^2 - dX_2^2 - dX_3^2 + dX_4^2 \qquad (1)]$$

by means of real quantities $dX_1 \ldots dX_4$. If we denote by $d\tau_0$ the "natural" element of volume dX_1, dX_2, dX_3, dX_4, then

$$d\tau_0 = \sqrt{(-g)}\, d\tau. \qquad (18a)$$

If $\sqrt{(-g)}$ were to vanish at a point of the four-dimensional continuum, it would mean that at this point an infinitely small "natural" volume would correspond to a finite volume in the co-ordinates. Let us assume that this is never the case. Then g cannot change sign. *We will assume that, in the sense of the special theory of relativity, g always has a finite negative value.* This is a hypothesis as to the physical nature of the continuum under consideration, and at the same time a convention as to the choice of coordinates.

But if $-g$ is always finite and positive, it is natural to settle the choice of coordinates a posteriori in such a way that this quantity is always equal to unity. We shall see later that by such a restriction of the choice of coordinates it is possible to achieve an important simplification of the laws of nature.

In place of (18), we then have simply $d\tau' = d\tau$, from which, in view of Jacobi's theorem, it follows that

$$|\, \partial x'_\sigma / \partial x_\mu \,| = 1. \qquad (19)$$

Thus, with this choice of coordinates, only substitutions for which the determinant is unity are permissible.

But it would be erroneous to believe that this step indicates a partial abandonment of the *general postulate of relativity*. We do not ask "What are the laws of nature which are covariant in face of all substitutions for which the determinant is unity?" but our question is "What are the generally covariant laws of nature?" It is not until we have formulated these that we simplify their expression by a particular choice of the system of reference.

The Formation of New Tensors by Means of the Fundamental Tensor.

...

Einstein defines the shortest line element ds between two points in space time in terms of the components of the fundamental tensor which describe the relationship between the square of the line element ds and the products of pairs of its contravariant vectors in 4-dimensional space time.

He then introduces the equation of the *geodetic line* in pseudo-Riemannian space in a frame moving with uniform acceleration relative to the reference frame, $d^2x_\tau/ds^2 = -\frac{1}{2} g^{\tau\alpha} (\delta g_{\mu\alpha}/\delta x_\nu + \delta g_{\nu\alpha}/\delta x_\mu - \delta g_{\mu\nu}/\delta x_\alpha) dx_\mu/ds\, dx_\nu/ds$.

§ 9. Equation of the geodetic line (or of point-motion). The motion of a particle.

As the "line element" ds is a definite magnitude independent of the co-ordinate system, we have also between two points P_1 and P_2 of a four-dimensional continuum a line for which $\int$ **ds** is an extremum (*geodetic line*), i.e., one which has got a significance independent of the choice of co-ordinates. Its equation is

$$\delta\{\int_{p1}^{p2} ds\} = 0 \, . \qquad (20)$$

[The noun *geodesic* and the adjective *geodetic* come from *geodesy*, the science of measuring the size and shape of Earth, though many of the underlying principles can be applied to any ellipsoidal geometry. In the original sense, a geodesic was the shortest route between two points on the Earth's surface. For a spherical Earth, it is a segment of a great circle. The term has since been generalized to more abstract mathematical spaces. The *geodetic line* is the shortest line that can be drawn between two points on any given curved surface so that the osculating plane of the curve at every point shall contain the normal to the surface. It is the minimum line that can be drawn on any surface between any two points.

A *geodesic* is a curve representing the shortest path between two points in a surface, or more generally in a Riemannian manifold. In a Riemannian manifold or submanifold, geodesics are characterized by the property of having vanishing geodesic curvature. A *manifold* is a topological space that locally resembles Euclidean space near each point. A *differentiable manifold* (also differential manifold) is a type of manifold that is locally similar enough to a vector space to allow one to apply calculus. In differential geometry, a *Riemannian manifold* is a real, smooth manifold equipped with a positive-definite *inner product* g_p on the tangent space at each point p. The family g_p of inner products is called a *Riemannian metric* (or Riemannian metric tensor).

An *inner product* space is a real vector space or a complex vector space with an operation called an inner product. The inner product of two vectors in the space is a scalar. Inner products allow formal definitions of intuitive geometric notions, such as lengths, angles, and orthogonality (zero inner product) of vectors. Inner product spaces generalize Euclidean vector spaces, in which the inner product is the dot product or scalar product of Cartesian coordinates.]

Carrying out the variation in the usual way, we can deduce four differential equations which define the *geodetic line*, this deduction is given here for the sake of completeness. Let λ be a function of the co-ordinates x_ν; this defines a series of surfaces which cut the required *geodetic line* as well as all neighboring lines drawn through the points P_1 and P_2. We can suppose that all such curves are given when the value of its co-ordinates x_ν are given in terms of λ. The sign δ corresponds to a passage from a point of the required geodetic to a point of the contiguous curve, both lying on the same surface λ. Then (20)

$$[\delta\{\int_{p1}^{p2} ds\} = 0 \qquad (20)]$$

can be replaced by

$$\int_{\lambda 1}^{\lambda 2} \delta w \, d\lambda = 0 \qquad (20a)$$

where

$$w^2 = g_{\mu\nu}\, dx_\mu/d\lambda \; dx_\nu/d\lambda$$

[= $ds^2/d\lambda^2$ where the square of the line element, $ds^2 = g_{\mu\nu}\, dx_\mu dx_\nu$, and $g_{\mu\nu}$ is the "fundamental tensor"]

But since

$$\delta w = 1/w \, \{½\, \partial g_{\mu\nu}/\partial x_\sigma \, dx_\mu/d\lambda \; dx_\nu/d\lambda \; \delta x_\sigma + g_{\mu\nu}\, dx_\mu/d\lambda \; \delta(dx_\nu/d\lambda)\}$$

and

$$\delta(dx_\mu/d\lambda) = d/d\lambda \, (\delta x_\nu),$$

we obtain from (20a),

$$[\int_{\lambda 1}^{\lambda 2} \delta w \, d\lambda = 0 \qquad (20a)]$$

after a partial integration

$$\int_{\lambda 1}^{\lambda 2} \kappa_\sigma \delta w \, d\lambda = 0,$$

where $\kappa_\sigma = d/d\lambda \, \{g_{\mu\nu}/w \, dx_\mu/\delta\lambda\} - ½w \, \partial g_{\mu\nu}/\partial x_\sigma \, dx_\mu/d\lambda \, dx_\nu/d\lambda$. (20b)

$[\int_{\lambda 1}^{\lambda 2} d\lambda \, [d/d\lambda \, \{g_{\mu\nu}/w \, dx_\mu/d\lambda\} - ½w \, \partial g_{\mu\nu}/\partial x_\sigma \, dx_\mu/d\lambda \, dx_\nu/d\lambda] \, \delta x_\sigma = 0$

so $\int_{\lambda 1}^{\lambda 2} \kappa_\sigma \, \delta x_\sigma \, d\lambda = 0]$

From which it follows, since the choice of δx_σ is perfectly arbitrary, that κ_σ should vanish; Then

$$\kappa_\sigma = 0 \qquad (20c)$$

are the equations of *geodetic line.*

If ds does not vanish along the *geodetic line* considered, we can choose the "length of the arc" s, measured along the geodetic line, for the parameter λ. Then w = 1, and in place of (20c) we obtain,

$$g_{\mu\nu} \, dx_\mu^2/ds^2 + \partial g_{\mu\nu}/\partial x_\sigma \, dx_\sigma/ds \, dx_\nu/ds - ½ \, \partial g_{\mu\nu}/\partial x_\sigma \, dx_\mu/ds \, dx_\nu/ds = 0$$

or by merely changing the notation suitably,

$$g_{\alpha\sigma} \, dx_\alpha^2/ds^2 + [{}_\sigma{}^{\mu\nu}] \, dx_\mu/ds \, dx_\nu/ds = 0 \qquad (20d)$$

where we have put, following Christoffel

$$[{}_\sigma{}^{\mu\nu}] = ½ \, (\partial g_{\mu\sigma}/\partial x_\nu + \partial g_{\nu\sigma}/\partial x_\mu - \partial g_{\mu\nu}/\partial x_\sigma) \qquad (21)$$

Finally, if we multiply (20d) by $g^{\sigma\tau}$ (outer multiplication with reference to τ, and inner with respect to σ) we get obtain the equations of the *geodetic line* in the form

$$d^2x_\tau/ds^2 + \{{}_\tau{}^{\mu\nu}\} dx_\mu/ds \, dx_\nu/ds = 0 \qquad (22)$$

where, following Christoffel, we have set

$$\{{}_\tau{}^{\mu\nu}\} = g^{\tau\alpha}[{}_\alpha{}^{\mu\nu}]. \qquad (23)$$

[Einstein also expresses the equation of the *geodetic line* in terms of the differential-quotient $\chi = d\psi/ds$ taken along any geodesic curve.]

§ 10. Formation of Tensors through Differentiation.

Based on the *equation of the geodetic line*, we can now easily derive the laws according to which new tensors can be formed from tensors by differentiation. This puts us in a position to establish *generally covariant differential equations*. We achieve this goal by repeatedly applying the following simple law: -

If a certain curve be given in our continuum whose points are characterized by the arc-distances as measured from a fixed point on the curve, and if further, φ be an invariant space function, then $d\varphi/ds$ is also an invariant.

The proof follows from the fact that $d\varphi$ as well as ds, are both invariants.

Since

$$d\varphi/ds = \partial\varphi/\partial x_\mu \, dx_\mu/ds$$

so that

$$\psi = \partial\varphi/\partial x_\mu \, dx_\mu/ds$$

is also an invariant for all curves which go out from a point in the continuum, i.e., for any choice of the vector dx_μ. From which follows immediately that

$$A_\mu = \partial\varphi/\partial x_\mu \qquad (24)$$

is a *covariant four-vector* (the "gradient" of φ).

According to our law, the differential-quotient

$$\chi = d\psi/ds$$

taken along any curve is likewise an invariant. Substituting the value of ψ,

$$[\chi = d(d\varphi/dx_\mu \, dx_\mu/ds)/ds]$$

we get

$$\chi = \partial^2\varphi/\partial x_\mu \partial x_\nu \, dx_\mu/ds \, dx_\nu/ds + \partial\varphi/\partial x_\mu \, d^2x_\mu/ds^2.$$

Here however we cannot at once deduce the existence of any tensor. If we however take the curve along which we have differentiated to be a *geodesic,* we obtain by substituting d^2x_ν/ds^2 from (22)

$$[d^2x_\nu/ds^2 + \{_\nu{}^{\mu\tau}\} dx_\mu/ds \, dx_\tau/ds = 0 \qquad (22)],$$

[so $\chi = \partial^2\varphi/\partial x_\mu \partial x_\nu \, dx_\mu/ds \, dx_\nu/ds - \partial\varphi/\partial x_\mu \, \{_\nu{}^{\mu\tau}\} dx_\mu/ds \, dx_\nu/ds$,
giving the equation of the *geodetic line*]

$$\chi = (\partial^2\varphi/\partial x_\mu \partial x_\nu - \partial\varphi/\partial x_\mu \, \{_\nu{}^{\mu\tau}\}) \, dx_\mu/ds \, dx_\nu/ds.$$

From the interchangeability of the differentiation with regard to μ and ν, and also according to (23)

$[\{_\tau^{\mu\nu}\} = g^{\tau\alpha}[_\alpha^{\mu\nu}]$ (23)]

and (21)

$[[_\sigma^{\mu\nu}] = \frac{1}{2}\,(\partial g_{\mu\sigma}/\partial x_\nu + \partial g_{\nu\sigma}/\partial x_\mu - \partial g_{\mu\nu}/\partial x_\sigma)$ (21)]

we see that the bracket $\{_\tau^{\mu\nu}\}$ is symmetrical with respect to μ and ν.

As we can draw a *geodetic line* in any direction from any point in the continuum, dx_μ/ds is thus a four-vector, with an arbitrary ratio of components, so that it follows from the results of § 7 [Multiplication of Tensors] that

$$A_{\mu\nu} = \partial^2\varphi/\partial x_\mu\partial x_\nu - \{_\tau^{\mu\nu}\}\ \partial\varphi/\partial x_\tau \qquad (25)$$

is a covariant tensor of the second rank. We have thus got the result that out of the covariant tensor of the first rank

$$A_\mu = \partial\varphi/\partial x_\mu \qquad (24)$$

we can get by differentiation a covariant tensor of second rank

$$A_{\mu\nu} = \partial A_\mu\delta x_\nu - \{_\tau^{\mu\nu}\}\ A_\tau. \qquad (26)$$

...

... By adding these, we have the tensor of the third rank

$$A_{\mu\nu\sigma} = \partial A_{\mu\nu}\partial x_\sigma - \{_\tau^{\sigma\mu}\}\ A_{\tau\nu} - \{_\tau^{\sigma\mu}\}\ A_{\mu\tau}. \qquad (27)$$

where we have put $A_{\mu\nu} = A_\mu\ B_\nu$.

... We call $A_{\mu\nu\sigma}$ the extension of the tensor $A_{\mu\nu}$.

...

Einstein substitutes $\Gamma^\tau_{\mu\nu} = -\ \{_\tau^{\mu\nu}\} = -\ g^{\tau\alpha}[_\alpha^{\mu\nu}] = -\ \frac{1}{2}\ g^{\tau\alpha}\ (\partial g_{\mu\alpha}/\partial x_\nu + \partial g_{\nu\alpha}/\partial x_\mu - \partial g_{\mu\nu}/\partial x_\alpha)$ in the equation of the *geodetic line* in pseudo-Riemannian space and equates the equation of the *geodetic line* with the *equation of motion* of a freely moving body in a frame moving with uniform acceleration relative to the reference frame. The negative sign appears to have been introduced in the definition of $\Gamma^\tau_{\mu\nu}$ in order to produce an equation in the form $d^2x_\tau/ds^2 = \Gamma^\tau_{\mu\nu}\ dx_\mu/ds\ dx_\nu/ds$.

§ 11 Some special cases of particular importance.

The Fundamental Tensor. We will first prove' some lemmas which will be useful hereafter. By the rule for the differentiation of determinants

$$dg = g^{\mu\nu}\, g\, dg_{\mu\nu} = -\, g_{\mu\nu}\, g\, dg^{\mu\nu} \qquad (28)$$

The last member is obtained from the last but one, if we bear in mind that

$g_{\mu\nu}\, g\, dg^{\mu'\nu} = \delta_{\mu}^{\mu'}$, so that $g_{\mu\nu}\, g^{\mu\nu} = 4$, and consequently
$g_{\mu\nu}\, g\, dg^{\mu\nu} + g^{\mu\nu}\, g\, dg_{\mu\nu} = 0$.

…

Further from $g_{\mu\nu}\, g_{\nu\sigma} = \delta_{\mu}^{\nu}$, it follows on differentiation that

$$g_{\mu\sigma}\, dg^{\nu\sigma} = -\, g^{\nu\sigma}\, dg_{\mu\sigma} \qquad (30)$$
$$g_{\mu\sigma}\, \partial g^{\nu\sigma}/\partial x_{\lambda} = -\, g^{\nu\sigma}\, \partial g_{\mu\sigma}/\partial x_{\lambda}.$$

…

The "Curl" (rotation) of a covariant four-vector. The second term in (26) is symmetrical in the indices μ and ν. Therefore $A_{\mu\nu} - A_{\nu\mu}$ is a particularly simple *anti-symmetrical* tensor. We obtain

$$B_{\mu\nu} = \partial A_{\mu}/\partial x_{\nu} - \partial A_{\nu}/\partial x_{\mu}. \qquad (36)$$

Antisymmetrical extension of a six-vector. If one applies (27) to an *antisymmetrical* tensor *of the second rank* $A_{\mu\nu}$, forming in addition the two equations created by cyclical permutations of the indices, and adding these three equations, we obtain the tensor *of the third rank*

$$B_{\mu\nu\sigma} = A_{\mu\nu\sigma} + A_{\nu\sigma\mu} + A_{\sigma\mu\nu} = \partial A_{\mu\nu}/\partial x_{\sigma} + \partial A_{\nu\sigma}/\partial x_{\mu} + \partial A_{\sigma\mu}/\partial x_{\nu} \qquad (37)$$

of which it is easy to prove that it is *antisymmetrical.*

…

The Divergence of a Six-vector. …

…

… Thus, we obtain

$$A^{\alpha} = \sqrt{(-g)}\, \partial\{\sqrt{(-g)}\, A^{\alpha\beta}\}/\partial x_{\beta} \qquad (40)$$

This is the expression for the divergence of a contravariant six-vector.

The Divergence of a Mixed Tensor of the Second Rank. …

…

$$\sqrt{(-g)}\, A_{\mu} = \partial\{\sqrt{(-g)}\, A_{\mu}^{\sigma}\}/\partial x_{\sigma} + \tfrac{1}{2}\, \partial g^{\rho\sigma}/\partial x_{\mu}\, \sqrt{(-g)}\, A_{\rho\sigma} \qquad (41b)$$

which we have to employ later on.

§ 12. The Riemann-Christoffel tensor.

Einstein introduces a new tensor, the *extension* of the *fundamental tensor* $g_{\mu\nu}$, known as the *Riemann-Christoffel Tensor*, $B^{\rho}{}_{\mu\sigma\tau} = -\,\partial/\partial x_{\tau}\,\{_{\rho}^{\mu\sigma}\} + \partial/\partial x_{\sigma}\,\{_{\rho}^{\mu\tau}\} - \{_{\alpha}^{\mu\sigma}\}\{_{\rho}^{\alpha\tau}\} + \{_{\alpha}^{\mu\tau}\}\{_{\rho}^{\alpha\sigma}\}$, where $\{_{\alpha}^{\mu\nu}\} = \frac{1}{2}\, g^{\alpha\tau}\,(\partial g_{\mu\tau}/\partial x_{\nu} + \partial g_{\nu\tau}/\partial x_{\mu} - \partial g_{\mu\nu}/\partial x_{\tau})$, by applying the rules for the formation of tensors to the *fundamental tensor*. He did this in order to focus on tensors which can be obtained from the *fundamental tensor* by differentiation alone and on account of its transformation properties, in particular that by a suitable choice of the coordinate system the components of the Riemann Tensor all vanish so that the $g_{\mu\nu}$ can be taken as constants, and *special relativity* holds.

> We now seek the tensor that can be obtained from the *fundamental tensor* of the $g_{\mu\nu}$ *alone*, by differentiation. At first sight the solution seems obvious. We place the fundamental tensor of the $g_{\mu\nu}$ in (27)
>
> $$[A_{\mu\nu\sigma} = \partial A_{\mu\nu}\partial x_{\sigma} - \{_{\tau}^{\sigma\mu}\}\, A_{\tau\nu} - \{_{\tau}^{\sigma\mu}\}\, A_{\mu\tau}. \qquad (27)]$$
>
> instead of any given tensor $A_{\mu\nu}$, and thus have a new tensor, namely, the extension of the fundamental tensor. But we easily convince ourselves that this extension vanishes identically. We reach our goal, however, in the following way. In (27) place
>
> $$A_{\mu\nu} = \partial A_{\mu}\partial x_{\nu} - \{_{\rho}^{\mu\nu}\}\, A_{\rho},$$
>
> i.e. *the extension of the four-vector* A_{μ}. Then (with a somewhat different naming of the indices) we get the tensor *of the third rank*
>
> $$A_{\mu\sigma\tau} = \ldots - \ldots - \ldots - \ldots + [- \ldots + \ldots + \ldots]\, A_{\rho}$$
>
> This expression suggests forming the tensor $A_{\mu\sigma\tau} - A_{\mu\tau\sigma}$. For, if we do so, the following terms of the expression for $A_{\mu\sigma\tau}$ cancel those of $A_{\mu\tau\sigma}$, the first, the fourth, and the member corresponding to the last term in square brackets; because all these are symmetrical in σ and τ. The same holds good for the sum of the second and third terms. Thus, we obtain
>
> $$A_{\mu\sigma\tau} - A_{\mu\tau\sigma} = B^{\rho}{}_{\mu\sigma\tau}\, A_{\alpha}. \qquad (42)$$
>
> where
>
> $$B^{\rho}{}_{\mu\sigma\tau} = -\,\partial/\partial x_{\tau}\,\{_{\rho}^{\mu\sigma}\} + \partial/\partial x_{\sigma}\,\{_{\rho}^{\mu\tau}\} - \{_{\alpha}^{\mu\sigma}\}\{_{\rho}^{\alpha\tau}\} + \{_{\alpha}^{\mu\tau}\}\{_{\rho}^{\alpha\sigma}\} \qquad (43)$$
>
> [where $\{_{\alpha}^{\mu\nu}\} = \frac{1}{2}\, g^{\alpha\tau}\,(\partial g_{\mu\tau}/\partial x_{\nu} + \partial g_{\nu\tau}/\partial x_{\mu} - \partial g_{\mu\nu}/\partial x_{\tau})$.]
>
> The essential feature of the result is that on the right side of (42) the A_{ρ} occur alone, without their derivatives. From the tensor character of $A_{\mu\sigma\tau} - A_{\mu\tau\sigma}$ in conjunction with the fact that A_{ρ} is an arbitrary vector, it follows, by reason of § 7 [Multiplication of Tensors], that $B^{\rho}{}_{\mu\sigma\tau}$ *is a tensor (the Riemann-Christoffel tensor).*

The mathematical importance of this tensor is as follows:

If the continuum is of such a nature that there is a co-ordinate system with reference to which the guv are constants, then all the $B^{\rho}_{\mu\sigma\tau}$ vanish.

If we choose any new system of coordinates in place of the original ones, the $g_{\mu\nu}$ referred thereto will not be constants, but in consequence of its tensor nature, the transformed components of $B^{\rho}_{\mu\sigma\tau}$ will still vanish in the new system. Thus, *the vanishing of the Riemann tensor is a necessary condition that, by an appropriate choice of the system of reference, the $g_{\mu\nu}$ may be constants.* In our problem this corresponds to the case in which*, with a suitable choice of the system of reference, the *special theory of relativity* holds good for a finite region of the continuum.

* The mathematicians have proved that this is also a sufficient condition.

Contracting (43) with respect to the indices τ and ρ we obtain the *covariant tensor of second rank*

$$G_{\mu\nu} = B^{\rho}_{\mu\sigma\tau} = R_{\mu\nu} + S_{\mu\nu}$$

where

$$R_{\mu\nu} = - \partial/\partial x_{\alpha} \{_{\alpha}{}^{\mu\alpha}\} + \{_{\beta}{}^{\mu\tau}\}\{_{\alpha}{}^{\nu\beta}\} \qquad (44)$$
$$S_{\mu\nu} = \partial^2 \log\sqrt{(-g)}/\partial x_{\mu}\partial x_{\nu} - \{_{\alpha}{}^{\mu\nu}\}\partial \log\sqrt{(-g)}/\partial x_{\alpha}.$$

[and g is the determinant $|g_{\mu\nu}|$ of $g_{\mu\nu}$, and $\{_{\alpha}{}^{\mu\nu}\} = \frac{1}{2} g^{\alpha\tau} (\partial g_{\mu\tau}/\partial x_{\nu} + \partial g_{\nu\tau}/\partial x_{\mu} - \partial g_{\mu\nu}/\delta x_{\tau})$].

Note on the Choice of Co-ordinates. It has already been observed in § 8 [*Some aspects of the Fundamental Tensor $g_{\mu\nu}$.*], in connection with equation (18a),

$$[d\tau_0 = \sqrt{(-g)}\, d\tau. \qquad (18a)]$$

that the choice of co-ordinates may with advantage be made so that $\sqrt{(-g)} = 1$. A glance at the equations obtained in the last two sections shows that by such a choice the laws of formation of tensors undergo an important simplification. This applies particularly to $G_{\mu\nu}$, the tensor just developed, which plays a fundamental part in the theory to be set forth. For this specialization of the choice of coordinates brings about the vanishing of $S_{\mu\nu}$, so that the tensor $G_{\mu\nu}$, reduces to $R_{\mu\nu}$.

On this account I shall hereafter give all relations in the simplified form which this specialization of the choice of coordinates brings with it. It will then be an easy matter to revert to the *generally* covariant equations, if this seems desirable in a special case.

C. Theory of the Gravitational Field.

§ 13. Equation of motion of a material point in a gravitation-field. Expression for the field-components of gravitation.

A freely moving body not acted on by external forces moves, according to the *special theory of relativity*, along a straight line and uniformly. This also holds for the *general theory of relativity* for any part of the four-dimensional region, in which the co-ordinates K_0 can be, and are, so chosen that $g_{\mu\nu}$ have special constant values of the expression (4).

$$\begin{bmatrix} -1 & 0 & 0 & 0 \\ 0 & -1 & 0 & 0 \\ 0 & 0 & -1 & 0 \\ 0 & 0 & 0 & +1 \end{bmatrix}. \qquad (4)$$

If we look at this motion from an arbitrarily chosen coordinate system K_1, the body, observed from K_1, moves, according to the considerations of § 2 *in a gravitational field.* The *law of motion* with reference to K_1 easily results from the following consideration. With reference to K_0, the *law of motion* is a four-dimensional straight line and thus a *geodesic* [the shortest line between two points on a mathematically defined surface]. As a *geodetic-line* is defined independently of the system of co-ordinates, it would also be the *law of motion* for the motion of the material-point with reference to K_1.

Einstein substitutes $\Gamma^{\tau}_{\mu\nu}$ for $-\{_{\tau}^{\mu\nu}\}$ in the covariant tensor of the second rank $R_{\mu\nu}$, obtained by the reduction of the Riemann-Christoffel Tensor $B^{\rho}_{\mu\sigma\tau}$, and describes these as the "components of the gravitational field". He proposes that $R_{\mu\nu} = 0$, resulting in the *field equations* for the force field *in a uniformly accelerated frame in the absence of matter* in the form $\partial\Gamma^{\alpha}_{\mu\nu}/\partial x_{\alpha} + \Gamma^{\alpha}_{\mu\beta}\,\Gamma^{\beta}_{\nu\alpha} = 0$, or $2\,\partial g^{\alpha\tau}\,(\partial g_{\mu\tau}/\partial x_{\nu} + \partial g_{\nu\tau}/\partial x_{\mu} - \partial g_{\mu\nu}/\partial x_{\tau})/\partial x_{\alpha} - g^{\alpha\tau}\,(\partial g_{\mu\tau}/\partial x_{\beta} + \partial g_{\beta\tau}/\partial x_{\mu} - \partial g_{\mu\beta}/\partial x_{\tau})\,(\partial g_{\nu\tau}/\partial x_{\alpha} + \partial g_{\alpha\tau}/\partial x_{\nu} - \partial g_{\nu\alpha}/\partial x_{\tau}) = 0$.

If we put

$$\Gamma^{\tau}_{\mu\nu} = -\,\{_{\tau}^{\mu\nu}\} \qquad (45),$$

[where $\{_{\tau}^{\mu\nu}\} = \{_{\tau}^{\mu\nu}\} = g^{\tau\alpha}[_{\alpha}^{\mu\nu}]$]

[in the final form of the equation of the *geodetic line*

$$d^2x_{\tau}/ds^2 + \{_{\tau}^{\mu\nu}\}dx_{\mu}/ds\; dx_{\nu}/ds = 0 \qquad (22)]$$

[$\quad d^2x_{\tau}/ds^2 = -\,\tfrac{1}{2}\,g^{\tau\alpha}\,(\partial g_{\mu\alpha}/\partial x_{\nu} + \partial g_{\nu\alpha}/\partial x_{\mu} - \partial g_{\mu\nu}/\partial x_{\alpha})\;dx_{\mu}/ds\; dx_{\nu}/ds$]

we get the motion of the point [*in a reference frame K_1 which is moving with uniform acceleration relative to K_0*] with reference to K_1 given by

$d^2x_\tau/ds^2 = \Gamma^\tau{}_{\mu\nu}\, dx_\mu/ds\, dx_\nu/ds$ (46)

[where $\Gamma^\tau{}_{\mu\nu} = -\, \{_\tau{}^{\mu\nu}\}$ (45),

$\{_\tau{}^{\mu\nu}\} = g^{\tau\alpha}[_\alpha{}^{\mu\nu}]$ (23),

and $[_\alpha{}^{\mu\nu}] = ½\, (\partial g_{\mu\alpha}/\partial x_\nu + \partial g_{\nu\alpha}/\partial x_\mu - \partial g_{\mu\nu}/\partial x_\alpha)$ (21),

so $\Gamma^\tau{}_{\mu\nu} = -\, ½\, g^{\tau\alpha}\, (\partial g_{\mu\alpha}/\partial x_\nu + \partial g_{\nu\alpha}/\partial x_\mu - \partial g_{\mu\nu}/\partial x_\alpha)$].

Einstein then equates the *equation of motion* of a freely moving body in a frame moving with uniform acceleration relative to the reference frame, i.e. along a *geodetic line* in space time, with the *equation of motion* of a *material-point* in a gravitational field. *He assumes that the $\Gamma^\tau{}_{\mu\nu}$ which arise in a uniformly accelerating frame are the components of the gravitational field.* Unlike Newton, Einstein, by equating acceleration of the frame of reference and the gravitational field, ignores the existence of acceleration of any body by any means other than gravitation, including by electrical or magnetic forces. In contrast, Newton goes to great lengths to discuss the effects of centripetal forces in general in Book I of *Principia* before applying this to gravitation in particular in Book 3.

We now make the very simple *assumption* that this general covariant system of equations defines also the *motion* of the point in the *gravitational field*, when there exists no reference-system K_0, with reference to which the special relativity theory holds throughout a finite region. The assumption seems to us to be all the more legitimate, as (46)

[$d^2x_\tau/ds^2 = \Gamma^\tau{}_{\mu\nu}\, dx_\mu/ds\, dx_\nu/ds$ (46)

or $d^2x_\tau/ds^2 = -\, ½\, g^{\tau\alpha}\, (\partial g_{\mu\alpha}/\partial x_\nu + \partial g_{\nu\alpha}/\partial x_\mu - \partial g_{\mu\nu}/\partial x_\alpha)\, dx_\mu/ds\, dx_\nu/ds$]

contains only the *first* derivatives of the $g_{\mu\nu}$, among which there is no relation in the special case when K_0 exists[1].

[1] According to § 12, it is only between the second (and first) derivatives that the relationships $B^\rho{}_{\mu\sigma\tau} = 0$ exist.

If $\Gamma^\tau{}_{\mu\nu}$ vanish, the point moves uniformly and in a straight line; *these magnitudes therefore determine the deviation from uniformity. They are the components of the gravitational field.*

§ 14. The Field-equation of Gravitation in the absence of matter.

In the following, we differentiate "gravitation-field" from "matter", in the sense that everything besides the gravitation-field will be signified as matter; therefore, *the term includes not only "matter" in the usual sense, but also the electro-dynamic field.*

Our next problem is to seek the field-equations of gravitation in the absence of matter. For this we apply the same method as employed in the foregoing paragraph for the deduction of the *equations of motion* for material points. A special case in which the field-equations sought-for are evidently satisfied is that of the *special theory of relativity* in which $g_{\mu\nu}$ have certain constant values. This would be the case in a certain finite region with reference to a definite co-ordinate system K_0. With reference to this system, all the components $B^{\rho}_{\mu\sigma\tau}$ of the Riemann's Tensor vanish. These vanish then also in the region considered, with reference to every other co-ordinate system.

The equations of the *gravitation-field* free from matter [field equations] must thus be in every case satisfied when all $B^{\rho}_{\mu\sigma\tau}$ vanish. But this condition is clearly one which goes too far. For it is clear that the *gravitation-field* generated by a material point in its own neighborhood can never be transformed away by any choice of axes, i.e., it cannot be transformed to a case of constant $g_{\mu\nu}$.

Therefore, it is clear that, for a gravitational field free from matter, we require that the symmetrical tensors $B^{\rho}_{\mu\sigma\tau}$ deduced from the tensors $B_{\mu\nu}$ should vanish. We thus get 10 equations for 10 quantities $g_{\mu\nu}$, which are fulfilled in the special case when $B^{\rho}_{\mu\sigma\tau}$ all vanish.

When referred to the special coordinate-system that we have chosen, and taking into consideration (44),

$$[B_{\mu\nu} = R_{\mu\nu} + S_{\mu\nu} \qquad (44)$$

$$R_{\mu\nu} = -\,\partial/\partial x_\alpha \{{}_\alpha{}^{\mu\nu}\} + \{{}_\beta{}^{\mu\alpha}\}\{{}_\alpha{}^{\nu\beta}\}$$

$$S_{\mu\nu} = \partial^2 \log(-g)^{1/2}/\partial x_\mu\, \partial x_\nu - \{{}_\alpha{}^{\mu\nu}\}\, \partial \log(-g)^{1/2}/\partial x_\alpha$$

where g is the determinant $|g_{\mu\nu}|$ of $g_{\mu\nu}$, and

$$\{{}_\alpha{}^{\mu\nu}\} = \tfrac{1}{2}\, g^{\alpha\tau}\,(\partial g_{\mu\tau}/\partial x_\nu + \partial g_{\nu\tau}/\partial x_\mu - \partial g_{\mu\nu}/\delta x_\tau)]$$

and substituting

$$\Gamma^{\tau}_{\mu\nu} = -\,\{{}_\tau{}^{\mu\nu}\} \qquad (45)$$

[so

$$\{{}_\alpha{}^{\mu\nu}\} = -\,\Gamma^{\alpha}_{\mu\nu}$$

$$\{{}_\beta{}^{\mu\alpha}\} = -\,\Gamma^{\beta}_{\mu\alpha}$$

$$\{{}_\alpha{}^{\nu\beta}\} = -\,\Gamma^{\alpha}_{\nu\beta}$$

in $R_{\mu\nu} = \delta/\delta x_\alpha\, \Gamma^{\alpha}_{\mu\nu} + \Gamma^{\beta}_{\mu\alpha}\, \Gamma^{\alpha}_{\nu\beta}$, where $R_{\mu\nu} = 0$],

we see that in the absence of matter the *field equations* [for the motion of the point *in a frame moving with uniform acceleration* relative to the reference frame] come out as follows;

$$\delta\Gamma^{\alpha}_{\mu\nu}/\delta x_\alpha + \Gamma^{\alpha}_{\mu\beta}\, \Gamma^{\beta}_{\nu\alpha} = 0 \qquad (47)$$

$$(-g)^{1/2} = 1$$

It can also be shown that there is only a minimum of arbitrariness in the choice of these equations. For besides $G_{\mu\nu}$, there is no tensor of the second rank, which can be formed from the $g_{\mu\nu}$ and its derivatives, which contains no derivatives higher than second, and is also linear in them[1].

[1] Actually, this can only be said of the tensor $G_{\mu\nu} + \lambda\, g_{\mu\nu}\, g^{\alpha\beta}\, G_{\alpha\beta}$, where λ is a constant. However, if you set this tensor = 0, you come back to the equations $G_{\mu\nu} = 0$.

It will be shown that these equations arising in a purely mathematical way out of the conditions of the *general relativity*, together with the equation of the motion [of a point in a frame, which is moving with uniform acceleration relative to the reference frame] (46)

$$[d^2x_\tau/ds^2 = \Gamma^\tau_{\mu\nu}\, dx_\mu/ds\, dx_\nu/ds \qquad (46)]$$

where the "invariant ds is the "line-element",

$$\Gamma^\tau_{\mu\nu} = -\tfrac{1}{2}\, g^{\tau\alpha}\, (\delta g_{\mu\alpha}/\delta x_\nu + \delta g_{\nu\alpha}/\delta x_\mu - \delta g_{\mu\nu}/\delta x_\alpha),$$

and $(-g)^{1/2} = 1]$

give us the *Newtonian law of attraction* as a first approximation, and *lead in the second approximation to the explanation of the perihelion-motion of mercury discovered by Leverrier* (as it remains after corrections for perturbation have been made). *My view is that these are convincing proofs of the physical correctness of the theory.*

§ 15. Hamiltonian Function for the Gravitation-field. Laws of Momentum and Energy.

Einstein then expressed the *field equations* of forces arising in an accelerated frame in the absence of matter in Hamiltonian form, $\delta\{\int H\, d\tau\} = 0$, where $H = g^{\mu\nu}\Gamma^\alpha_{\mu\beta}\, \Gamma^\beta_{\nu\alpha}$.

In order to show that the *field equations* [*of forces arising in an accelerated frame in the absence of matter*] correspond to the *laws of momentum and energy*, it is most convenient to write them [the *field equations*] in the following Hamiltonian form: —

$$\delta\{\int H\, d\tau\} = 0 \qquad (47a)$$

$$H = g^{\mu\nu}\Gamma^\alpha_{\mu\beta}\, \Gamma^\beta_{\nu\alpha}$$

$$(-g)^{1/2} = 1,$$

where, on the boundary of the finite four-dimensional region of integration-space, which we have in view, the variations vanish.

It is first necessary to show that the form (47a) is equivalent to (47).

$$[\qquad \partial\Gamma^\alpha_{\mu\nu}/\partial x_\alpha + \Gamma^\alpha_{\mu\beta}\, \Gamma^\beta_{\nu\alpha} = 0 \qquad (47)$$

where the motion of the point in a frame which is moving with uniform acceleration relative to the reference frame is given by

$$d^2x_\tau/ds^2 = \Gamma^\tau_{\mu\nu}\, dx_\mu/ds\, dx_\nu/ds \qquad (46)$$

and $\Gamma^{\tau}{}_{\mu\nu} = -\frac{1}{2} g^{\tau\alpha} (\partial g_{\mu\alpha}/\partial x_\nu + \partial g_{\nu\alpha}/\partial x_\mu - \partial g_{\mu\nu}/\partial x_\alpha)$ and $(-g)^{1/2} = 1$].

For this purpose, we regard H as a function of the $g^{\mu\nu}$ and the $g_\sigma{}^{\mu\nu}$ ($= \partial g^{\mu\nu}/\partial x_\sigma$). Then in the first place

$$\delta H = \Gamma^{\alpha}{}_{\mu\beta}\, \Gamma^{\beta}{}_{\nu\alpha}\, \delta g^{\mu\nu} + 2g^{\mu\nu}\, \Gamma^{\alpha}{}_{\mu\beta}\, \delta\Gamma^{\beta}{}_{\nu\alpha}$$
$$= -\Gamma^{\alpha}{}_{\mu\beta}\, \Gamma^{\beta}{}_{\nu\alpha}\, \delta g^{\mu\nu} + 2\Gamma^{\alpha}{}_{\mu\beta}\, \delta(g^{\mu\nu}\Gamma^{\beta}{}_{\nu\alpha}).$$

But

$$\delta(g^{\mu\nu}\Gamma^{\beta}{}_{\nu\alpha}) = -\tfrac{1}{2}\, [g^{\mu\nu}\, g^{\beta\lambda}\, (\partial g_{\nu\lambda}/\partial x_\alpha + \partial g_{\alpha\lambda}/\partial x_\nu - \partial g_{\alpha\nu}/\partial x_\lambda)].$$

The terms arising from the last two terms in round brackets are of different sign, and result from each other (since the denomination of the summation indices is immaterial) through interchange of the indices μ and β. They cancel each other in the expression for δH, because they are multiplied by the quantity $\Gamma^{\alpha}{}_{\mu\beta}$, which is *symmetrical* with respect to the indices μ and β. Thus, there remains only the first term in round brackets to be considered, so that, taking (31)

$$[dg^{\mu\nu} = -g^{\mu\alpha}\, g^{\nu\beta}\, dg_{\alpha\beta} \qquad (31)$$
$$\partial g^{\mu\nu}/\partial x_\sigma = -g^{\mu\alpha}\, g^{\nu\beta}\, \partial g_{\alpha\beta}/\partial x_\sigma]$$

into account, we obtain

$$\delta H = -\Gamma^{\alpha}{}_{\mu\beta}\, \Gamma^{\beta}{}_{\nu\alpha}\, \delta g^{\mu\nu} + \Gamma^{\alpha}{}_{\mu\beta}\, \delta g_\alpha{}^{\mu\beta}.$$

Therefore

$$\partial H/\partial g^{\mu\nu} = -\Gamma^{\alpha}{}_{\mu\beta}\, \Gamma^{\beta}{}_{\nu\alpha} \qquad (48)$$
$$\partial H/\partial g_\sigma{}^{\mu\nu} = -\Gamma^{\sigma}{}_{\mu\nu}$$

If we now carry out the variations in (47a)

$$[\delta\{\textstyle\int H\, d\tau\} = 0 \qquad (47a)$$
$$H = g^{\mu\nu}\Gamma^{\alpha}{}_{\mu\beta}\, \Gamma^{\beta}{}_{\nu\alpha}\,]$$

we obtain the system of equations,

$$\partial/\partial x_\alpha\, (\partial H/\partial g_\sigma{}^{\mu\nu}) - \partial H/\partial g^{\mu\nu} = 0 \qquad (47b)$$

which, owing to the relations (48) agrees with (47)

$$[\partial\Gamma^{\alpha}{}_{\mu\nu}/\partial x_\alpha + \Gamma^{\alpha}{}_{\mu\beta}\, \Gamma^{\beta}{}_{\nu\alpha} = 0 \qquad (47)]$$

as was required to be proved.

Einstein uses the *field equations* of forces arising in an accelerated frame in the absence of matter expressed in terms of the Hamiltonian to obtain an equation $\partial t_\sigma{}^\alpha/\partial x_\alpha = 0$ corresponding to the laws of conservation of momentum and energy, in terms of the *energy components* $t_\sigma{}^\alpha$ of the force field, where $t_\sigma{}^\alpha = g_\sigma{}^{\mu\nu}\, \partial H/\partial g_\alpha{}^{\mu\nu} - \delta_\sigma{}^\alpha H = \frac{1}{2}\, g^{\mu\nu}\Gamma^{\alpha}{}_{\mu\beta}\, \Gamma^{\beta}{}_{\nu\sigma} - \frac{1}{4}\, \delta_\sigma{}^\alpha g^{\mu\nu}\Gamma^{\lambda}{}_{\mu\beta}\, \Gamma^{\beta}$. Einstein denoted the

magnitudes t_σ^α to be the energy components of the *gravitation field, and introduced an arbitrary factor (– 2κ) to this expression* for the energy-components resulting in
$-2\kappa t_\sigma^\alpha = g_\sigma^{\mu\nu}\, \partial H/\partial g_\alpha^{\mu\nu} - \delta_\sigma^\alpha H$.

If we multiply (47b)

$$[\partial/\partial x_\alpha\, (\partial H/\partial g_\sigma^{\mu\nu}) - \partial H/\partial g^{\mu\nu} = 0 \qquad (47b)]$$

by $g_\sigma^{\mu\nu}$, then since

$$\partial g_\sigma^{\mu\nu}/\partial x_\alpha = \partial g_\alpha^{\mu\nu}/\partial x_\sigma$$

and, consequently,

$$g_\sigma^{\mu\nu}\, \partial/\partial x_\alpha\, (\partial H/\partial g_\sigma^{\mu\nu}) = \partial/\partial x_\alpha\, (g_\sigma^{\mu\nu}\, \partial H/\partial g_\sigma^{\mu\nu}) - \partial H/\partial g_\alpha^{\mu\nu}\, \partial g_\alpha^{\mu\nu}/\partial x_\sigma$$

we obtain the equation

$$\partial/\partial x_\alpha\, (g_\sigma^{\mu\nu}\, \partial H/\partial g_\alpha^{\mu\nu}) - \partial H/\partial x_\sigma = 0$$

or[1] $\partial t_\sigma^\alpha/\partial x_\alpha = 0$ (49)

$$-2\kappa t_\sigma^\alpha = g_\sigma^{\mu\nu}\, \partial H/\partial g_\alpha^{\mu\nu} - \delta_\sigma^\alpha H$$

[where $t_\sigma^\alpha = g_\sigma^{\mu\nu}\, \partial H/\partial g_\alpha^{\mu\nu} - \delta_\sigma^\alpha H$]

[or removing the factor – 2κ,

$$t_\sigma^\alpha = g_\sigma^{\mu\nu}\, \partial H/\partial g_\alpha^{\mu\nu} - \delta_\sigma^\alpha H \qquad (49^*)]$$

[1] The reason for the introduction of the factor – 2κ will become clear later.

where, on account of (48)

$$[\partial H/\partial g^{\mu\nu} = -\Gamma^\alpha_{\mu\beta}\, \Gamma^\beta_{\nu\alpha} \qquad (48)$$

$$\partial H/\partial g_\sigma^{\mu\nu} = -\Gamma^\sigma_{\mu\nu}]$$

and

$$H = g^{\mu\nu}\Gamma^\alpha_{\mu\beta}\, \Gamma^\beta_{\nu\alpha}]$$

and the second equation of (47),

$$[\partial \Gamma^\alpha_{\mu\nu}/\partial x_\alpha + \Gamma^\alpha_{\mu\beta}\, \Gamma^\beta_{\nu\alpha} = 0 \qquad (47)$$

$$\text{so } \partial g_{\mu\nu}/\partial x_\sigma = g^{\mu\tau}\, \Gamma^\nu_{\tau\sigma} + g^{\nu\tau}\, \Gamma^\mu_{\tau\sigma}]$$

and (34)

$$[\partial g_{\mu\nu}/\partial x_\sigma = -(g^{\mu\tau}\{_\nu{}^{\tau\sigma}\} + g^{\nu\tau}\{_\mu{}^{\tau\sigma}\}), \qquad (34)$$

$$\text{where } \Gamma^\tau_{\mu\nu} = -\{_\tau{}^{\mu\nu}\} \qquad (45)$$

$$\text{so} \quad \{_\nu{}^{\tau\sigma}\} = -\Gamma^\nu_{\tau\sigma} \text{ and } \{_\mu{}^{\tau\sigma}\} = -\Gamma^\mu_{\tau\sigma}$$

$$\text{so} \quad \partial g_{\mu\nu}/\partial x_\sigma = g^{\mu\tau}\, \Gamma^\nu_{\tau\sigma} + g^{\nu\tau}\, \Gamma^\mu_{\tau\sigma},$$

$$\text{and} \quad \partial t_\sigma^\alpha/\partial x_\alpha = 0 \qquad (49)]$$

$$\kappa t_\sigma{}^\alpha = \tfrac{1}{2}\, \delta_\sigma{}^\alpha g^{\mu\nu}\Gamma^\lambda{}_{\mu\beta}\, \Gamma^\beta{}_{\nu\lambda} - g^{\mu\nu}\Gamma^\alpha{}_{\mu\beta}\, \Gamma^\beta{}_{\nu\sigma} \qquad (50)$$

[or removing the factor -2κ,

$$t_\sigma{}^\alpha = \tfrac{1}{2}\, g^{\mu\nu}\Gamma^\alpha{}_{\mu\beta}\, \Gamma^\beta{}_{\nu\sigma} - \tfrac{1}{4}\, \delta_\sigma{}^\alpha g^{\mu\nu}\Gamma^\lambda{}_{\mu\beta}\, \Gamma^\beta{}_{\nu\lambda}. \qquad (50^*)]$$

It is to be noticed that $t_\sigma{}^\alpha$ is not a tensor, so that the equation (49) holds only for systems for which [the units are chosen so that] $(-g)^{1/2} = 1$.

This equation expresses the laws of conservation of momentum and energy in a gravitation-field [in an accelerated frame in the absence of matter]. In fact, the integration of this equation over a *three-dimensional* volume V leads to the four equations

$$d/dx_4\, \{\int t_\sigma{}^4\, dV\} = \int (l t_\sigma{}^1 + m t_\sigma{}^2 + n t_\sigma{}^3)\, dS \qquad (49a)$$

where *l*, m, n are the direction-cosines of the inward drawn normal to the surface-element dS in the sense of Euclidean geometry. *We recognize in this the usual expression for the laws of conservation.* We denote the magnitudes $t_\sigma{}^\alpha$ as the *energy-components* of the gravitation-field [*force field arising in an accelerated frame in the absence of matter*].

Einstein then expresses the *field equations* of forces arising in an accelerated frame in the absence of matter in terms of the energy components $t_\mu{}^\sigma$ of the force field arising in an accelerated frame in the absence of matter, $\partial/\partial x_\alpha\, (g^{\sigma\beta}\Gamma^\alpha{}_{\mu\beta}) = -\kappa\, (t_\mu{}^\sigma - \tfrac{1}{2}\, \delta_\mu{}^\sigma\, t)$, where $t = t^\alpha{}_\alpha$.

I will now put [the *field equations*] (47) in a third form which will be very serviceable for a quick realization of our object.

By multiplying the field-equations (47)

$$[\delta\Gamma^\alpha{}_{\mu\nu}/\delta x_\alpha + \Gamma^\alpha{}_{\mu\beta}\, \Gamma^\beta{}_{\nu\alpha} = 0 \qquad (47)$$
$$(-g)^{1/2} = 1]$$

by $g^{\nu\sigma}$, these are obtained in the "mixed" form. If we remember that

$$g^{\nu\sigma}\, \partial\Gamma^\alpha{}_{\mu\nu}/\partial x_\alpha = \partial/\partial x_\alpha\, (g^{\nu\sigma}\Gamma^\alpha{}_{\mu\nu}) - \partial g^{\nu\sigma}/\partial x_\alpha\, \Gamma^\alpha{}_{\mu\nu}$$

which owing to (34),

$$[\quad \partial g_{\mu\nu}/\partial x_\sigma = -\, (g^{\mu\tau}\{_\nu{}^{\tau\sigma}\} + g^{\nu\tau}\{_\mu{}^{\tau\sigma}\}) \qquad (34)$$
$$\text{or} \quad \partial g^{\nu\sigma}/\partial x_\alpha = -\, (g^{\nu\tau}\{_\sigma{}^{\tau\alpha}\} + g^{\sigma\tau}\{_\nu{}^{\tau\alpha}\}),]$$

is equal to

$$\partial/\partial x_\alpha\, (g^{\nu\sigma}\Gamma^\alpha{}_{\mu\nu}) - g^{\nu\beta}\, \Gamma^\sigma{}_{\alpha\beta}\, \Gamma^\alpha{}_{\mu\nu} - g^{\alpha\beta}\, \Gamma^\nu{}_{\beta\alpha}\, \Gamma^\alpha{}_{\mu\nu}$$

or (with different symbols for the summation indices)

$$\partial/\partial x_\alpha (g^{\alpha\beta}\Gamma^\alpha_{\mu\nu}) - g^{mn}\,\Gamma^\sigma_{m\beta}\,\Gamma^\beta_{n\mu} - g^{\nu\sigma}\,\Gamma^\alpha_{\mu\beta}\,\Gamma^\beta_{\nu\alpha}.$$

The third member of this expression [$- g^{\sigma\tau}\,\Gamma^\nu_{\tau\alpha}\Gamma^\alpha_{\mu\nu}$] cancels with that arising from the second member of the field-equations (47) [$g^{\nu\sigma}\,\Gamma^\alpha_{\mu\beta}\,\Gamma^\beta_{\nu\alpha}$]. In place of the second term of this expression [$- g^{mn}\,\Gamma^\sigma_{m\beta}\,\Gamma^\beta_{n\mu}$], we can, on account of the relations (50)

[$\kappa t_\sigma^{\ \alpha} = \tfrac{1}{2}\,\delta_\sigma^{\ \alpha} g^{\mu\nu}\Gamma^\lambda_{\mu\beta}\,\Gamma^\beta_{\nu\lambda} - g^{\mu\nu}\Gamma^\alpha_{\mu\beta}\,\Gamma^\beta_{\nu\sigma}$ (50)]

put

$$\kappa(t^\sigma_{\ \nu} - \tfrac{1}{2}\,\delta_\mu^{\ \sigma}\, t),$$

where $t = t^\alpha_{\ \alpha}$.

[or removing the factor $- 2\kappa$,
$t_\sigma^{\ \alpha} = \tfrac{1}{2}\, g^{\mu\nu}\Gamma^\alpha_{\mu\beta}\,\Gamma^\beta_{\nu\sigma} - \tfrac{1}{4}\,\delta_\sigma^{\ \alpha} g^{\mu\nu}\Gamma^\lambda_{\mu\beta}\,\Gamma^\beta_{\nu\lambda}$ (50*)
put $-\tfrac{1}{2}\, t^\sigma_{\ \nu} + \tfrac{1}{4}\,\delta_\mu^{\ \sigma}\, t$]

Therefore, in the place of [the *field equations*] (47)
[$\delta\Gamma^\alpha_{\mu\nu}/\delta x_\alpha + \Gamma^\alpha_{\mu\beta}\,\Gamma^\beta_{\nu\alpha} = 0$ (47)
$(-g)^{1/2} = 1$]
we obtain [the *field equations* in a third form]

$$\partial/\partial x_\alpha (g^{\sigma\beta}\Gamma^\alpha_{\mu\beta}) = - \kappa(t_\mu^{\ \sigma} - \tfrac{1}{2}\,\delta_\mu^{\ \sigma}\, t) \qquad (51)$$
$$(-g)^{1/2} = 1$$

where the $t_\mu^{\ \sigma}$ are the *energy-components* of the gravitation-field (*force field arising in an accelerated frame in the absence of matter*), and $t = t^\alpha_{\ \alpha}$.

[or removing the factor $- 2\kappa$,
$\partial/\partial x_\alpha (g^{\sigma\beta}\Gamma^\alpha_{\mu\beta}) = \tfrac{1}{2}\, t^\sigma_{\ \nu} - \tfrac{1}{4}\,\delta_\mu^{\ \sigma}\, t$ (51*)
where $\Gamma^\tau_{\mu\nu} = - \tfrac{1}{2}\, g^{\tau\alpha}\,(\partial g_{\mu\alpha}/\partial x_\nu + \partial g_{\nu\alpha}/\partial x_\mu - \partial g_{\mu\nu}/\partial x_\alpha)$,
and $(-g)^{1/2} = 1$.]

§ 16. The general formulation of the Field Equations of Gravitation.

Einstein claims to introduce matter into the field equations by adding an *energy-tensor* $T_\sigma^{\ \alpha}$ *associated with matter*, "corresponding to the density ρ of Poisson's equation $\Delta\varphi = 4\pi\kappa\rho$, where φ is the gravitational potential, and ρ is the density of matter, to the energy components $t_\mu^{\ \sigma}$ of the force field in an accelerated frame in the absence of matter, which like the energy components $t_\sigma^{\ \alpha}$ can be connected with symmetrical covariant tensors.

The *field-equations* established in the preceding paragraph for spaces free from matter are to be compared with the equation

$\nabla^2\varphi = 0$

of the *Newtonian theory* [where φ is the *gravitational potential*].

We have now to find the equations which correspond to *Poisson's Equation*

$\nabla^2\varphi = 4\pi\kappa\rho$

where ρ signifies the *density of matter*.

The *theory of special relativity* has led to the conception that the *inertial mass* is no other than *energy*, which finds its complete mathematical expression in a symmetrical tensor of the second rank, the *energy-tensor*.

We have therefore to introduce in our *generalized theory* an *energy-tensor* $T_\sigma{}^\alpha$ *associated with matter*, which like the *energy components* $t_\sigma{}^\alpha$ *of the gravitational field* [*the force field in an accelerated frame in the absence of matter*] (equations 49,

$$[\partial t_\sigma{}^\alpha/\partial x_\alpha = 0 \qquad (49)$$
$$-2\kappa t_\sigma{}^\alpha = g_\sigma{}^{\mu\nu}\,\partial H/\partial g_\alpha{}^{\mu\nu} - \delta_\sigma{}^\alpha H$$
$$\text{where } t_\sigma{}^\alpha = g_\sigma{}^{\mu\nu}\,\partial H/\partial g_\alpha{}^{\mu\nu} - \delta_\sigma{}^\alpha H]$$

and 50)

$$[\kappa t_\sigma{}^\alpha = \tfrac{1}{2}\,\delta_\sigma{}^\alpha g^{\mu\nu}\Gamma^\lambda{}_{\mu\beta}\,\Gamma^\beta{}_{\nu\lambda} - g^{\mu\nu}\Gamma^\alpha{}_{\mu\beta}\,\Gamma^\beta{}_{\nu\sigma} \qquad (50)]$$

[or removing the factor – 2κ,

$$t_\sigma{}^\alpha = g_\sigma{}^{\mu\nu}\,\partial H/\partial g_\alpha{}^{\mu\nu} - \delta_\sigma{}^\alpha H \qquad (49^*)$$
$$t_\sigma{}^\alpha = \tfrac{1}{2}\,g^{\mu\nu}\Gamma^\alpha{}_{\mu\beta}\,\Gamma^\beta{}_{\nu\sigma} - \tfrac{1}{4}\,\delta_\sigma{}^\alpha g^{\mu\nu}\Gamma^\lambda{}_{\mu\beta}\,\Gamma^\beta{}_{\nu\lambda} \qquad (50^*)]$$

have a mixed character but will pertain to a *symmetrical covariant* tensor[1].

[1] $g_{\sigma\tau}T_\sigma{}^\alpha = T_{\sigma\tau}$ and $g^{\sigma\beta}T_\sigma{}^\alpha = T^{\alpha\beta}$ *shall be symmetrical tensors.*

Einstein claims that the total mass depends on the total energy of the system, presumably including the energy associated with the force field in an accelerated frame in the absence of matter. It is yet to be seen how the energy-tensor $T_\sigma{}^\alpha$ associated with matter, corresponds to the density ρ of Poisson's equation.

The *field equation* (51)

$$[\partial/\partial x_\alpha\,(g^{\sigma\beta}\Gamma^\alpha{}_{\mu\beta}) = -\,\kappa(t_\mu{}^\sigma - \tfrac{1}{2}\,\delta_\mu{}^\sigma\, t) \qquad (51)$$
$$(-g)^{1/2} = 1$$

where $t = t^\alpha{}_\alpha$],

[or removing the factor – 2χ,

$$\partial/\partial x_\alpha (g^{\sigma\beta}\Gamma^\alpha_{\mu\beta}) = \tfrac{1}{2} t^\sigma_\nu - \tfrac{1}{4} \delta_\mu^\sigma t \qquad (51^*)]$$

shows us how to introduce the *energy-tensor* (corresponding to the energy tensor in Poisson's equation) in the *field equations of gravitation*. If we consider a complete system (for example the Solar-system) its total mass, as also its total gravitating action, will depend on the total energy of the system, ponderable as well as gravitational. *This can be expressed, by introducing into [the field equations] (51), in place of energy components t_μ^σ of gravitation-field alone, the sum of the energy components of matter and gravitation*, i. e. $t_\mu^\sigma + T_\mu^\sigma$. We thus get instead of (51)

$$[\partial/\partial x_\alpha (g^{\sigma\beta}\Gamma^\alpha_{\mu\beta}) = -\kappa(t_\mu^\sigma - \tfrac{1}{2} \delta_\mu^\sigma t) \qquad (51)]$$

[or removing the factor -2κ,

$$\partial/\partial x_\alpha (g^{\sigma\beta}\Gamma^\alpha_{\mu\beta}) = \tfrac{1}{2} t^\sigma_\nu - \tfrac{1}{4} \delta_\mu^\sigma t \qquad (51^*)]$$

the tensor-equation

$$\partial/\partial x_\alpha (g^{\sigma\beta}\Gamma^\alpha_{\mu\beta}) = -\kappa\{(t_\mu^\sigma + T_\mu^\sigma) - \tfrac{1}{2} \delta_\mu^\sigma (t + T)\} \qquad (52)$$

$$(-g)^{1/2} = 1,$$

[or removing the factor -2κ,

$$\partial/\partial x_\alpha (g^{\sigma\beta}\Gamma^\alpha_{\mu\beta}) = \tfrac{1}{2} (t_\mu^\sigma + T_\mu^\sigma) - \tfrac{1}{4} \delta_\mu^\sigma (t + T), \qquad (52^*)]$$

where we have set $T = T_\mu^\mu$ (Laue's scalar),

$$[t = t^\alpha_\alpha, \text{ and } \Gamma^\tau_{\mu\nu} = -\tfrac{1}{2} g^{\tau\alpha} (\partial g_{\mu\alpha}/\partial x_\nu + \partial g_{\nu\alpha}/\partial x_\mu - \partial g_{\mu\nu}/\partial x_\alpha)].$$

These are the *general field-equations of gravitation* in the mixed form.

Working back from these, we have in place of (47) [the *field equations* in the form in which in the absence of matter, when referred to the special coordinate-system we have chosen]

$$[\partial\Gamma^\alpha_{\mu\nu}/\partial x_\alpha + \Gamma^\alpha_{\mu\beta}\, \Gamma^\beta_{\nu\alpha} = 0 \qquad (47)$$

the system [the *field equations* with the sum of the energy components of matter and gravitation, i. e. $t_\mu^\sigma + T_\mu^\sigma$ in place of the energy components t_μ^σ of *the force field in an accelerated frame in the absence of matter* alone],

$$\partial\Gamma^\alpha_{\mu\nu}/\partial x_\alpha + \Gamma^\alpha_{\mu\beta}\, \Gamma^\beta_{\nu\alpha} = -\kappa(T_{\mu\nu} - \tfrac{1}{2} g_{\mu\nu} T) \qquad (53)$$

$$(-g)^{1/2} = 1.$$

[or removing the factor -2κ,

$$\partial\Gamma^\alpha_{\mu\nu}/\partial x_\alpha + \Gamma^\alpha_{\mu\beta}\, \Gamma^\beta_{\nu\alpha} = \tfrac{1}{2} T_{\mu\nu} - \tfrac{1}{4} g_{\mu\nu} T). \qquad (53^*)]$$

Einstein claims that the strongest ground for the introduction of the *energy-tensor* of matter in the *field equations* in an accelerated frame in the absence of matter is that they lead to equations expressing the *conservation of the momentum and energy components of total energy*. Einstein has not yet managed to get mass into the components of the *energy-tensor* T_σ^α associated with matter; or to generate a flux from which an inverse square law can be derived.

> It must be admitted, that *this introduction of the energy-tensor of matter cannot be justified by means of the relativity-postulate alone.* For this reason, we have in the foregoing analysis deduced it from the condition that the energy of the gravitation-field [*of an accelerated frame in the absence of matter*] should exert gravitating action in the same way as every other kind of energy. The strongest ground for the choice of the above equation however lies in this, that they lead, as their consequences, to equations expressing the *conservation of the components of total energy* (momentum and energy) which exactly correspond to the equations (49) and (49a).
>
> [$\partial t_\sigma^\alpha / \partial x_\alpha = 0$ (49),
>
> and
>
> $d/dx_4 \{\int t_\sigma^4 \, dV\} = \int (l t_\sigma^1 + m t_\sigma^2 + n t_\sigma^3) \, dS$ (49a)
>
> where *l*, m, n are the direction-cosines of the inward drawn normal to the surface-element dS in the Euclidean sense.]
>
> This shall be shown in § 17.
>
> ***§ 17. The Laws of Conservation in the general case.***
>
> Equation (52)
>
> [$\partial/\partial x_\alpha (g^{\sigma\beta} \Gamma^\alpha_{\mu\beta}) = -\kappa\{(t_\mu^\sigma + T_\mu^\sigma) - \tfrac{1}{2} \delta_\mu^\sigma (t + T)\}$ (52)]
>
> is easy to transform so that the second term on the right-hand side is omitted. Contract (52) with respect to the indices μ and σ, and after multiplying the resulting expression by $\tfrac{1}{2} \delta_\mu^\sigma$, subtract it from the equation (52) thus obtained. This gives
>
> $$\partial/\partial x_\alpha (g^{\sigma\beta} \Gamma^\alpha_{\mu\beta}) - \tfrac{1}{2} \delta_\mu^\sigma g^{\lambda\beta} \Gamma^\alpha_{\lambda\beta}) = -\kappa\{(t_\mu^\sigma + T_\mu^\sigma) \qquad (52a)$$
>
> …
>
> From (55) and (52a), it follows that
>
> $$\partial(t_\mu^\sigma + T_\mu^\sigma)/\partial x_\sigma = 0. \qquad (56)$$
>
> Thus, it results from our *field equations of gravitation* that the laws of conservation of momentum and energy are satisfied. This may be seen most easily from the consideration which leads to equation (49a);
>
> $$[d/dx_4 \{\int t_\sigma^4 \, dV\} = \int (l t_\sigma^1 + m t_\sigma^2 + n t_\sigma^3) \, dS \qquad (49a)]$$
>
> except that here, instead of the energy components t^σ of the gravitational field, *we have to introduce the totality of the energy components of matter and gravitational field.*

§ 18. The Laws of Momentum and Energy for Matter, as a consequence of the Field Equations.

Multiplying (53)

$$[\partial\Gamma^{\alpha}_{\mu\nu}/\partial x_\alpha + \Gamma^{\alpha}_{\mu\beta}\,\Gamma^{\beta}_{\nu\alpha} = -\,\kappa(T_{\mu\nu} - \tfrac{1}{2}\,g_{\mu\nu}T) \qquad (53)$$

$$(-g)^{1/2} = 1]$$

by $\partial g^{\mu\nu}/\partial x_\sigma$, we obtain, by the method adopted in § 15, in view of the vanishing of

$$g_{\mu\nu}\,\partial g^{\mu\nu}/\partial x_\sigma,$$

the equation

$$\partial t_\sigma{}^\alpha/\partial x_\alpha + \tfrac{1}{2}\,\partial g^{\mu\nu}/\partial x_\sigma\,T_{\mu\nu} = 0,$$

or, in view of (56),

$$[\partial(t_\mu{}^\sigma + T_\mu{}^\sigma)/\partial x_\sigma = 0, \qquad (56)]$$

$$\partial T_\sigma{}^\alpha/\partial x_\alpha + \tfrac{1}{2}\,\partial g^{\mu\nu}/\partial x_\sigma\,T_{\mu\nu} = 0. \qquad (57)$$

Comparison with (41b)

$$[\sqrt{(-g)}\,A_\mu = \partial\{\sqrt{(-g)}\,A_\mu{}^\sigma\}/\partial x_\sigma + \tfrac{1}{2}\,\partial g^{\rho\sigma}/\partial x_\mu\,\sqrt{(-g)}\,A_{\rho\sigma} \qquad (41b)]$$

shows that with the choice of system of co-ordinates which we have made, this equation predicates nothing more or less than *the vanishing of the divergence of the material energy-tensor*. Physically, the occurrence of the second term on the left-hand side shows that *laws of conservation of momentum and energy do not apply in the strict sense for matter alone*, or else that *they apply only when the $g^{\mu\nu}$ are constant, i.e. when the field intensities of gravitation vanish*. This second term is an expression for momentum, and for energy, as transferred per unit of volume and time from the gravitational field to matter. This is brought out still more clearly by re-writing (57) in the sense of (41) as

$$\partial T_\sigma{}^\alpha/\partial x_\alpha = -\,T_{\alpha\sigma}{}^\beta\,T_\beta{}^\alpha \qquad (57a)$$

The right side expresses *the energetic effect of the gravitational field on matter*.
Thus, the *field equations of gravitation* contain four conditions which govern the course of material phenomena. They give the equations of material phenomena completely, if the latter is capable of being characterized by four differential equations independent of one another*.

* On this question cf. Hilbert, H. (1915). *Nachr. d. K. Gesellsch. d. Wiss. zu Gottingen, Math.-phys. Klasse*, p. 3.

D. Material Phenomena.

The mathematical tools developed under B enable us to generalize the physical laws of matter (hydrodynamics, Maxwell's electrodynamics), as formulated in the *special theory of relativity*, in such a way that they fit into the *general theory of relativity*. The general *principle of relativity* does not result in any further restriction of possibilities; but it teaches us to know exactly the influence of the *gravitational field* on all processes, without the need to introduce any new hypothesis.

This state of affairs means that the physical nature of matter (in the strict sense) does not necessarily require the introduction of definite assumptons. In particular, *the question may remain open as to whether or not the theory of the electromagnetic field and the gravitational field together provide a sufficient basis for the theory of matter*. The *general postulate of relativity* cannot, in principle, teach us anything about this. As the theory is expanded, it remains to be seen whether *electromagnetics* and *gravitational theory* together can achieve what the former alone is unable to do.

§ 19. Euler's equations for frictionless adiabatic liquid.

Einstein introduces *Euler's equation of motion for a frictionless adiabatic liquid* in a *relativistic* form in an abortive attempt to provide a link between the *stress-energy tensor* defined in his *field equations* and matter. The problem with this is that by omitting $-\ g_{\mu\nu}p$ (see ***§ 21*** below, in the paragraph following equation 67) in his approximation, Einstein removed the connection to body *accelerations* [force fields] (per unit mass) acting on the continuum, including that to the weak *attractive* gravitational force. The remaining force on matter in Euler's equation is much stronger, has nothing to do with the weak force of gravitational attraction between matter, and is of opposite sign.

Let p and ρ, be two scalars, of which the first denotes the "pressure" and the last the "density" of the liquid; between them there is a relation. Let the contravariant symmetrical tensor

$$T^{\alpha\beta} = -\ g^{\alpha\beta}p + \rho\ dx_\alpha/ds\ dx_\beta/ds \qquad (58)$$

be the contravariant *energy-tensor* of the liquid. To it also belongs the covariant tensor

$$T_{\mu\nu} = -\ g_{\mu\nu}p + g_{\mu\alpha}\ dx_\alpha/ds\ g_{\mu\beta}\ dx_\beta/ds\ \rho \qquad (58a)$$

as well as the mixed tensor[1]

$$T_\sigma{}^\alpha = -\ \delta_\sigma{}^\alpha p + g_{\alpha\beta}\ dx_\beta/ds\ dx_\alpha/ds\ \rho \qquad (58b)$$

[1] For a co-moving observer, who uses for an infinitely small region a reference frame in the sense of *special theory of relativity*, the energy-density T_4^4 is equal to ρ – p. This gives the definition of ρ. Thus, ρ is not constant for an incompressible fluid.

[In fluid dynamics, the *Euler equations* are a set of quasilinear hyperbolic equations governing adiabatic and inviscid flow. They are named after Leonhard Euler. The equations represent Cauchy equations of conservation of mass (continuity), and balance of momentum and energy, and can be seen as particular Navier–Stokes equations with zero viscosity and zero thermal conductivity.

... the Euler equations originally formulated in convective form (also called "Lagrangian form") can also be put in the "conservation form" (also called "Eulerian form"). The *convective form* emphasizes changes to the state in a frame of reference moving with the fluid. The *conservation form* emphasizes the mathematical interpretation of the equations as conservation equations through a control volume fixed in space, and is the most important for these equations also from a numerical point of view.

The *Euler equations* first appeared in published form in Euler's article "Principes généraux du mouvement des fluides", published in *Mémoires de l'Académie des Sciences de Berlin* in 1757 (in this article Euler actually published only the general form of the *continuity equation* and the *momentum equation*; the *energy balance equation* would be obtained a century later). They were among the first partial differential equations to be written down. At the time Euler published his work, the system of equations consisted of the momentum and continuity equations, and thus was underdetermined except in the case of an incompressible fluid. An additional equation, which was later to be called the adiabatic condition, was supplied by Pierre-Simon Laplace in 1816. During the second half of the 19th century, it was found that the equation related to the balance of energy must at all times be kept, while the adiabatic condition is a consequence of the fundamental laws in the case of smooth solutions.

Euler equations.

In differential *convective form* (i.e., the form with the convective operator made explicit in the *momentum equation*), the compressible (and most general) Euler equations can be written with the *material derivative notation* (the *material derivative* describes the time rate of change of some physical quantity of a material element that is subjected to a space-and-time-dependent macroscopic velocity field):

$D\rho/Dt = -\, \nabla . \mathbf{u}$
$D\mathbf{u}/Dt = -\, \nabla p/\rho + \mathbf{g}$
$De/Dt = -\, p/\rho\, \nabla . \mathbf{u}$

where:

- ρ is the *density* of the fluid,
- p is the mechanical *pressure*,
- e is the specific internal *energy* (internal energy per unit mass),
- **u** is the *flow velocity* vector, with components in an N-dimensional space $u_1, u_2, \ldots, u_N$,
- w is the specific (with the sense of per *unit mass*) *thermodynamic work*, the internal source term,
- **g** represents body *accelerations* [force fields] (per unit mass) acting on the continuum, for example gravity, inertial accelerations, electric field acceleration, and so on,
- **∇** denotes the gradient with respect to space,
- **.** denotes the scalar product,
- **∇** is the nabla operator, used in the second equation to represent the specific thermodynamic work gradient, and
- **∇ . u** is the *flow velocity divergence.*

The equations above thus represent conservation of mass, momentum, and energy. *Mass density, flow velocity* and *pressure* are the so-called convective variables (or physical variables, or Lagrangian variables), while *mass density, momentum density* and *total energy density* are the so-called conserved variables (also called Eulerian, or mathematical variables).

With the discovery of the *special theory of relativity*, the concepts of *energy density, momentum density, and stress* were unified into the concept of the *stress–energy tensor*, and *energy and momentum* were likewise unified into a single concept, the *energy–momentum vector*.

For a *perfect fluid in thermodynamic equilibrium*, the *stress–energy tensor* takes on a particularly simple form

$$T^{\alpha\beta} = g^{\alpha\beta}p + (\rho + p/c^2)\, dx_\alpha/ds\, dx_\beta/ds$$

where ρ is the mass–energy density (kilograms per cubic meter), p is the hydrostatic pressure (pascals), dx_α/ds is the fluid's four velocity, and $g^{\alpha\beta}$ is the reciprocal of the *metric tensor*. The four-velocity satisfies

$$dx_\alpha/ds\ dx_\beta/ds\ g_{\alpha\beta} = -c^2.$$

(A four-velocity is a four-vector in four-dimensional spacetime that represents the relativistic counterpart of velocity, which is a three-dimensional vector in space.)]

If we put the right-hand side of (58b)

$$[T_\sigma{}^\alpha = -\delta_\sigma{}^\alpha p + g_{\alpha\beta}\ dx_\beta/ds\ dx_\alpha/ds\ \rho \qquad (58b)]$$

in (57a) [the *energy-momentum law* for matter]

$$[\partial T_\sigma{}^\alpha/\partial x_\alpha = -\Gamma^\alpha{}_{\sigma\beta}\ \Gamma^\beta{}_\alpha \qquad (57a)]$$

we get the *general hydrodynamical equations* of Euler according to the *general theory of relativity*. This in principle completely solves the problem of motion; for the four equations (57a) together with the given equation between p and ρ, and the equation

$$g_{\alpha\beta}\ dx_\alpha/ds\ dx_\beta/ds = 1$$

are sufficient, with the given values of $g_{\alpha\beta}$, for finding out the six unknowns p, ρ, dx_1/ds, dx_2/ds, dx_3/ds, dx_4/ds.

If the $g_{\mu\nu}$ are unknown we have also to bring in the equations (53) [the *field equations* with the sum of the energy components of matter and gravitation, i. e. $t_\mu{}^\sigma + T_\mu{}^\sigma$ in place of the energy components $t_\mu{}^\sigma$ of *the force field in an accelerated frame in the absence of matter* alone,

$$\partial\Gamma^\alpha{}_{\mu\nu}/\partial x_\alpha + \Gamma^\alpha{}_{\mu\beta}\ \Gamma^\beta{}_{\nu\alpha} = -\kappa(T_{\mu\nu} - \tfrac{1}{2}\ g_{\mu\nu}T) \qquad (53)$$

[or removing the factor -2κ,

$$\partial\Gamma^\alpha{}_{\mu\nu}/\partial x_\alpha + \Gamma^\alpha{}_{\mu\beta}\ \Gamma^\beta{}_{\nu\alpha} = \tfrac{1}{2}\ T_{\mu\nu} - \tfrac{1}{4}\ g_{\mu\nu}T) \qquad (53^*)]$$

[where $\Gamma^\alpha{}_{\mu\nu} = -\{_\alpha{}^{\mu\nu}\}$, $\Gamma^\alpha{}_{\mu\beta} = -\{_\alpha{}^{\mu\beta}\}$, $\Gamma^\beta{}_{\nu\alpha} = -\{_\beta{}^{\nu\alpha}\}$, $T_{\mu\nu}$ is the *stress–energy tensor*, $T = T_{\mu\mu}$ (Laue's scalar),

$$\Gamma^\tau{}_{\mu\nu} = -\{_\tau{}^{\mu\nu}\}, \qquad (45)$$

$$\{_\tau{}^{\mu\nu}\} = g^{\tau\alpha}[_\alpha{}^{\mu\nu}], \qquad (23)$$

$$[_\alpha{}^{\mu\nu}] = \tfrac{1}{2}\ (\partial g_{\mu\alpha}/\partial x_\nu + \partial g_{\nu\alpha}/\partial x_\mu - \partial g_{\mu\nu}/\partial x_\alpha), \qquad (21)$$

and $(-g)^{1/2} = 1$].

[Substituting

$$T_{\mu\nu} = -g_{\mu\nu}p + g_{\mu\alpha}\ dx_\alpha/ds\ g_{\mu\beta}\ dx_\beta/ds\ \rho \qquad (58a)$$

in the *field equations*

$$\partial\Gamma^\alpha{}_{\mu\nu}/\partial x_\alpha + \Gamma^\alpha{}_{\mu\beta}\ \Gamma^\beta{}_{\nu\alpha} = -\kappa(T_{\mu\nu} - \tfrac{1}{2}\ g_{\mu\nu}T), \qquad (53)$$

[or removing the factor -2κ,

$$\partial\Gamma^\alpha{}_{\mu\nu}/\partial x_\alpha + \Gamma^\alpha{}_{\mu\beta}\ \Gamma^\beta{}_{\nu\alpha} = \tfrac{1}{2}\ T_{\mu\nu} - \tfrac{1}{4}\ g_{\mu\nu}T), \qquad (53^*)]$$

where Laue's scalar, $T = T_{\mu\mu} = - g_{\mu\mu}p + g_{\mu\alpha}\, dx_\alpha/ds\, g_{\mu\beta}\, dx_\beta/ds\, \rho$
gives

$$\partial\Gamma^\alpha{}_{\mu\nu}/\partial x_\alpha + \Gamma^\alpha{}_{\mu\beta}\,\Gamma^\beta{}_{\nu\alpha} = \tfrac{1}{2}\,(- g_{\mu\nu}p + g_{\mu\alpha}g_{\mu\beta}\, dx_\alpha/ds\, dx_\beta/ds\, \rho$$
$$- \tfrac{1}{2}\, g_{\mu\nu}\{- g_{\mu\mu}p + g_{\mu\alpha}\, dx_\alpha/ds\, g_{\mu\beta}\, dx_\beta/ds\, \rho\})$$
$$= (- \tfrac{1}{2}\, g_{\mu\nu}p + \tfrac{1}{2}\, g_{\mu\alpha}g_{\mu\beta}\, dx_\alpha/ds\, dx_\beta/ds\, \rho$$
$$+ \tfrac{1}{4}\, g_{\mu\nu}g_{\mu\mu}p - \tfrac{1}{4}\, g_{\mu\nu}g_{\mu\alpha}g_{\mu\beta}\, dx_\alpha/ds\, dx_\beta/ds\, \rho)$$

or $$\partial\Gamma^\alpha{}_{\mu\nu}/\partial x_\alpha + \Gamma^\alpha{}_{\mu\beta}\,\Gamma^\beta{}_{\nu\alpha} = \{(\tfrac{1}{4}\, g_{\mu\mu} - \tfrac{1}{2})g_{\mu\nu}\, p + (\tfrac{1}{4}\, g_{\mu\nu} - \tfrac{1}{2})g_{\mu\alpha}g_{\mu\beta}\, dx_\alpha/ds\, dx_\beta/ds\, \rho\}.]$$

There are now 11 equations for defining the 10 functions $g_{\mu\nu}$, so that the number is more than sufficient. We must remember, however, that the equations (57a)

$$[\partial T_\sigma{}^\alpha/\partial x_\alpha = - \Gamma^\alpha{}_{\sigma\beta}\,\Gamma^\beta{}_\alpha \qquad (57a)]$$

are already contained in (53), so that the latter only represents (7) independent equations. This indefiniteness is due to the wide freedom in the choice of co-ordinates, so that mathematically the problem is indefinite in the sense that three of the space-functions can be arbitrarily chosen[1].

[1] On abandonment of the choice of coordinates with $g = - 1$, there remain *four* functions of space free to choose, corresponding to the four arbitrary functions, which we can freely use in the choice of coordinates.

§ 20 Maxwell's electromagnetic field equations for the vacuum.

Let φ_ν be the components of a *covariant four-vector*, the four-vector of the *electromagnetic potential*. From them, in accordance with (36),

$$[B_{\mu\nu} = \partial A_\mu/\partial x_\nu - \partial A_\nu/\partial x_\mu \qquad (36)]$$

we form the components $F_{\rho\sigma}$ of the *covariant six-vector* of the *electromagnetic field* according to the system of equations

$$F_{\rho\sigma} = \partial\varphi_\rho/\partial x_\sigma - \partial A_\sigma/\partial x_\rho. \qquad (59)$$

It follows from (59) that the system of equations

$$\partial F_{\rho\sigma}/\partial x_\tau + \partial F_{\sigma\tau}/\partial x_\rho + \partial F_{\tau\sigma}/\partial x_\sigma = 0 \qquad (60)$$

is satisfied; whose left side, according to (37),

$$[B_{\mu\nu\sigma} = A_{\mu\nu\sigma} + A_{\nu\sigma\mu} + A_{\sigma\mu\nu} = \partial A_{\mu\nu}/\partial x_\sigma + \partial A_{\nu\sigma}/\partial x_\mu + \partial A_{\sigma\mu}/\partial x_\nu \qquad (37)]$$

is an *antisymmetric tensor of the third order*. Thus, the system (60) essentially contains four equations, which are written out as follows:

$$\partial F_{23}/\partial x_4 + \partial F_{34}/\partial x_2 + \partial F_{42}/\partial x_3 = 0 \qquad (60a)$$
$$\partial F_{34}/\partial x_1 + \partial F_{41}/\partial x_3 + \partial F_{13}/\partial x_4 = 0$$
$$\partial F_{41}/\partial x_2 + \partial F_{12}/\partial x_4 + \partial F_{24}/\partial x_1 = 0$$
$$\partial F_{12}/\partial x_3 + \partial F_{23}/\partial x_1 + \partial F_{31}/\partial x_2 = 0.$$

This system of equations corresponds to Maxwell's second system of equations. You can see this immediately by placing

$$F_{23} = H_x, \qquad F_{14} = E_x \qquad (61)$$
$$F_{31} = H_y, \qquad F_{24} = E_y$$
$$F_{12} = H_z, \qquad F_{34} = E_z.$$

Then, instead of (60a) in the usual notation of three-dimensional vector analysis, one can set

$$-\,\partial H/\partial t = \text{curl } E \qquad (60b)$$
$$\text{div } H = 0.$$

Maxwell's first system is obtained by generalizing the form given by Minkowski. We introduce the *contravariant six-vector* associated with $F^{\alpha\beta}$

$$F^{\mu\nu} = g^{\mu\alpha} g^{\nu\beta} F_{\alpha\beta} \qquad (62)$$

as well as the *contravariant four-vector* J^μ of the *electric vacuum current density*. Then, taking (40)

[*The Divergence of a Six-vector*. … "Thus, we obtain

$$A^\alpha = \sqrt{(-g)}\, \partial\{\sqrt{(-g)}\, A^{\alpha\beta}\}/\partial x_\beta \qquad (40)$$

This is the expression for the divergence of a contravariant six-vector."]

into consideration, the following equations will be invariant with regard to *arbitrary substitutions of determinant 1* (according to the chosen coordinates):

$$\partial/\partial x_\nu\, F^{\mu\nu} = J^\mu. \qquad (63)$$

Let

$$F_{23} = H'_x, \qquad F_{14} = -\,E'_z \qquad (64)$$
$$F_{31} = H'_y, \qquad F_{24} = -\,E'_y$$
$$F_{12} = H'_z, \qquad F_{34} = -\,E'_z.$$

which quantities in the special case of the restricted *theory of special relativity* are equal to the quantities H_x... E_z, and in addition

$$J^1 = j_x,\ J^2 = j_y,\ J^3 = j_z,\ J^4 = \rho,$$

thus, instead of (63) we obtain

$$\partial E'/\partial t + j = \text{curl } H' \qquad (63a)$$
$$\text{div } E' = \rho.$$

Equations (60), (62) and (63) thus form the generalization of *Maxwell's field equations of vacuum*, with the convention which we have made with regard to the choice of coordinates. *The energy components of the electromagnetic field.* We form the inner product

$$\kappa_\sigma = F_{\sigma\mu} J^\mu \qquad (65)$$

Its components are according to (61)

$$[F_{23} = H_x, \quad F_{14} = E_x \qquad (61)$$
$$F_{31} = H_y, \quad F_{24} = E_y$$
$$F_{12} = H_z, \quad F_{34} = E_z]$$

in three-dimensional notation

$$\kappa_1 = \rho E_x + [j \,.\, H]^x \qquad (65a)$$
...
...
$$\kappa_4 = -\,(jE)$$

κ_σ is a *covariant four-vector* of whose components are equal to the negative *momentum* or, respectively, the *energy* transmitted from the electromagnetic field by the electric masses per unit of time and volume. If the electric masses are free, i.e. under the sole influence of the electromagnetic field, the *covariant four-vector* κ_σ will disappear.

To obtain the *energy* components T_σ^{ν} of the electromagnetic field, we only need to give the equation $\kappa_\sigma = 0$ the form of equation (57)

$$[\partial T_\sigma^{\alpha}/\partial x_\alpha + \tfrac{1}{2}\, \partial g^{\mu\nu}/\partial x_\sigma \, T_{\mu\nu} = 0. \qquad (57)]$$

From (63) and (65)

$$[\partial/\partial x_\nu \, F^{\mu\nu} = J^\mu. \qquad (63)$$
$$\kappa_\sigma = F_{\sigma\mu} J^\mu \qquad (65)]$$

it follows first:

$$\kappa_\sigma = F_{\sigma\mu}\, \partial F^{\mu\nu}/\partial x_\nu = \partial/\partial x_\nu \,(F_{\sigma\mu} F^{\mu\nu}) - F^{\mu\nu}\, \partial F_{\sigma\mu}/\partial x_\nu.$$

The second term of the right-hand side, by reason of (60),

$$[\partial F_{\rho\sigma}/\partial x_\tau + \partial F_{\sigma\tau}/\partial x_\rho + \partial F_{\tau\sigma}/\partial x_\sigma = 0 \qquad (60)]$$

permits the forming the transformation

$$F^{\mu\nu}\, \partial F_{\sigma\mu}/\partial x_\nu = -\,\tfrac{1}{2}\, F^{\mu\nu}\, \partial F_{\mu\nu}/\partial x_\sigma = -\,\tfrac{1}{2}\, g^{\mu\alpha} g^{\nu\beta} F_{\alpha\beta}\, \partial F_{\mu\nu}/\partial x_\sigma,$$

which, for reasons of symmetry, the latter expression can also be written in the form

$$-\frac{1}{4}\left[g^{\mu\alpha} g^{\nu\beta} F_{\alpha\beta} \partial F_{\mu\nu}/\partial x_\sigma + g^{\mu\alpha} g^{\nu\beta} \partial F_{\mu\nu}/\partial x_\sigma F_{\alpha\beta}\right].$$

But for this we may set

$$-\frac{1}{4} \partial/\partial x_\sigma (g^{\mu\alpha} g^{\nu\beta} F_{\alpha\beta} F_{\mu\nu}) + \frac{1}{4} F_{\alpha\beta} F_{\mu\nu} \partial/\partial x_\sigma (g^{\mu\alpha} g^{\nu\beta}).$$

The first of these terms can be written in shorter notation

$$-\frac{1}{4} \partial/\partial x_\sigma (F^{\mu\nu} F_{\mu\nu});$$

the second, after carrying out the differentiation, after some reduction, results in

$$-\frac{1}{2} F^{\mu\tau} F_{\mu\nu} g^{\nu\rho} \partial g_{\sigma\tau}/\partial x_\sigma.$$

If one takes all three terms together, one obtains the relation

$$\kappa_\sigma = \partial T_\sigma{}^\nu/\partial x_\nu - \frac{1}{2} g^{\tau\mu} \partial g_{\mu\nu}/\partial x_\sigma T_\tau{}^\nu \qquad (66)$$

where

$$T_\sigma{}^\nu = -F_{\sigma\alpha} F^{\sigma\alpha} + \frac{1}{4} \delta_\sigma{}^\nu F_{\alpha\beta} F^{\alpha\beta}.$$

Equation (66), if κ_σ vanishes, is equivalent to (57) or (57a)

$$[\partial T_\sigma{}^\alpha/\partial x_\alpha + \frac{1}{2} \partial g^{\mu\nu}/\partial x_\sigma T_{\mu\nu} = 0. \qquad (57)$$

$$\partial T_\sigma{}^\alpha/\partial x_\alpha = -\Gamma^\alpha{}_{\sigma\beta} \Gamma^\beta{}_\alpha \qquad (57a)]$$

respectively, on account of (30)

$$[g_{\mu\sigma} dg^{\nu\sigma} = -g^{\nu\sigma} dg_{\mu\sigma} \qquad (30)$$

$$g_{\mu\sigma} \partial g^{\nu\sigma}/\partial x_\lambda = -g^{\nu\sigma} \partial g_{\mu\sigma}/\partial x_\lambda].$$

Therefore, the $T_\sigma{}^\nu$ are the energy components of the electromagnetic field. With the help of (61) and (64) it is easy to show that these *energy components of the electromagnetic field* give the well-known *Maxwell-Pointing* expressions in the case of *special relativity. We have now derived the most general laws that the gravitational field and matter satisfy by consistently using a coordinate system for which $\sqrt{(-g)} = 1$.* We thus achieved a considerable simplification of the formulas and calculations, without abandoning the requirement of *general covariance*: for we found our equations by specializing the coordinate system from generally covariant equations.

Still, the question is not without formal interest as to whether, with a correspondingly generalized definition of the *energy* components of the *gravitational field* and *matter*, even without specialization of the coordinate system, it is possible to formulate laws of conservation in the form of equation (56),

$$[\partial(t_\mu{}^\sigma + T_\mu{}^\sigma)/\partial x_\sigma = 0. \qquad (56)]$$

and even field equations of gravitation of the of the same nature as equations (52)

$$[\partial/\partial x_\alpha (g^{\sigma\beta}\Gamma^\alpha{}_{\mu\beta}) = - \kappa\{(t_\mu{}^\sigma + T_\mu{}^\sigma) - \tfrac{1}{2} \delta_\mu{}^\sigma (t + T)\} \qquad (52)]$$

or (52a),

$$[\partial/\partial x_\alpha (g^{\sigma\beta}\Gamma^\alpha{}_{\mu\beta}) - \tfrac{1}{2} \delta_\mu{}^\sigma g^{\lambda\beta} \Gamma^\alpha{}_{\lambda\beta}) = - \kappa\{(t_\mu{}^\sigma + T_\mu{}^\sigma) \qquad (52a)]$$

in such a way that there is a divergence (in the ordinary sense) on the left, and the sum of the energy components of matter and gravitation on the right. *I have found that both are indeed the case*. However, I do not believe that it would be worthwhile to share my rather extensive observations on this subject, since nothing objectively new comes out of it.

§ 21. Newton's Theory as a First Approximation.

Einstein introduces a series of "approximations" to his *equation of motion* of a particle in a frame, which is moving with uniform acceleration relative to the reference frame, along a geodesic, $d^2x_\tau/ds^2 = \Gamma^\tau{}_{\mu\nu}\ dx_\mu/ds\ dx_\nu/ds$, and to the contravariant *energy-tensor* of a frictionless adiabatic liquid, $T^{\alpha\beta} = - g^{\alpha\beta}p + \rho\ dx_\alpha/ds\ dx_\beta/ds$, in his *field equation* $\partial\Gamma^\alpha{}_{\mu\nu}/\partial x_\alpha + \Gamma^\alpha{}_{\mu\beta}\,\Gamma^\beta{}_{\nu\alpha} = - \kappa(T_{\mu\nu} - \tfrac{1}{2}\, g_{\mu\nu}T)$, in order to create an equation in a similar form to that of Newton's law of gravitation.

Einstein first assumes that "we get a more realistic approximation [of the influence of gravitation] if we consider the case when $g_{\mu\nu}$ differ from the values in an inertial frame under special relativity (constant velocity of light in all inertial frames) only by small magnitudes (compared to 1) where we can neglect small quantities of the second and higher orders (his "first aspect of the approximation") and only consider the case *when the velocity of the particle is very small compared with the speed of light* so dx_1/ds, dx_2/ds, dx_3/ds can be treated as small quantities, whereas dx_4/ds is equal to 1 (his "second point of view for approximation). This reduces the *equation of motion* of a particle in a frame, which is moving with uniform acceleration relative to the reference frame, from $d^2x_\tau/ds^2 = \Gamma^\tau{}_{\mu\nu}\ dx_\mu/ds\ dx_\nu/ds$, where
$\Gamma^\tau{}_{\mu\nu} = - \tfrac{1}{2}\, g^{\tau\alpha} (\partial g_{\mu\alpha}/\partial x_\nu + \partial g_{\nu\alpha}/\partial x_\mu - \partial g_{\mu\nu}/\partial x_\alpha)$ to $d^2x_\tau/dt^2 = - \tfrac{1}{2}\, \partial g_{44}/\partial x_\tau$ ($\tau = 1, 2, 3$).

We have already mentioned several times that the *special theory of relativity* is to be looked upon as a special case of the *general theory of relativity*, in which $g_{\mu\nu}$ have constant values

$$\begin{matrix} -1 & 0 & 0 & 0 \\ 0 & -1 & 0 & 0 \\ 0 & 0 & -1 & 0 \\ 0 & 0 & 0 & +1. \end{matrix} \qquad (4)$$

This signifies, according to what has been said before, a total neglect of the influence of gravitation. We get a more realistic approximation *if we consider the case when* $g_{\mu\nu}$ *differ from (4) only by small magnitudes (compared to 1)* where we can neglect small quantities of the second and higher orders (*first point of view of the approximation.*)

Further it should be assumed that within the *space-time* region considered, $g_{\mu\nu}$ at infinite distances (using the word infinite in a spatial sense) can, by a suitable choice of coordinates, tend to the limiting values (4); i.e., we consider only those *gravitational fields* which can be regarded as produced by masses distributed over finite regions.

It might be thought that this approximation should lead to Newton's theory. For it *however, it is necessary to approximate the fundamental equations from a second point of view*. Let us consider the motion of a material point according to the equation (16).

$[g_{\mu\nu} g^{\nu\mu} = \delta_\mu^\nu$ (16)

where the symbol δ_μ^ν means 1 or 0, depending on $\mu = \nu$ or $\mu \neq \nu$, and $g^{\nu\mu}$ is a *contravariant tensor*].

In the case of the *special theory of relativity*, the components dx_1/ds, dx_2/ds, dx_3/ds [where$ds^2 = \sum_{\mu\nu} g_{\mu\nu}\, dx_\mu dx_\nu$; $\mu,\nu = 1,..,4$] can take any values. This signifies that any velocity

$$\upsilon = (dx_1^2/dx_4 + dx_2^2/dx_4 + dx_3^2/dx_4)^{1/2}$$

can appear which is less than the velocity of light in vacuum. If we limit ourselves to the consideration of the case, which almost exclusively offers itself to our experience, *when* υ *is small compared to the velocity of light*, it signifies that the components

$$dx_1/ds,\ dx_2/ds,\ dx_3/ds$$

can be treated as small quantities, whereas dx_4/ds is equal to 1, up to the second-order magnitudes (*the second point of view for approximation*).

Now we see that, according to the first view of approximation, the magnitudes $\Gamma^\tau_{\mu\nu}$ are all small quantities of at least the first order. A glance at (46)

[the equation of motion of a point in a frame, which is moving with uniform acceleration relative to the reference frame

$d^2x_\tau/ds^2 = \Gamma^\tau_{\mu\nu}\, dx_\mu/ds\, dx_\nu/ds$ (46)]

[where the invariant ds is the "line-element",

$\Gamma^\tau_{\mu\nu} = -\ \{_\tau{}^{\mu\nu}\}$ (45)

$\{_\tau{}^{\mu\nu}\} = g^{\tau\alpha}[_\alpha{}^{\mu\nu}]$ (23)

$[_\sigma{}^{\mu\nu}] = \frac{1}{2}\,(\partial g_{\mu\sigma}/\partial x_\nu + \partial g_{\nu\sigma}/\partial x_\mu - \partial g_{\mu\nu}/\partial x_\sigma)$ (21),

and $(-g)^{1/2} = 1$
so $\Gamma^{\tau}_{\mu\nu} = - g^{\tau\alpha}[_{\alpha}{}^{\mu\nu}]$
and $\Gamma^{\tau}_{\mu\nu} = - \frac{1}{2} g^{\tau\alpha} (\partial g_{\mu\alpha}/\partial x_{\nu} + \partial g_{\nu\alpha}/\partial x_{\mu} - \partial g_{\mu\nu}/\partial x_{\alpha})]$

will also show, that in this equation according to *the second view of approximation*, we are only to take into account those terms for which $\mu = \nu = 4$. By limiting ourselves only to terms of the lowest order [and only taking into account those terms for which $\mu = \nu = 4$] we get instead of (46) the equations

$$d^2x_{\tau}/dt^2 = \Gamma^{\tau}_{44}, [(\tau = 1, 2, 3, 4)$$

where we have set $ds = dx_4 = dt$; or, by limiting ourselves only to those terms which according to the *first point of view of approximations* are of the first order,

$$d^2x_{\tau}/dt^2 = [_{\tau}{}^{44}] \ (\tau = 1, 2, 3),$$
$$d^2x_4/dt^2 = - [_4{}^{44}].$$

If we further assume that the gravitation-field is quasi-static, i.e., it is limited only to the case when the *matter producing the gravitation-field is moving slowly (relative to the velocity of light)* we can neglect on the right-hand side differentiations with respect to the time in comparison with those with respect to the positional coordinates, so that we get

$$d^2x_{\tau}/dt^2 = - \frac{1}{2} \partial g_{44}/\partial x_{\tau} \qquad (\tau = 1, 2, 3) \qquad (67)$$

This is the equation of motion of a material point according to Newton's theory, where ½ g_{44} plays the part of the gravitational potential. The remarkable thing in the result is that in the *first approximation* of motion of the material point *only the component g_{44} of the fundamental tensor appears.*

Einstein then "*makes the necessary approximations*" to the contravariant *energy-tensor* $T_{\mu\nu}$ of a frictionless adiabatic liquid $T^{\alpha\beta}$ so that all of the components of the righthand side of the *field equations* for a frictionless adiabatic liquid *in a uniformly accelerated frame*, vanish apart from $T_{44} = \rho = T$.

Let us now turn to the *field equations* (53) [with the sum of the energy components of matter and gravitation, i. e. $t_{\mu}{}^{\sigma} + T_{\mu}{}^{\sigma}$ in place of the energy components $t_{\mu}{}^{\sigma}$ of *the force field in an accelerated frame in the absence of matter* alone,]

$$[\partial\Gamma^{\alpha}_{\mu\nu}/\partial x_{\alpha} + \Gamma^{\alpha}_{\mu\beta}\,\Gamma^{\beta}_{\nu\alpha} = - \kappa(T_{\mu\nu} - \tfrac{1}{2} g_{\mu\nu}T) \qquad (53)]$$

[or removing the factor $- 2\kappa$,

$$\partial\Gamma^{\alpha}_{\mu\nu}/\partial x_{\alpha} + \Gamma^{\alpha}_{\mu\beta}\,\Gamma^{\beta}_{\nu\alpha} = \tfrac{1}{2} T_{\mu\nu} - \tfrac{1}{4} g_{\mu\nu}T) \qquad (53^*)]$$

[where $\Gamma^{\alpha}{}_{\mu\nu} = -\{_{\alpha}{}^{\mu\nu}\}$, $\Gamma^{\alpha}{}_{\mu\beta} = -\{_{\alpha}{}^{\mu\beta}\}$, $\Gamma^{\beta}{}_{\nu\alpha} = -\{_{\beta}{}^{\nu\alpha}\}$, $T_{\mu\nu}$ is the *stress–energy tensor*, $T = T_{\mu\mu}$ (Laue's scalar),

$$\Gamma^{\tau}{}_{\mu\nu} = -\{_{\tau}{}^{\mu\nu}\} \qquad (45),$$

$$\{_{\tau}{}^{\mu\nu}\} = g^{\tau\alpha}[_{\alpha}{}^{\mu\nu}] \qquad (23),$$

$$[_{\alpha}{}^{\mu\nu}] = \tfrac{1}{2}(\partial g_{\mu\alpha}/\partial x_{\nu} + \partial g_{\nu\alpha}/\partial x_{\mu} - \partial g_{\mu\nu}/\partial x_{\alpha}) \qquad (21)$$

and $(-g)^{1/2} = 1$.]

In this case, we have to remember that the *energy-tensor of matter is exclusively defined by the density ρ of matter in a narrow sense*, i.e., by the second term [$\rho\, dx_{\alpha}/ds\, dx_{\beta}/ds$] on the right-hand side of equation (58) [or respectively by (58a) or (58b)] [defining the *contravariant energy-tensor* of a liquid];

[$$T^{\alpha\beta} = -g^{\alpha\beta}p + \rho\, dx_{\alpha}/ds\, dx_{\beta}/ds \qquad (58),$$

or

$$T_{\mu\nu} = -g_{\mu\nu}p + g_{\mu\alpha}\, dx_{\alpha}/ds\, g_{\mu\beta}\, dx_{\beta}/ds\, \rho \qquad (58a),$$

$$T_{\sigma}{}^{\alpha} = -\delta_{\sigma}{}^{\alpha}p + g_{\alpha\beta}\, dx_{\beta}/ds\, dx_{\alpha}/ds\, \rho \qquad (58b)$$

where p and ρ denote the "*pressure*" and the "*density*" of the liquid.]

If we make the "necessary approximations" then all components vanish except

$$T_{44} = \rho = T.$$

[Omitting $- g_{\mu\nu}p$ removes the connection to body *accelerations* [force fields] (per unit mass) acting on the continuum, for example gravity, inertial accelerations, electric field acceleration.]

On the left-hand side of (53) [the *field equations* for a frictionless adiabatic liquid *in a uniformly accelerated frame*]

[$$\partial\Gamma^{\alpha}{}_{\mu\nu}/\partial x_{\alpha} + \Gamma^{\alpha}{}_{\mu\beta}\,\Gamma^{\beta}{}_{\nu\alpha} = -\kappa(T_{\mu\nu} - \tfrac{1}{2} g_{\mu\nu}T) \qquad (53),$$

[or removing the factor -2κ,

$$\partial\Gamma^{\alpha}{}_{\mu\nu}/\partial x_{\alpha} + \Gamma^{\alpha}{}_{\mu\beta}\,\Gamma^{\beta}{}_{\nu\alpha} = \tfrac{1}{2} T_{\mu\nu} - \tfrac{1}{4} g_{\mu\nu}T \qquad (53^*)],$$

the *second term*

[$$\Gamma^{\alpha}{}_{\mu\beta}\,\Gamma^{\beta}{}_{\nu\alpha} = g^{\alpha\tau}g^{\beta\tau}\,[_{\tau}{}^{\mu\beta}]\,[_{\tau}{}^{\nu\alpha}] = -\tfrac{1}{4} g^{\alpha\tau}(\partial g_{\mu\tau}/\partial x_{\beta} + \partial g_{\beta\tau}/\partial x_{\mu} - \partial g_{\mu\beta}/\partial x_{\tau})(\partial g_{\nu\tau}/\partial x_{\alpha} + \partial g_{\alpha\tau}/\partial x_{\nu} - \partial g_{\nu\alpha}/\partial x_{\tau})]$$

is an infinitesimal of the second order, so that the *first term*

[$$\partial\Gamma^{\alpha}{}_{\mu\nu}/\partial x_{\alpha} = -\partial g^{\alpha\tau}[_{\tau}{}^{\mu\nu}]/\partial x_{\alpha} = \tfrac{1}{2}\partial g^{\alpha\tau}(\partial g_{\mu\tau}/\partial x_{\nu} + \partial g_{\nu\tau}/\partial x_{\mu} - \partial g_{\mu\nu}/\partial x_{\tau})/\partial x_{\alpha}]$$

leads to the approximation in question

$$\partial/\partial x_1\,[_1{}^{\mu\nu}] + \partial/\partial x_2\,[_2{}^{\mu\nu}] + \partial/\partial x_3\,[_3{}^{\mu\nu}] - \partial/\partial x_4\,[_4{}^{\mu\nu}].$$

[or $+ ½\, δ/δx_1\, (δg_{μ1}/δx_ν + δg_{ν1}/δx_μ − δg_{μν}/δx_1)$
$+ ½\, δ/δx_2\, (δg_{μ2}/δx_ν + δg_{ν2}/δx_μ − δg_{μν}/δx_2)$
$+ ½\, δ/δx_3\, (δg_{μ3}/δx_ν + δg_{ν3}/δx_μ − δg_{μν}/δx_3)$
$− ½\, δ/δx_4\, (δg_{μ4}/δx_ν + δg_{ν4}/δx_μ − δg_{μν}/δx_4)$
where $[_α{}^{μν}] = ½\, (δg_{μα}/δx_ν + δg_{να}/δx_μ − δg_{μν}/δx_α)$,
and $α = τ = 1, 2, 3, 4$ (21)

or $[_1{}^{μν}] = ½\, (δg_{μ1}/δx_ν + δg_{ν1}/δx_μ − δg_{μν}/δx_1)$
$[_2{}^{μν}] = ½\, (δg_{μ2}/δx_ν + δg_{ν2}/δx_μ − δg_{μν}/δx_2)$
$[_3{}^{μν}] = ½\, (δg_{μ3}/δx_ν + δg_{ν3}/δx_μ − δg_{μν}/δx_3)$
$[_4{}^{μν}] = ½\, (δg_{μ4}/δx_ν + δg_{ν4}/δx_μ − δg_{μν}/δx_4)$

or setting $μ = ν = 4$
$\{+ ½\, δ/δx_1\, (δg_{41}/δx_4 + δg_{41}/δx_4)\} − ½\, δ/δx_1\, δg_{44}/δx_1$
$+ ½\, δ/δx_2\, (δg_{42}/δx_4 + δg_{42}/δx_4)\} − ½\, δ/δx_2\, δg_{44}/δx_2$
$+ ½\, δ/δx_3\, (δg_{43}/δx_4 + δg_{43}/δx_4)\} − ½\, δ/δx_3\, δg_{44}/δx_3$
$\{− ½\, δ/δx_4\, (δg_{44}/δx_4 + δg_{44}/δx_4 − δg_{44}/δx_4)\}$.]

For $μ = ν = 4$, *neglecting all differentiations with regard to time* [$/∂x_4$], this leads to the expression

$$− ½\, (δg^2_{44}/\, δx^2_1 + ½\, δg^2_{44}/\, δx^2_2 + ½\, δg^2_{44}/\, δx^2_3) = − ½\, Δg_{44},$$

[assuming $g^{11} = g^{22} = g^{33} = 1$,
where $Δg_{44} = (δg^2_{44}/δx^2_1 + δg^2_{44}/δx^2_2 + δg^2_{44}/δx^2_3)$.]

The last of the *field equations* (53)

$$[∂Γ^α_{μν}/∂x_α + Γ^α_{μβ}\, Γ^β_{να} = − κ(T_{μν} − ½\, g_{μν}T) \quad (53)]$$

[or removing the factor − 2κ,

$$∂Γ^α_{μν}/∂x_α + Γ^α_{μβ}\, Γ^β_{να} = ½\, T_{μν} − ¼\, g_{μν}T \quad (53^*)]$$

thus leads to

$$Δg_{44} = ∇^2g_{44} = κρ. \quad (68)$$

[or removing the factor − 2κ,

$$Δg_{44} = − ½\, ρ, \quad (68^*)$$

or $(∂g^2_{44}/∂x^2_1 + ∂g^2_{44}/∂x^2_2 + ∂g^2_{44}/∂x^2_3) = − ½\, ρ$,
where $Δg_{44} = ∂g^2_{44}/∂x^2_1 + ∂g^2_{44}/∂x^2_2 + ∂g^2_{44}/∂x^2_3$,
and $ρ$ is the *density* of the liquid in Euler's equation for a frictionless adiabatic liquid.]

Einstein asserts that these approximations of the *equation of motion* of a particle in a frame, which is moving with uniform acceleration relative to the reference frame along a geodesic, and the *field equations*, for a frictionless adiabatic liquid in a uniformly accelerated frame, together, are equivalent to *Newton's law of gravitation.*

The equations (67)

[$d^2x_\tau/dt^2 = -\frac{1}{2}\,\delta g_{44}/\delta x_\tau$ ($\tau = 1, 2, 3$) (67)]

[the *equation of motion* of a material point which is moving with uniform acceleration relative to the reference frame along a geodesic, *where $g_{44}/2$ plays the part of gravitational potential φ*]

and (68)

[$\Delta g_{44} = \nabla^2 g_{44} = \kappa\rho$ (68),

[or removing the factor -2κ,

$\Delta g_{44} = -\frac{1}{2}\rho$ (68*)]

the *field equations*, for a frictionless adiabatic liquid in a uniformly accelerated frame]

together, are equivalent to Newton's law of gravitation.

Under a series of assumptions Einstein manages to reduce his *equation of motion* and *field equation* to form an equation for the *gravitational potential* in terms of the integral of density of matter divide by the distance from the center of the matter similar to Newton's law of gravitation. He then equates the two equations to determine the differential-quotient taken along any geodesic curve the differential-quotient taken along any geodesic curve κ, which is then treated as a *gravitational constant.* On his assumption that "$g_{44}/2$ plays the part of *gravitational potential*", Einstein asserts that his *field theory* reduces to the Newtonian law of gravitation as a first approximation, and that under his *theory of general relativity*, the *gravitational potential* at radius r, $\varphi(r) = -\kappa/8\pi \int \rho d\tau/r$. Comparing this with the *gravitational potential* under the Newtonian theory for the chosen unit of time, $\varphi(r) = -K/c^2 \int \rho d\tau/r$ where K denotes the gravitation-constant 6.7×10^{-8}, he obtains $\kappa = 8\pi K/c^2 = 1.87 \times 10^{-27}$.

From (67)

[$d^2x_\tau/dt^2 = -\frac{1}{2}\,\partial g_{44}/\partial x_\tau$ ($\tau = 1, 2, 3$) (67)]

and (68)

[$\Delta g_{44} = \kappa\rho$ (68),

[or removing the factor -2κ,

$\Delta g_{44} = -\frac{1}{2}\rho$ (68*)

or $(\partial g^2_{44}/\partial x^2_1 + \partial g^2_{44}/\partial x^2_2 + \partial g^2_{44}/\partial x^2_3) = -\frac{1}{2}\rho$,

or $\partial g^2_{44}/\partial x^2_\tau = -\frac{1}{2}\rho$, ($\tau = 1, 2, 3$)]

and integrating (68) over a sphere,

$$\int(\partial g^2{}_{44}/\partial x^2{}_\tau)dx_\tau = (\partial g_{44}/\partial x_\tau) = \kappa \int\rho\ dx_\tau, \quad (\tau = 1, 2, 3)$$

so $\quad \partial g_{44}/\partial x_\tau = \kappa \int\rho\ dx_\tau,$

and $\quad g_{44} = \kappa \iint\rho\ dx_\tau dx_\sigma, \qquad (\tau, \sigma = 1, 2, 3)$

so the *gravitation-potential*, represented by $\varphi = g_{44}/2$,

$$\varphi = \tfrac{1}{2}\kappa \iint\rho\ dx_\tau dx_\sigma, \ldots.$$

and $\quad d^2x_\tau/dt^2 = -\tfrac{1}{2}\,\partial g_{44}/\partial x_\tau, \qquad (\tau = 1, 2, 3),$

we get the expression for the *gravitation-potential*:

$$[\varphi(r) =] - \kappa/8\pi \int\rho d\tau/r \qquad (68a)$$

[or removing the factor -2κ,

$$[\varphi(r) =]\ 1/16\pi \int\rho d\tau/r \qquad (68a^*)]$$

whereas the Newtonian theory *for the chosen unit of time* gives

$$[\varphi(r) =] - K/c^2 \int\rho d\tau/r$$

where K denotes the *gravitation-constant* 6.7×10^{-8} [m g^{-1}]; *equating them* we get

$$\kappa = 8\pi K/c^2 = 1.87 \times 10^{-27}\ [???]. \qquad (69)$$

[This is incorrect by a factor of 10^{-4} if measured in m g^{-1} or by a factor of 10^2 if measured in km kg^{-1}: substituting $\pi = 3.14159$, $K = 6.7 \times 10^{-8}$ m g^{-1}, and $c = 299{,}792{,}000$ m s^{-2},

$$\kappa = 8 \times 3.14159 \times 6.7 \times 10^{-8}/2.99792^2 \times 10^{16} = 18.736 \times 10^{-24}$$
$$\kappa = 1.8736 \times 10^{-23}\ \text{m g}^{-1} = 1.8736 \times 10^{-29}\ \text{km kg}^{-1}]$$

[Compared with the empirically determined Newtonian coefficient in $\varphi(r) = -K/c^2 \int\rho d\tau/r$ of $-K/c^2 = -6.7 \times 10^{-8}/(2.99 \times 10^{10})^2 = -0.7494323 \times 10^{-28}$, this results in a coefficient in $\varphi(r) = 1/16\pi \int\rho d\tau/r$ of $1/16\pi = 0.0198943 = 1.98943 \times 10^{-2}$, *which is of opposite sign (repulsion) and 2.67×10^{26} greater*. It is now evident why Einstein added the factor -2κ in (49):

$$[-2\kappa t_\sigma{}^\alpha = g_\sigma{}^{\mu\nu}\,\partial H/\partial g_\alpha{}^{\mu\nu} - \delta_\sigma{}^\alpha H. \qquad (49)]$$

[1] The reason for the introduction of the factor -2κ will become clear later.]

[As note above, it is impossible to calculate a "gravitational" constant from Einstein's equations on their own as *they include nothing representing the weak gravitational attraction between matter*. Omitting $- g_{\mu\nu}p$ from the Euler equation removed the only connection to body *accelerations* [force fields] (per unit mass) acting on the continuum, including that to the weak *attractive* gravitational force. *The remaining force on matter in Euler's equation is much stronger, has nothing to do with the weak force of gravitational attraction between matter, and is of opposite sign.*]

§ 22. Behavior of measuring rods and clocks in a statical gravitation-field. Curvature of light-rays. Perihelion-motion of the paths of the Planets.

In order to obtain Newton's theory as a *first approximation* we had to calculate only g_{44} out of the 10 components of the gravitation potential $g_{\mu\nu}$, for that is the only component which enters into the *first approximation* (67),

$$[d^2x_\tau/dt^2 = -\tfrac{1}{2}\,\partial g_{44}/\partial x_\tau \qquad (\tau = 1, 2, 3), \qquad (67)]$$

of the *equations of motion of a material point in a gravitational field.* We see however, that the other components of $g_{\mu\nu}$ should also differ from the values given in (4)

$$\left[\begin{matrix} -1 & 0 & 0 & 0 \\ 0 & -1 & 0 & 0 \\ 0 & 0 & -1 & 0 \\ 0 & 0 & 0 & +1 \end{matrix}\right] \qquad (4)$$

as required by the condition $g = -1$.

For a mass-point at the origin of co-ordinates generating the *gravitational field*, we get as a *first approximation* the radially symmetrical solution of the equation

$$g_{\rho\sigma} = -\delta_{\rho\sigma} - \alpha\, x_\rho x_\sigma/r^3 \text{ (}\rho\text{ and }\sigma\text{ between 1 and 3)} \qquad (70)$$
$$g_{\rho 4} = g_{4\rho} = 0 \text{ (}\rho\text{ between 1 and 3)}$$
$$g_{44} = 1 - \alpha/r$$

where $\delta_{\rho\sigma}$ is 1 or 0, according as $\rho = \sigma$ or $\rho \neq \sigma$, and r is the quantity

$$+ (x_1^2 + x_2^2 + x_3^2)^{1/2}$$

[The last of the *field equations* (53)

$$\partial\Gamma^\alpha_{\mu\nu}/\partial x_\alpha + \Gamma^\alpha_{\mu\beta}\,\Gamma^\beta_{\nu\alpha} = -\kappa(T_{\mu\nu} - \tfrac{1}{2}\, g_{\mu\nu}T) \qquad (53)$$

thus leads to

$$\Delta g_{44} = \nabla^2 g_{44} = \kappa\rho. \qquad (68)$$

From (67)

$$[d^2x_\tau/dt^2 = -\tfrac{1}{2}\,\partial g_{44}/\partial x_\tau \qquad (\tau = 1, 2, 3) \qquad (67)]$$

and (68)

$[\Delta g_{44} = \kappa\rho,$ (68)]

and integrating (68) over a sphere,

$\int(\partial g^2{}_{44}/\partial x^2{}_\tau)dx_\tau = (\partial g_{44}/\partial x_\tau) = \kappa \int\rho\, dx_\tau,\quad (\tau = 1, 2, 3)$

so $\partial g_{44}/\partial x_\tau = \kappa \int\rho\, dx_\tau,$

and $g_{44} = \kappa \iint\rho\, dx_\tau dx_\sigma,\quad (\tau, \sigma = 1, 2, 3),$

the *gravitation-potential*, represented by $\varphi = g_{44}/2$,

$\varphi = \frac{1}{2} \kappa \iint\rho\, dx_\tau dx_{\sigma}, \ldots.$

and $d^2x_\tau/dt^2 = -\frac{1}{2}\, \partial g_{44}/\partial x_\tau \quad (\tau = 1, 2, 3),$

with $g_{\rho\sigma} = -\delta_{\rho\sigma} - \alpha\, x_\rho x_\sigma/r^3$ (ρ and σ between 1 and 3) (70)

$g_{\rho 4} = g_{4\rho} = 0$ (ρ between 1 and 3)

$g_{44} = 1 - \alpha/r$

then $\varphi = g_{44}/2 = 1/2 - \alpha/2r$.]

On account of (68a)

$[\varphi(r) = -\kappa/8\pi \int\rho d\tau/r$ (68a)

with $M = \int\rho d\tau$,

$\varphi(r) = -\kappa M/8\pi r$

$\varphi(r) = -\frac{1}{2}\, \alpha/r$]

we have

$$\alpha = \kappa M/4\pi. \quad (70a)$$

where M denotes the mass generating the field.

[or removing the factor -2κ,

we have $\alpha = -M/8\pi$, (70a*)

so from (70),

$g_{\rho\sigma} = -\delta_{\rho\sigma} + Mx_\rho x_\sigma/8\pi r^3$ (ρ and σ between 1 and 3)

$g_{\rho 4} = g_{4\rho} = 0$ (ρ between 1 and 3)

$g_{44} = 1 + M/8\pi r$]

It is easy to verify that this solution satisfies to the first order of small quantities the field-equation outside the mass.

Let us now investigate the influences which the field of the mass M will have upon the metrical properties of space. Between the lengths and times ds measured "locally" on the one hand, and the differences in co-ordinates dx on the other, we have the relation

$$ds^2 = \sum_{\mu\nu} g_{\mu\nu}\, dx_\mu dx_\nu;\quad \mu, \nu = 1,..,4$$

For a unit measuring rod placed "parallel" to the x-axis, for example, we have to set $ds^2 = -1$; $dx_2 = dx_3 = dx_4 = 0$. Then $-1 = g_{11}\, dx_1^2$. If, in addition, the unit measuring rod lies on the x-axis, the first of the equations (70)

[$g_{\rho\sigma} = -\delta_{\rho\sigma} + Mx_\rho x_\sigma/8\pi r^3$ (ρ and σ between 1 and 3) (70)
where $\delta_{\rho\sigma}$ is 1 or 0, according as $\rho = \sigma$ or $\rho \neq \sigma$]

gives

$g_{11} = -(1+\alpha/r)$.

[so $g_{11} = -(1 - Mx_1^2/8\pi r^3)$.]

From both these relations

[$-1 = g_{11}\, dx_1^2$
$g_{11} = -(1+\alpha/r)$]

it follows as a first approximation that

$$dx = 1 - \alpha/2r, \qquad (71)$$

[$dx_1^2 = -1/g_{11} = 1/(1+\alpha/r)$],
or with $\alpha = -M/8\pi$, or $g_{11} = -(1 - Mx_1^2/8\pi r^3)$
$dx_1^2 = 1/(1 - Mx_1^2/8\pi r^3)$
$dx = 1 + Mx_1^2/16\pi r^3$.]

The unit measuring rod thus appears, when referred to the co-ordinate-system, shortened [lengthened] *by the calculated magnitude through the presence of the gravitational field,* when we place it radially in the field.

[The factor -2κ was required to make it appear shortened.]

Similarly, we can get its co-ordinate-length in a tangential position, if we put for example, $ds^2 = -1$; $dx_2 = dx_3 = dx_4 = 0$; $x_1 = r$, $x_2 = x_3 = 0$

[where $ds^2 = \sum_{\mu\nu} g_{\mu\nu}\, dx_\mu dx_\nu$]

we then get

$$-1 = g_{22}\, dx_2^2 = -dx_2^2. \qquad (71a)$$

[assuming that $g_{22} = -1$.]

The gravitational field has no influence upon the length of the rod, when we put it tangentially in the field.

Thus, [according to Einstein's relativistic calculation] *Euclidean geometry does not hold in the gravitational field even in the first approximation*, if we conceive that one and the same rod independent of its position and its orientation can serve as the measure of the same extension. But a glance at (70a)

[$\alpha = \kappa M/4\pi$. (70a)]

and (69)

[$\kappa = 8\pi K/c^2 = 1.87 \times 10^{-27}$, (69)]

shows that the expected difference is much too small to be noticeable in the measurement of earth's surface.

> [The latter is only true if *Newton's universal law of gravitation* is substituted in place of Einstein's relativistic calculation.]

We would further investigate *the rate of going of a unit-clock which is placed in a static gravitational field.* Here we have for a period of the clock

$$ds = 1;\ dx_1 = dx_2 = dx_3 = 0$$

then we have [assuming that $ds^2 = \sum_{\mu\nu} g_{\mu\nu}\, dx_\mu dx_\nu$]

$$1 = g_{44}dx_4^2;$$

$$dx_4 = 1/(g_{44})^{1/2} = 1/[1 + (g_{44} - 1)]^{1/2} = 1 - \tfrac{1}{2}(g_{44} - 1),$$

or $$dx_4 = 1 + \kappa/8\pi \int \rho d\tau/r \quad (72)$$

[$dx_4 = 1 - \tfrac{1}{2}(g_{44} - 1)$,

$g_{44} = 1 - \alpha/r$ (70)

$\alpha = \kappa M/4\pi$ (70a)

$g_{44} = 1 - \kappa M/4\pi r$

$M = \int \rho d\tau$

so $g_{44} = 1 - \kappa \int \rho d\tau/4\pi r$

$g_{44} - 1 = -\kappa \int \rho d\tau/4\pi r$

and $dx_4 = 1 - \tfrac{1}{2}(g_{44} - 1) = 1 - \tfrac{1}{2}(-\kappa \int \rho d\tau/4\pi r) = 1 + \kappa \int \rho d\tau/8\pi r$.]

> [or removing the factor $- 2\kappa$,
>
> $dx_4 = 1 - 1/16\pi \int \rho d\tau/r$ (72*).]]

Therefore, the clock goes more slowly [faster] when it is placed in the neighborhood of ponderable masses. *It follows from this that the spectral lines in the light coming to us from the surfaces of big stars should appear shifted towards the red end of the spectrum*[1].

[1] In support of the existence of such an effect we can allude to the spectral observations on fix-stars according to E. Freundlich. However, a concluding examination of that consequence is still missing.

[The factor -2κ was required to make it appear to go more slowly.]

Bending of light-rays by the sun.

Let us further investigate the path of light-rays in a statical gravitational field. According to the *special theory of relativity*, the velocity of light is given by the equation

$$-dx_1^2 - dx_2^2 - dx_3^2 - dx_4^2 = 0 \quad (1)$$

and therefore, according to the *general theory of relativity* it is given by the equation

$$ds^2 = \sum_{\mu\nu} g_{\mu\nu} dx_\mu dx_\nu = 0, \quad (73)$$

[where ds is the magnitude of the line-element belonging to two infinitely near points in the four-dimensional region, and $g_{\mu\nu}$ is a symmetric covariant tensor of the second rank called the "*fundamental tensor*". From the physical stand-point the quantities $g_{\mu\nu}$ are to be looked upon as magnitudes which describe the gravitation-field with reference to the chosen system of axes].

If the direction, i.e., the ratio $dx_1 : dx_2 : dx_3$, is given, equation (73) gives the magnitudes

$$dx_1/dx_4,\ dx_2/dx_4,\ dx_3/dx_4,$$

and with it the velocity,

$$[(dx_1/dx_4)^2 + (dx_2/dx_4)^2 + (dx_3/dx_4)^2]^{1/2} = \gamma$$

defined in the sense of the Euclidean geometry. *We can easily see that, with reference to the co-ordinate system, the rays of light must appear curved in the case in which the $g_{\mu\nu}$ are not constant.*

If n be the direction perpendicular to the direction of propagation, we have, from Huygen's principle [that all points of a wave front of light in a vacuum or transparent medium may be regarded as new sources of wavelets that expand in every direction at a rate depending on their velocities], shows that light-rays, taken in the plane (γ, n), must have the curvature

$$-d\gamma/\delta n.$$

Let us find out the curvature which a light-ray undergoes when it passes a mass M at a distance Δ from it. If we use the co-ordinate system according to the above scheme, then the total bending B of light-rays (*reckoned positive when it is concave to the origin*) is given in sufficient approximation by

$$B = \int_{-\infty}^{+\infty} \partial\gamma/\partial x_1 \, dx_2$$

while (73)

$$[ds^2 = \sum_{\mu\nu} g_{\mu\nu} d\mathbf{x}_\mu d\mathbf{x}_\nu = 0,] \qquad (73)$$

and (70)

$$[g_{\rho\sigma} = -\,\delta_{\rho\sigma} - \alpha\, x_\rho x_\sigma / r^3 \text{ (}\rho\text{ and }\sigma\text{ between 1 and 3)} \qquad (70)$$

$$g_{\rho 4} = g_{4\rho} = 0 \text{ (}\rho\text{ between 1 and 3)}$$

$$g_{44} = 1 - \alpha/r$$

$\delta_{\rho\sigma}$ is 1 or 0, according as $\rho = \sigma$ or $\rho \neq \sigma$, and r is the quantity $+ (x_1^2 + x_2^2 + x_3^2)^{1/2}$]

give

$$\gamma = (-\,g_{44}/g_{22})^{1/2} = 1 - \alpha/2r\,(1 + x_2^2/r^2).$$

Einstein obtained the bending of light which a light-ray undergoes when it passes a mass M at a distance Δ by integrating the expression for the curvature to give the approximation, $B = 2\alpha/\Delta$, where α was derived by introducing the Newtonian expression for the weak gravitational *attraction* of matter $-\,\alpha\, x_\rho x_\sigma/r^3$ in $g_{\rho\sigma} = -\,\delta_{\rho\sigma} - \alpha\, x_\rho x_\sigma/r^3$ (ρ and σ between 1 and 3). He then equated his expression for the *gravitational potential* derived from Euler's equation, with the addition of the factor $-\,2k$, $\varphi(r) = -\,\kappa M/8\pi r$, with the Newtonian expression, $\varphi(r) = -\,½\,\alpha/r$, to obtain $\alpha = \kappa M/4\pi$, which he substituted for α in the equation for the bending of light B, so $B = 2\alpha/\Delta = \kappa M/2\pi\Delta$. Then he substituted for his expression by equating it again to the Newtonian expression in terms of the gravitational constant K, $\varphi(r) = -\,\kappa M/8\pi r = -\,KM/rc^2$, so that the final expression was based solely on equations derived from Newtonian theory. It is consequently not surprising that it agrees with Soldner's 1801 calculation, also based on Newton's theory. *This is the Newtonian calculation.*

The calculation gives [total bending B of light-rays]

$$[B = \int_{-\infty}^{+\infty} \partial\gamma/\partial x_1 \, dx_2 = \int_{-\infty}^{+\infty} \partial\{1 - \alpha/2r\,(1 + x_2^2/r^2)\}/\partial x_1 \, dx_2]$$

$$B = 2\alpha/\Delta = \kappa M/2\pi\Delta \qquad (74)$$

[by substituting $\alpha = \kappa M/4\pi$.]

According to this, a ray of light just grazing the sun would suffer a bending of 1.7" [towards the sun] whereas one coming by Jupiter would have a deviation of about 0.02".

[In Einstein (November 18, 1915) this calculation was also based on Einstein's *first approximation* $\sum_{\alpha} \partial\Gamma^{\alpha}_{\mu\nu}/\partial x^{\alpha} + \sum_{\alpha\beta} \Gamma^{\alpha}_{\mu\beta} \Gamma^{\beta}_{\nu\alpha} = 0$ *in which the link to the weak gravitational force was provided by importing Newton's law of gravitation* in the form $-\alpha\, x_{\rho}x_{\sigma}/r^3$ into his *assumed solution* of his field equations *so that in the first approximation these reduced to Newton's equations.* It was given as follows: "As we perceive from equation (4b),

$$[g_{\rho\sigma} = -\delta_{\rho\sigma} + \alpha(\partial^2 r/\partial x_{\rho}x_{\sigma} - \delta_{\rho\sigma}/r) = -\delta_{\rho\sigma} - \alpha\, x_{\rho}x_{\sigma}/r^3 \quad (4b)]$$

my theory implies that in the case of a resting mass, the components g_{11} up to g_{33} are in quantities of first order already different from 0. Therefore, as we shall see later, *no disagreement with Newton's law arises in the first approximation.* This theory, however, produces an influence of the gravitational field on a light ray somewhat different from that given in my earlier work, *because the velocity of light is determined by the equation*

$$\sum g_{\mu\nu}dx_{\mu}dx_{\nu} = 0. \qquad (5)$$

Upon the application of Huygen's principle, we find from equations (5) and (4b), *after a simple calculation*, that a light ray passing at a distance Δ suffers an angular deflection of magnitude $2\alpha/\Delta$, while the earlier calculation, which was not based upon the hypothesis $\sum \Gamma^{\mu}_{\mu} = 0$, had produced the value α/Δ. A light ray grazing the surface of the sun should experience a deflection of 1.7 sec of arc instead of 0.85 sec of arc."]

[To calculate this, Einstein must have substituted the value for κ obtained by equating his expression for the gravitational potential for Newton's expression: "we get the expression for the *gravitation-potential*:

$$[\varphi(r) =] - \kappa/8\pi \int \rho d\tau/r \qquad (68a)$$

whereas the Newtonian theory *for the chosen unit of time* gives

$$[\varphi(r) =] - K/c^2 \int \rho d\tau/r$$

where K denotes the *gravitation-constant* 6.7×10^{-8} [$m^3\, g^{-1}\, s^{-2}$]; equating them we get

$$\kappa = 8\pi K/c^2 = 1.87 \times 10^{-27} \text{ [cm } g^{-1}.] \qquad (69)"]$$

[Substituting $\pi = 3.14159$, $K = 6.7 \times 10^{-8}$ m g^{-1} s^{-2}, and
$c = 3.0 \times 10^{10}$ cm s^{-2},
$\kappa = 8 \times 3.14159 \times 6.7 \times 10^{-8}/(3.0 \times 10^{10})^2 = 18.7 \times 10^{-28}$
$\kappa = 1.87 \times 10^{-27}$ cm $g^{-1} = 1.87 \times 10^{-29}$ km kg^{-1}.]

[Compared with the empirically determined Newtonian coefficient in $\varphi(r) = -K/c^2 \int\rho d\tau/r$ of $-K/c^2 = -6.7 \times 10^{-8}/(3.0 \times 10^{10})^2 = -0.74 \times 10^{-28}$, Einstein's equation results in a coefficient in $\varphi(r) = 1/16\pi \int\rho d\tau/r$ of $1/16\pi = 0.020 = 2.0 \times 10^{-2}$, *which is of opposite sign (repulsion) and* 2.7×10^{26} *greater*. It is evident why Einstein added the factor -2κ in (49) $-2\kappa t_\sigma^{\alpha} = g_\sigma^{\mu\nu}\, \partial H/\partial g_\alpha^{\mu\nu} - \delta_\sigma^{\alpha} H$.]

Then $B = 2\alpha/\Delta = \kappa M/2\pi\Delta = 1.87 \times 10^{-29}\ M/2\pi\Delta$.

Substituting

$\kappa = 1.87 \times 10^{-29}$ km kg^{-1},

M = mass of the sun = 1.988×10^{30} kg,

$\Delta = r$ is the radius of the sun = 695,700 km

gives $B = 1.87 \times 10^{-29} \times 1.988 \times 10^{30}/2 \times 3.14159 \times 6.957 \times 10^5$

or $B = 0.085 \times 10^{-4} = 8.5 \times 10^{-6}$ radians

$B = 8.5 \times 10^{-6} \times 57.296 = 4.87 \times 10^{-4}$ degrees

$B = 4.87 \times 10^{-4} \times 3.6 \times 10^3 = 17.53 \times 10^{-1} = 1.75$ arcseconds).]

[If this were calculated directly from Einstein's equation with the link to matter derived from Euler's equation, (74*), $B = -M/4\pi\Delta$ obtained by removing the factor -2κ, with M, the mass of the sun = 1.988×10^{30} kilograms, and Δ, the radius of the sun = 700,000 kilometers, the total bending of light rays at distance Δ from it, according to Einstein's equation is $B = -1.988\times10^{30}/4\pi \times 700{,}000$, i.e. *away from the sun.*]

The addition to the precession of the perihelion of Mercury.

If we calculate the gravitation-field to a higher degree of approximation and with corresponding accuracy the orbital motion *of a material point of a relatively small (infinitesimal) mass* we find a deviation of the following kind from the Kepler-Newtonian Laws of Planetary motion. The orbital ellipse of a planet undergoes a slow rotation in the direction of motion, of amount per revolution

$$\varepsilon = 24\pi^3 a^2/T^2c^2(1 - e^2). \qquad (75)$$

In this formula *a* signifies the major semi-axis, c the velocity of light, measured in the usual way, e the eccentricity, T the time of revolution in seconds[1].

[1] As regards the calculation I allude to the original papers, Einstein, A. (November 18, 1915). Erklärung der Perihelbewegung des Merkur aus der allgemeinen Relativitätstheorie. (Explanation of the Perihelion Motion of Mercury from the General Theory of Relativity.)

Sitzungsber. d. Preuß. Akad. d. Wiss., 47, 831-9; Schwarzschild, K. (1916) Über das Gravitationsfeld eines Massenpunktes nach der Einstein'schen Theorie. (On the Gravitational Field of a Point-Mass, according to Einstein's Theory.) *Sitzungsber. d. Preuß. Akad. d. Wiss*., 189-96.

[Einstein does not provide the units of ε in this paper or in Einstein, A. (November 18, 1915). Erklärung der Perihelbewegung des Merkur aus der allgemeinen Relativitätstheorie. (Explanation of the Perihelion Motion of Mercury from the General Theory of Relativity.) where the formula was derived, but he does on the 97th page of his 1921 Princeton lectures [Einstein, A. (1922). *Vier Vorlesungen über Relativitätstheorie: gehalten im Mai 1921 an der Universität Princeton* (*The Meaning of Relativity: Four Lectures Delivered at Princeton University, May 1921.*), Doc. 71, Lecture 4, page 357. Einstein claimed that "The most important result is a *secular rotation of the elliptic orbit of the planet in the same sense as the orbital revolution of the plane,* amounting in *radians per revolution* to

$$24\pi^3 a^2/(1 - e^2)c^2T^2, \qquad (113)$$

where

a is the semi-major axis of the planetary orbit in centimeters,
e is the numerical eccentricity in centimeters,
$c = 3.\ 10^{10}$ [cm sec^{-1}] is the speed of vacuum light,
T is the orbital period in seconds."

Examination of the units show that distance and time cancel, leaving a number:
units: $km^2.sec.^{-2}.\ km^{-2}.sec^2$.

Reference to the origin of the assumed 43" per Julian century discrepancy [Le Verrier, U. -J. (1941). *Développements sur Plusieurs Points de la Théorie des Perturbations des Planètes*. (Developments on several points of the Theory of Perturbations of Planets.) Bachelier, Imprimeur-Libraire de L'École Polytechnique, du Bureau des Longitudes, Paris, p. 45; Le Verrier, U. -J. (1845). *Théorie du mouvement de Mecure*. (The Theory of the Movement of Mercury.) Bachelier, Imprimeur-Libraire du Bureau des Longitudes de l'École royale Polytechnique, Paris, p. 7; see Underwood, T. G. (2021). *Urbain Le Verrier on the Movement of Mercury - annotated translations*, pp. 193, 275, 358], show that the natural units of theoretically derived precession of the perihelion are *orbits per rotation.* In order to calculate ε in arcseconds per Julian year, Le Verrier multiplied ε by 360*60*60*365/365.00637 = 1,295,977.383 = 1.296 x 10^6, to obtain ε in arcseconds per Julian year, then by a further 100, to obtain ε in arcseconds per

Julian century. See table of "Contributions to the precession of perihelion of Mercury" below.

This means that in order to represent his equations as $\varepsilon = 3\pi\ \alpha/a(1 - e^2)$ and $\varepsilon = 24\pi^3\ \alpha^2/T^2c^2(1 - e^2)$ in *radians per revolution* Einstein multiplied them by 2π. (1 turn = 2π radians.) *In orbits per rotation* of these equations were

$\varepsilon = 3/2\ \alpha/a(1 - e^2)$ and
$\varepsilon = 12\pi^2\ \alpha^2/T^2c^2(1 - e^2)$, respectively.]

The calculation gives for the planet Mercury, a rotation of path of amount 43" per century, corresponding sufficiently to what has been found by astronomical observation (Leverrier). For the astronomers have discovered in the *perihelion* motion of this planet a residual of the given magnitude which cannot be explained by the perturbation of the other planets. [*End of text.*]

[It is notable that Einstein did not include the derivation of his calculation of the *addition to the precession of the perihelion of Mercury* in this paper. This gave the impression that this was based on his theory in which the sum of the energy components of matter and gravitation, i. e. $t_\mu{}^\sigma + T_\mu{}^\sigma$ in place of the energy components $t_\mu{}^\sigma$ of the force field in an accelerated frame in the absence of matter alone,

$$\sum_\alpha \partial\Gamma^\alpha{}_{\mu\nu}/\partial x_\alpha + \sum_{\alpha\beta} \Gamma^\alpha{}_{\mu\beta}\, \Gamma^\beta{}_{\nu\alpha} = -\ \kappa(T_{\mu\nu} - \tfrac{1}{2}\, g_{\mu\nu}T) \qquad (53)$$
and $(-g)^{1/2} = 1$,

in which Einstein introduced the link to matter based on Euler's equation, and then arbitrarily introduced the factor $-\ 2\kappa$.

However, the starting assumption in both Einstein's (November 18, 1915), and Schwartzchild's (1916), calculations of the addition to the *precession of the perihelion of Mercury* resulting from Einstein's theory of general relativity was Einstein's *first approximation* of his field equations $\sum_\alpha \partial\Gamma^\alpha{}_{\mu\nu}/\partial x^\alpha + \sum_{\alpha\beta} \Gamma^\alpha{}_{\mu\beta}\, \Gamma^\beta{}_{\nu\alpha} = 0$, in which there was no link to matter. *The link to the weak attractive gravitational force* was provided by including the sum of the energy components of *matter* and *gravitation*, defined by Newton's equation of gravitation, in place of the energy components of *the force field in an accelerated frame in the absence of matter*, in Einstein's *assumed solution* of his field equations $g_{\rho\sigma} = -\ \delta_{\rho\sigma} - \alpha\ x_\rho x_\sigma/r^3$, so that in the *first approximation* these reduced to Newton's equations. After obtaining $g_{\mu\nu}$ in the *first approximation*, Einstein calculated the *components of the gravitational field* $\Gamma^\alpha{}_{\mu\nu}$ to the *first approximation*, resulting in $\Gamma^\sigma{}_{44} = \Gamma^4{}_{4\sigma} = -\ \alpha/2\ x_\sigma/r^3$, from which he obtained his *second approximation* $\Gamma^\sigma{}_{44} = -\ \alpha/2\ x_\sigma/r^3\ (1 - \alpha/r)$ by adding

$+ \alpha^2/2\ x_\sigma/r^4$ "*based on the symmetry properties" of his assumed solution.* Einstein then applied his expression for the *components of the gravitational field* to the *equation of motion of the point mass* in the *gravitational field*, and developed his *equations of motion* to the *second approximation*, $d^2x_\nu/ds^2 = - \alpha/2\ x_\nu/r^3\ \{1 + \alpha/r + 2u^2 - 3\ (dr/ds)^2\}$, $\nu = 1, 2, 3$. Einstein then applied this *equation of motion* to the secular rotation of the orbital ellipse to obtain the angle described by the radius vector between the perihelion and the aphelion. Corresponding to the addition in his *second approximation*, this included an additional term $+ \alpha/r^3$ ($+ \alpha\ x^3$ where $x = 1/r$), which is not included in the classic Newtonian theory. By integration of the elliptical integral Einstein obtained the *contribution of the additional term*, from which he claimed that after a complete orbit, the perihelion of Mercury advances by an additional amount $\varepsilon = 3\pi\ \alpha/a(1 - e^2)$.]

Einstein's calculations of the bending of light, and the addition to the precession of the perihelion of Mercury, which provided Einstein's main evidence for his *theory of general relativity*, were obtained from approximations for his equation of the *geodetic line*

$$\sum_\alpha \partial\Gamma^\alpha{}_{\mu\nu}/\partial x^\alpha + \sum_{\alpha\beta} \Gamma^\alpha{}_{\mu\beta}\,\Gamma^\beta{}_{\nu\alpha} = 0,$$

where $\Gamma^\alpha{}_{\mu\nu} = - \frac{1}{2} \sum_\beta g^{\alpha\beta}\ (\delta g_{\mu\beta}/\delta x_\nu + \delta g_{\nu\beta}/\delta x_\mu - \delta g_{\mu\nu}/\delta x_\alpha)$, in which the link to the weak attractive force of gravitation was provided by *Newton's law of gravitation.* The computed value for the bending of light was equal to the Newtonian value; the value for the addition to the precession of the perihelion of Mercury is 42.9" (arcseconds) per Julian century, apparently confirming Einstein's claim.

There is no evidence, nor could there be, for the versions of Einstein's theory of general relativity, in which the link to matter was provided by Euler's equation, described in this paper; or by "the Entwurf theory", in Einstein, A. & Grossmann. M. (June, 1913), [Entwurf einer verallgemeinerten Relativitätstheorie und eine Theorie der Gravitation. (Outline of a generalized theory of relativity and a theory of gravitation.); or in Einstein, A. (November 4, 1915). Zur allgemeinen Relativitätstheorie. (On the General Theory of Relativity); or Einstein, A. (November 25, 1915). Die Feldgleichungen der Gravitation. (The Field Equations of Gravitation.), two of the three November 1915 papers.

Consequently, this version of Einstein's *theory of general relativity*, must be rejected for lack of any supporting evidence.

Why anyone gave credence to this is a mystery. By 1921 Einstein was already moving his research interests into superseding general relativity. He attempted to generalize his theory of gravitation to include electromagnetism in a unified theory but his efforts were ultimately unsuccessful. In 1922, he was awarded the Nobel Prize in Physics "for his services to theoretical physics, and especially for his discovery of the law of the photoelectric effect" with no mention of either his theory of special relativity of his theory of general relativity.

Einstein, A. (February, 1917). Kosmologische Betrachtungen zur allgemeinen Relativitätstheorie. (Cosmological Considerations in the General Theory of Relativity.)

Koniglich Preußische Akademie der Wissenschaften, Sitzungsberichte (Berlin), 142–52; translation by W. Perrett & G. B. Jeffery in A. Engel (translator), E. Schuckling (consultant). (1997). *The Collected Papers of Albert Einstein*, Volume 6: The Berlin Years: Writings, 1914-1917, Princeton University Press, Princeton, Doc. 43, 421-32; https://einsteinpapers.press.princeton.edu/vol6-trans/433.

Describes Einstein's struggles with supplementing the *relativistic differential equations* by *limiting conditions* at *spatial infinity* in order to regard the universe as being of infinite spatial extent. In his treatment of the planetary problem, he chose these limiting conditions on the basis of the assumption that it is possible to select a system of reference so that at spatial infinity all the gravitational potentials $g_{\mu\nu}$ become constant, but it was by no means evident that the same limiting conditions could be applied to larger portions of the physical universe. Einstein attempts to resolve this using a method analogous to the extension of Poisson's equation used in the *non-relativistic case*, by adding to the left-hand side of field equation, the fundamental tensor $g_{\mu\nu}$, multiplied by a universal constant, $-\lambda$. As he noted, "*we admittedly had to introduce an extension of the field equations of gravitation which is not justified by our actual knowledge of gravitation*".

It is well known that Poisson's equation

$$\nabla^2\varphi = 4\pi K\rho \qquad (1)$$

in combination with the *equations of motion of a material point* is not as yet a perfect substitute for *Newton's theory of action at a distance*. There is still to be taken into account the condition that at spatial infinity the potential φ tends toward a fixed limiting value. There is an analogous state of things in the *theory of gravitation* in *general relativity*. Here, too, we must supplement the differential equations by limiting conditions at spatial infinity, if we really have to regard the universe as being of infinite spatial extent.

In my treatment of the planetary problem, I chose these limiting conditions in the form of the following assumption: it is possible to select a system of reference so that at spatial infinity all the gravitational potentials $g_{\mu\nu}$ become constant. But *it is by no means evident a priori that we may lay down the same limiting conditions when we wish to take larger portions of the physical universe into consideration*. In the following pages the reflections will be given which, up to the present, I have made on this fundamentally important question.

§ 1. The Newtonian Theory

It is well known that Newton's limiting condition of the constant limit for φ at spatial infinity leads to the view that *the density of matter becomes zero at infinity*. For we imagine that there may be a place in universal space round about which the *gravitational field of matter*, viewed on a large scale, possesses spherical symmetry. It then follows from Poisson's equation that, in order that φ may tend to a limit at infinity, the mean density ρ must decrease toward zero more rapidly than $1/r^2$ as the distance r from the center increases.*

> * ρ is the mean density of matter, calculated for a region which is large as compared with the distance between neighboring fixed stars, but small in comparison with the dimensions of the whole stellar system.

In this sense, therefore, *the universe according to Newton is finite, although it may possess an infinitely great total mass.*

From this it follows in the first place that *the radiation emitted by the heavenly bodies* will, in part, *leave the Newtonian system of the universe*, passing radially outwards, to become ineffective and lost in the infinite. *May not entire heavenly bodies fare likewise*? It is hardly possible to give a negative answer to this question. For it follows from the assumption of a finite limit for φ at spatial infinity that a heavenly body with finite kinetic energy is able to reach spatial infinity by overcoming the Newtonian forces of attraction. By statistical mechanics this case must occur from time to time, as long as the total energy of the stellar system-transferred to one single star-is great enough to send that star on its journey to infinity, whence it never can return.

We might try to avoid this peculiar difficulty by assuming a very high value for the limiting potential at infinity. That would be a possible way, if the value of the gravitational potential were not itself necessarily conditioned by the heavenly bodies. The truth is that we are compelled to regard the occurrence of any great differences of potential of the gravitational field as contradicting the facts. These differences must really be of so low an order of magnitude that the stellar velocities generated by them do not exceed the velocities actually observed.

If we apply Boltzmann's law of distribution for gas molecules to the stars, by comparing the stellar system with a gas in thermal equilibrium, we find that the Newtonian stellar system cannot exist at all. For there is a finite ratio of densities corresponding to the finite difference of potential between the center and spatial infinity. A vanishing of the density at infinity thus implies a vanishing of the density at the center.

It seems hardly possible to surmount these difficulties on the basis of the Newtonian theory. We may ask ourselves the question whether they can be removed by a modification of the Newtonian theory. First of all, we will indicate a method which does not in itself claim to be taken seriously; it merely serves as a foil for what is to follow. In place of Poisson's equation, we write

$$\nabla^2\varphi - \lambda\varphi = 4\pi\kappa\rho \qquad (2)$$

where λ denotes a universal constant. If ρ_0 be the *uniform density of a distribution of mass*, then

$$\varphi = -\,4\pi\kappa/\lambda\ \rho_0 \qquad (3)$$

is a solution of equation (2). This solution would correspond to the case *in which the matter of the fixed stars was distributed uniformly through space*, if the density ρ_0 is equal to the actual mean density of the matter in the universe. The solution then corresponds to an infinite extension of the central space, filled uniformly with matter. If, without making any change in the mean density, we imagine matter to be non-uniformly distributed locally, there will be, over and above the φ with the constant value of equation (3), an additional φ, which in the neighborhood of denser masses will so much the more resemble the Newtonian field as $\lambda\varphi$ is smaller in comparison with $4\pi\kappa\rho$.

A universe so constituted would have, with respect to its gravitational field, no center. A decrease of density in spatial infinity would not have to be assumed, but both the mean potential and mean density would remain constant to infinity. The conflict with statistical mechanics which we found in the case of the Newtonian theory is not repeated. With a definite but extremely small density, matter is in equilibrium, without any internal material forces (pressures) being required to maintain equilibrium.

§ 2. The Boundary Conditions According to the General Theory of Relativity

In the present paragraph I shall conduct the reader over the road that I have myself travelled, rather a rough and winding road, because otherwise I cannot hope that he will take much interest in the result at the end of the journey. The conclusion I shall arrive at is that *the field equations of gravitation which I have championed hitherto still need a slight modification, so that on the basis of the general theory of relativity those fundamental difficulties may be avoided which have been set forth in § 1 as confronting the Newtonian theory*. This modification corresponds perfectly to the transition from Poisson's equation (1) to equation (2) of § 1. *We finally infer that boundary conditions in spatial infinity fall away altogether, because the universal continuum in respect of its spatial dimensions is to be viewed as a self-contained continuum of finite spatial (three-dimensional) volume*.

The opinion which I entertained until recently, as to the limiting conditions to be laid down in spatial infinity, took its stand on the following considerations. In a consistent theory of relativity there can be no inertia *relatively to "space,"* but only an inertia of masses *relatively to one another*. If, therefore, I have a mass at a sufficient distance from all other masses in the universe, its inertia must fall to zero. We will try to formulate this condition mathematically.

According to the *general theory of relativity* the negative momentum is given by the first three components, the energy by the last component of the covariant tensor multiplied by $\sqrt{(-g)}$

$$m \sqrt{(-g)}\, g_{\mu\alpha}\, dx_\alpha/ds \qquad (4)$$

where, as always, we set

$$ds^2 = g_{\mu\nu}\, dx_\mu dx_\nu \qquad (5)$$

In the particularly perspicuous case of the possibility of choosing the system of coordinates so that the gravitational field at every point is spatially isotropic, we have more simply

$$ds^2 = -A(dx_1^2 + dx_2^2 + dx_3^2) + Bdx_4^2.$$

If, moreover, at the same time

$$\sqrt{(-g)} = 1 = \sqrt{(A^3B)}$$

we obtain from (4), to a first approximation for small velocities,

$$mA/\sqrt{(B)}\, dx_1/dx_4, \qquad mA/\sqrt{(B)}\, dx_2/dx_4, \qquad mA/\sqrt{(B)}\, dx_3/dx_4,$$

for the components of *momentum*, and for the *energy* (in the static case)

$$m\sqrt{(B)}.$$

From the expressions for the momentum, it follows that $mA/\sqrt{(B)}$ plays the part of the *rest mass*. As m is a constant peculiar to the point of mass, independently of its position, this expression, *if we retain the condition $\sqrt{(-g)} = 1$ at spatial infinity*, can vanish only when A diminishes to zero, while B increases to infinity. *It seems, therefore, that such a degeneration of the coefficients $g_{\mu\nu}$ is required by the postulate of relativity of all inertia.* This requirement implies that *the potential energy $m\sqrt{(B)}$ becomes infinitely great at infinity*. Thus, a point of mass can never leave the system; and a more detailed investigation shows that the same thing applies to light-rays. A system of the universe with such behavior of the *gravitational potentials* at infinity would not therefore run the risk of wasting away which was mooted just now in connection with the Newtonian theory.

I wish to point out that the simplifying assumptions as to the *gravitational potentials* on which this reasoning is based, have been introduced merely for the sake of lucidity. It is possible to find general formulations for the behavior of the $g_{\mu\nu}$ at infinity which express the essentials of the question without further restrictive assumptions.

At this stage, with the kind assistance of the mathematician J. Grommer, I investigated *centrally symmetrical, static gravitational fields, degenerating at infinity* in the way mentioned. The gravitational potentials $g_{\mu\nu}$ were applied, and from them the *energy-tensor $T_{\mu\nu}$ of matter was calculated on the basis of the field equations of gravitation*. But here it proved that for the system of the fixed stars no boundary conditions of the kind can come into question at all, as was also rightly emphasized by the astronomer de Sitter recently.

For the contravariant energy-tensor $T_{\mu\nu}$ of ponderable matter is given by

$$T^{\mu\nu} = \rho\, dx_\mu/ds\, dx_\nu/ds,$$

where ρ is the *density of matter* in natural measure. With an appropriate choice of the system of co-ordinates the stellar velocities are very small in comparison with that of light. We may, therefore, substitute $\sqrt{(g_{44})}\, dx4$ for ds. This shows us that all components of $T^{\mu\nu}$ must be very small in comparison with the last component T^{44}. *But it was quite impossible to reconcile this condition with the chosen boundary conditions*. In the retrospect this result does not appear astonishing. The fact of the small velocities of the stars allows the conclusion that wherever there are fixed stars, the gravitational potential (in our case $\sqrt{B}$) can never be much greater than here on earth. *This follows from statistical reasoning, exactly as in the case of the Newtonian theory*. At any rate, *our calculations have convinced me that such conditions of degeneration for the $g_{\mu\nu}$ in spatial infinity may not be postulated.*

After the failure of this attempt, two possibilities next present themselves.

(a) We may require, as in the problem of the planets, that, with a suitable choice of the system of reference, the $g_{\mu\nu}$ in spatial infinity approximate to the values

$$\begin{matrix} -1 & 0 & 0 & 0 \\ 0 & -1 & 0 & 0 \\ 0 & 0 & -1 & 0 \\ 0 & 0 & 0 & 1 \end{matrix}$$

(b) We may refrain entirely from laying down boundary conditions for spatial infinity claiming general validity; but at the spatial limit of the domain under consideration we have to give the $g_{\mu\nu}$ separately in each individual case, as hitherto we were accustomed to give the initial conditions for time separately.

The possibility (b) holds out no hope of solving the problem, but amounts to giving it up. This is an incontestable position, which is taken up at the present time by de Sitter.*

*de Sitter, *Akad. van Wetensch. te Amsterdam*, Nov. 8, 1916.

But I must confess that such a complete resignation in this fundamental question is for me a difficult thing. I should not make up my mind to it until every effort to make headway toward a satisfactory view had proved to be vain.

Possibility (a) is unsatisfactory in more respects than one. In the first place those boundary conditions presuppose a definite choice of the system of reference, *which is contrary to the spirit of the relativity principle.* Secondly, if we adopt this view, *we fail to comply with the requirement of the relativity of inertia.* For the inertia of a material point of mass m (in natural measure) depends upon the $g_{\mu\nu}$; but these differ but little from their postulated values, as given above, for spatial infinity. Thus, inertia would indeed be influenced, but would not be conditioned by matter (present in finite space). If only one single point of mass were present, according to this view, it would possess inertia, and in fact an inertia almost as great as when it is surrounded by the other masses of the actual universe. Finally, those statistical objections must be raised against this view which were mentioned in respect of the Newtonian theory.

From what has now been said it will be seen that I have not succeeded in formulating boundary conditions for spatial infinity. Nevertheless, there is still a possible way out, without resigning as suggested under (b). *For if it were possible to regard the universe as a continuum which is finite (closed) with respect to its spatial dimensions, we should have no need at all of any such boundary conditions.* We shall proceed to show that both the *general postulate of relativity* and the fact of the small stellar velocities are compatible with the hypothesis of a spatially finite universe; though certainly, in order to carry through this idea, *we need a generalizing modification of the field equations of gravitation.*

§ 3. The Spatially Finite Universe with a Uniform Distribution of Matter

According to the general theory of relativity the metrical character (curvature) of the four-dimensional space-time continuum is defined at every point by the matter at that point and the state of that matter. Therefore, on account of the lack of uniformity in the distribution of matter, the metrical structure of this continuum must necessarily be extremely complicated.

But if we are concerned with the structure only on a large scale, we may represent matter to ourselves as being uniformly distributed over enormous spaces, so that its density of distribution is a variable function which varies extremely slowly. Thus, our procedure will somewhat resemble that of the geodesists who, by means of an ellipsoid, approximate to the shape of the earth's surface, which on a small scale is extremely complicated.

The most important fact that we draw from experience as to the distribution of matter is that the relative velocities of the stars are very small as compared with the velocity of light. So, I think that for the present we may base our reasoning upon the following approximative assumption. *There is a system of reference relatively to which matter may be looked upon as being permanently at rest.* With respect to this system, therefore, the contravariant energy-tensor $T^{\mu\nu}$ of matter is, by reason of (5), of the simple form

$$\begin{matrix} 0 & 0 & 0 & 0 \\ 0 & 0 & 0 & 0 \\ 0 & 0 & 0 & 0 \\ 0 & 0 & 0 & \rho \end{matrix} \qquad (6)$$

The scalar ρ of the (mean) *density of distribution* may be a priori a function of the space co-ordinates. But if we assume the universe to be spatially finite, we are prompted to the hypothesis that ρ is to be independent of locality. On this hypothesis we base the following considerations.

As concerns the *gravitational field*, it follows from the equation of motion of the material point

$$d^2x_\nu/ds^2 + \{\alpha\beta, \nu\}\, dx_\alpha/ds\, dx_\beta/ds = 0$$

that a material point in a static gravitational field can remain at rest only when g_{44} is independent of locality. Since, further, we presuppose independence of the time coordinate x_4 for all magnitudes, we may demand for the required solution that, for all x_ν,

$$g_{44} = 1. \qquad (7)$$

Further, as always with static problems, we shall have to set

$$g_{14} = g_{24} = g_{34} = 0 \qquad (8)$$

It remains now to determine those components of the gravitational potential which define the purely spatial-geometrical relations of our continuum (g_{11}, g_{12}, … g_{33}). *From our assumption as to the uniformity of distribution of the masses generating the field, it follows that the curvature of the required space must be constant.* With this distribution of mass, therefore, the required finite continuum of the x_1, x_2, x_3, with constant x_4, will be a spherical space.

We arrive at such a space, for example, in the following way. We start from a Euclidean space of four dimensions, ξ_1, ξ_2, ξ_3, ξ_4, with a linear element $d\sigma$; let, therefore,

$$d\sigma^2 = d\xi_1{}^2 + d\xi_2{}^2 + d\xi_3{}^2 + d\xi_4{}^2 \qquad (9)$$

In this space we consider the hyper-surface

$$R^2 = \xi_1{}^2 + \xi_2{}^2 + \xi_3{}^2 + \xi_4{}^2 \qquad (10)$$

where R denotes a constant. The points of this hyper-surface form a three-dimensional continuum, a spherical space of radius of curvature R.

The four-dimensional Euclidean space with which we started serves only for a convenient definition of our hypersurface. Only those points of the hypersurface are of interest to us which have metrical properties in agreement with those of physical space with a uniform distribution of matter. For the description of this three-dimensional continuum, we may employ the co-ordinates ξ_1, ξ_2, ξ_3 (the projection upon the hyperplane $\xi_4 = 0$) since, by reason of (10), ξ_4 can be expressed in terms of ξ_1, ξ_2, ξ_3. Eliminating ξ_4 from (9), we obtain for the linear element of the spherical space the expression

$$d\sigma^2 = \gamma_{\mu\nu}\, d\xi_\mu d\xi_\nu \qquad (11)$$
$$\gamma_{\mu\nu} = \delta_{\mu\nu} + \xi_\mu\xi_\nu/(R^2 - \rho^2)$$

where $\delta_{\mu\nu} = 1$, if $\mu = \nu$; $\delta_{\mu\nu} = 0$, if $\mu \# \nu$, and $\rho^2 = \xi_1{}^2 + \xi_2{}^2 + \xi_3{}^2$.

The co-ordinates chosen are convenient when it is a question of examining the environment of one of the two points $\xi_1 = \xi_2 = \xi_3 = 0$.

Now the linear element of the required four-dimensional space-time universe is also given us. For the potential $g_{\mu\nu}$, both indices of which differ from 4, we have to set

$$g_{\mu\nu} = -\,[\delta_{\mu\nu} + x_\mu x_\nu/\{R2 - (x_1{}^2 + x_2{}^2 + x_3{}^2)\}] \qquad (12)$$

which equation, in combination with (7) and (8), perfectly defines the behavior of measuring-rods, clocks, and light-rays.

§ 4. On an Additional Term for the Field Equations of Gravitation

My proposed *field equations of gravitation* for any chosen system of co-ordinates run as follows: -

$$G_{\mu\nu} = -\kappa(T_{\mu\nu} - \tfrac{1}{2} g_{\mu\nu} T), \qquad (13)$$

$$G_{\mu\nu} = -\partial/\partial x_\alpha \{\mu\nu, \alpha\} + \{\mu\alpha, \beta\}\{\nu\beta, \alpha\} + \partial^2 \log \sqrt{(-g)}/\partial x_\mu \partial x_\nu - \{\mu\nu, \alpha\}\partial \log \sqrt{(-g)}/\partial x_\alpha$$

The system of equations (13) *is by no means satisfied when we insert for the $g_{\mu\nu}$ the values given in (7), (8), and (12), and for the (contravariant) energy-tensor of matter the values indicated in (6).* It will be shown in the next paragraph how this calculation may conveniently be made. So that, if it were certain that the field equations (13) which I have hitherto employed were the only ones compatible with the postulate of general relativity, *we should probably have to conclude that the theory of relativity does not admit the hypothesis of a spatially finite universe.*

However, the system of equations (14) allows a readily suggested extension which is compatible with the relativity postulate, and is perfectly analogous to the extension of Poisson's equation given by equation (2). For on the left-hand side of field equation (13) *we may add the fundamental tensor $g_{\mu\nu}$, multiplied by a universal constant, $-\lambda$, at present unknown, without destroying the general covariance.* In place of field equation (13)

$$[G_{\mu\nu} = -\kappa(T_{\mu\nu} - \tfrac{1}{2} g_{\mu\nu} T), \qquad (13)]$$

we write

$$G_{\mu\nu} - \lambda g_{\mu} = -\kappa(T_{\mu\nu} - \tfrac{1}{2} g_{\mu\nu} T). \qquad (13a)$$

This field equation, with λ sufficiently small, is in any case also compatible with the facts of experience derived from the solar system. It also satisfies laws of conservation of momentum and energy, because we arrive at (13a) in place of (13) by introducing into Hamilton's principle, instead of the scalar of Riemann's tensor, this scalar increased by a universal constant; and Hamilton's principle, of course, guarantees the validity of laws of conservation. It will be shown in § 5 that field equation (13a) is compatible with our conjectures on field and matter.

§ 5. Calculation and Result

Since all points of our continuum are on an equal footing, it is sufficient to carry through the calculation for one point, e.g. for one of the two points with the co-ordinates

$$x_1 = x_2 = x_3 = x_4 = 0.$$

Then for the $g_{\mu\nu}$ (13a) we have to insert the values

$$\begin{matrix} -1 & 0 & 0 & 0 \\ 0 & -1 & 0 & 0 \\ 0 & 0 & -1 & 0 \\ 0 & 0 & 0 & 1 \end{matrix}$$

wherever they appear differentiated only once or not at all. We thus obtain in the first place

$$G_{\mu\nu} = \partial/\partial x_1[\mu\nu,1] + \partial/\partial x_2[\mu\nu,2] + \partial/\partial x_3[\mu\nu,3] + \partial^2 \log \sqrt{(-g)}/\partial x_\mu \partial x_\nu.$$

From this we readily discover, taking (7), (8), and (13)

[$g_{44} = 1$ (7)

$g_{14} = g_{24} = g_{34} = 0$ (8)

$G_{\mu\nu} = -\kappa(T_{\mu\nu} - \frac{1}{2} g_{\mu\nu} T)$, (13)]

into account, that all equations (13a) are satisfied if the two relations

$$-2/R^2 + \lambda = -\kappa\rho/2, \quad -\lambda = -\kappa\rho/2,$$

or

$$\lambda = \kappa\rho/2 = 1/R^2 \qquad (14)$$

are fulfilled.

Thus, the newly introduced universal constant λ defines both the *mean density of distribution* ρ which can remain in equilibrium and also the *radius* R and the *volume $2\pi^2R^3$ of spherical space*. The total mass M of the universe, according to our view, is finite, and is in fact

$$M = \rho \,.\, 2\pi^2 R^3 = 4\pi^2 R/\kappa = \pi^2\sqrt{(32/\kappa^3\rho)}. \qquad (15)$$

Thus, the theoretical view of the actual universe, if it is in correspondence with our reasoning, is the following. *The curvature of space is variable in time and place, according to the distribution of matter, but we may roughly approximate to it by means of a spherical space.* At any rate, this view is logically consistent, and from the standpoint of the *general theory of relativity* lies nearest at hand; whether, from the standpoint of present astronomical knowledge, it is tenable, will not here be discussed. In order to arrive at this consistent view, *we admittedly had to introduce an extension of the field equations of gravitation which is not justified by our actual knowledge of gravitation.* It is to be emphasized, however, that a positive curvature of space is given by our results, even if the supplementary term is not introduced. *That term is necessary only for the purpose of making possible a quasi-static distribution of matter, as required by the fact of the small velocities of the stars.*

[Peebles, P. J. E. & Ratra, B. (April 2003). The Cosmological Constant and Dark Energy. *Reviews of Modern Physics*, 75, 2, 559–606: "The record shows Einstein never liked the Λ term. His view of how general relativity might fit Mach's principle was disturbed by de Sitter's (1917) solution to Eq. (34) for empty space ($T_{\mu\nu} = 0$) with $\Lambda > 0$. Pais, A., 1982, *Subtle is the Lord* ... (Oxford University, New York, p. 288) points out that Einstein in a letter to Weyl in 1923 comments on the effect of Λ in Eq. (24): "According to De Sitter two material points that are sufficiently far apart, continue to be accelerated and move apart. If there is no quasistatic world, then away with the cosmological term." We do not know whether at this time Einstein was influenced by Slipher's redshifts or Friedmann's expanding world model. ... Further to this point, in the appendix of the second edition of his book, *The Meaning of Relativity*, Einstein (1945, p. 127) states that the "introduction of the 'cosmologic member' " — Einstein's terminology for Λ — "into the equations of gravity, though possible from the point of view of relativity, *is to be rejected from the point of view of logical economy*", and that if "Hubble's expansion had been discovered at the time of the creation of the general theory of relativity, the cosmologic member would never have been introduced. It seems now so much less justified to introduce such a member into the field equations, since its introduction loses its sole original justification, — that of leading to a natural solution of the cosmologic problem." Einstein knew that without the cosmological constant the expansion time derived from Hubble's estimate of H_0 is uncomfortably short compared to estimates of the ages of the stars, and opined that that might be a problem with the star ages. The big error, the value of H_0, was corrected by 1960 [Sandage, A. (1958). *Astrophys. J.* 127, 513; (1962). *Problems of Extragalactic Research*, edited by G. C. McVittie (McMillan, New York), p. 359.

Gamow, G. (1970). [*My World Line*, Viking, New York, p. 44] recalls that "when I was discussing cosmological problems with Einstein, he remarked that the introduction of the cosmological term was the biggest blunder he ever made in his life." This certainly is consistent with all of Einstein's written comments we have seen on the cosmological constant per se; we do not know whether Einstein was also referring to the missed chance to predict the evolution of the universe."]

Part II. What is Gravity?

The five fundamental interactions of physics.

"*Gravity* is the weakest of the four fundamental interactions of physics, approximately 10^{38} times weaker than the *strong interaction*, 10^{36} times weaker than the *electromagnetic force* and 10^{29} times weaker than the *weak interaction*."

> [It is impossible to compare the strength of different forces that depend on different properties and vary differently with distance. Although electric and magnetic forces and gravity vary inversely with the square of the distance, the former depend on the strength of the electric charge and magnetic pole strength, whilst the latter varies with the mass.]

As a consequence, gravity has no significant influence at the level of subatomic particles."

In contrast, it is the dominant interaction at the macroscopic scale, and is the cause of the formation, shape and trajectory (orbit) of astronomical bodies. Gravitational attraction occurs between any two particles of *matter, which comes in one type* (referred to as *mass*). Matter exists in five phases based on increasing temperature, Bose-Einstein condensates, solids, liquids, gases, and plasmas.

The *weak interaction*, which is also often called the *weak force* or *weak nuclear force*, is the second weakest of the four fundamental interactions. It is the mechanism of interaction between subatomic particles that is responsible for the radioactive decay of atoms: The weak interaction participates in nuclear fission and nuclear fusion. The theory describing its behavior and effects is sometimes called *quantum flavordynamics* (QFD); however, the term QFD is rarely used, because the weak force is better understood by *electroweak theory* (EWT). The effective range of the weak force is limited to subatomic distances and is less than the diameter of a proton. At high energy, the *weak force* and *electromagnetic force* are unified as a single *electroweak force*.

Electroweak theory describes both the *electromagnetic force* and the *weak force*. Superficially, these forces appear quite different. The *weak force* acts only across distances smaller than the atomic nucleus, while the *electromagnetic force* can extend for great distances (as observed in the light of stars reaching across entire galaxies), weakening only with the square of the distance. Moreover, comparison of the strength of these two fundamental interactions *between two protons*, for instance, reveals that *the weak force is some 10 million times weaker than the electromagnetic force*. Yet one of the major discoveries of the 20th century has been that these two forces are different facets of a single, more-fundamental *electroweak force*.

The *electroweak theory* arose principally out of attempts to produce a self-consistent gauge theory for the weak force, in analogy with *quantum electrodynamics* (QED), the successful modern theory of the *electromagnetic force* developed during the 1940s. There are two basic requirements for the gauge theory of the *weak force*. First, it should exhibit an underlying mathematical symmetry, called gauge invariance, such that the effects of the force are the same at different points in space and time. Second, the theory should be renormalizable; i.e., it should not contain nonphysical infinite quantities.

The *electrostatic force* is the third weakest force. It is the dominant force in the interactions of atoms and molecules. *Electrostatic forces* occur between any two particles of matter with an *electric charge, which comes with in two types* (referred to as *positive* and *negative signs*), causing an attraction between particles with opposite charges and repulsion between particles with the same charge. The *electrostatic forces* is responsible for many of the chemical and physical phenomena observed in daily life. The *electrostatic attraction* between atomic nuclei and their electrons holds atoms together. Electric forces also allow different atoms to combine into molecules, including the macromolecules such as proteins that form the basis of life.

The *magnetic force*, which occurs between charged particles *in relative motion* and *which comes with in two types* (referred to as *north* and *south poles*), causing an attraction between particles with opposite poles and repulsion between particles with the same poles. The *magnetic interactions* between the spin and angular momentum magnetic moments of electrons also play a role in chemical reactivity; such relationships are studied in spin chemistry.

These two effects combine to create *electromagnetic fields* in the vicinity of charged particles, which can accelerate other charged particles via the Lorentz force. Electromagnetism can be thought of as a combination of electrostatics and magnetism, two distinct but closely intertwined phenomena. *Electromagnetism* also plays a crucial role in modern technology: electrical energy production, transformation and distribution; light, heat, and sound production and detection; fiber optic and wireless communication; sensors; computation; electrolysis; electroplating; and mechanical motors and actuators.

The *strong interaction*, which is also often called the *strong force* or *strong nuclear force*, is the strongest of the four fundamental interactions. It confines quarks into protons, neutrons, and other hadron particles. The strong interaction also binds neutrons and protons to create atomic nuclei, where it is called the *nuclear force*.

Most of the mass of a proton or neutron is the result of the strong interaction energy; the individual quarks provide only about 1% of the mass of a proton. *At the range of* 10^{-15} *m* (1 femtometer, slightly more than the radius of a nucleon), the *strong force* is approximately 100 times as strong as *electromagnetism*, 106 times as strong as the *weak interaction*, and 10^{38} times as strong as *gravitation.*

The *strong interaction* is observable at two ranges, and mediated by *different force carriers* in each one. On a scale less than about 0.8 fm (roughly the radius of a nucleon), *the force is carried by gluons* and holds quarks together to form protons, neutrons, and other hadrons. This is often known as the *color force*, and is so strong that if hadrons are struck by high-energy particles, they produce jets of massive particles instead of emitting their constituents (quarks and gluons) as freely moving particles.

This property of the *strong force* is called *color confinement*. Gluons are thought to interact with quarks and other gluons by way of a type of charge called *color charge*. *Color charge* is analogous to *electrostatic charge (and magnetic poles)*, but it comes in *three six types, three colors, each with two signs, (± red, ± green, and ± blue)* rather than one, which results in different rules of behavior. These rules are described by *quantum chromodynamics* (QCD), the theory of quark–gluon interactions.

On a larger scale, up to about 3 fm, *the force is carried by mesons* and binds nucleons (protons and neutrons) together to form the nuclei of an atom. In the context of atomic nuclei, the force binds protons and neutrons together to form a nucleus and is called the *nuclear force* (or residual strong force). Because the force is mediated by massive, short-lived mesons on this scale, the residual strong interaction obeys a distance-dependent behavior between nucleons that is quite different from when it is acting to bind quarks within hadrons. There are also differences in the binding energies of the nuclear force with regard to nuclear fusion vs nuclear fission.

Nuclear fusion accounts for most energy production in the Sun and other stars. Nuclear fission allows for decay of radioactive elements and isotopes, although it is often mediated by the weak interaction. Artificially, the energy associated with the nuclear force is partially released in nuclear power and nuclear weapons, both in uranium or plutonium-based fission weapons and in fusion weapons like the hydrogen bomb.

Current models of particle physics imply that the earliest instance of *gravity* in the Universe, possibly in the form of quantum gravity, supergravity or a gravitational singularity, along with ordinary space and time, developed during the Planck epoch (up to 10^{-43} seconds after the birth of the Universe), possibly from a primeval state, such as a false vacuum, quantum vacuum or virtual particle, in a currently unknown manner.

Attempts to develop a theory of gravity consistent with quantum mechanics, a *quantum gravity theory*, which would allow gravity to be united in a common mathematical framework (a theory of everything) with the other three fundamental interactions of physics, are a current area of research.

Einstein, A. & Grommer, J. (1923). Proof of the Non-Existence of an Everywhere Regular, Centrally Symmetric Field According to the Field Theory of Kaluza.

Scripta Universitatis atque Bibliothecae Hierosolymitanarum. Mathematica et Physica 1 (1923), VII: 1–5 Kitvei ha-Universita ve-Beth-ha-Sfarim bi-Yerushalayim. Mathematica u'Fisica. A (5684), VII: 1–4 (Hebrew); translation in *The Collected Papers of Albert Einstein*, Volume 13: The Berlin Years: Writings & Correspondence January 1922 - March 1923 (English translation supplement); edited by Diana Kormos Buchwald, József Illy, Ze'ev Rosenkranz, & Tilman Sauer; Princeton University Press, Princeton, Doc. 12, 30-3; https://einsteinpapers.press.princeton.edu/vol13-trans/60.

Received 10 January 1922.

Between 1923 until he died in 1955, Einstein published 31 papers on *classical unified field theory*, a unified theory of electromagnetism and gravity, without any success. This is the first; which he published jointly with Jacob Grommer.

Surely the most important current issue of the general theory of relativity today is the essential unity of the gravitational field and the electromagnetic field. Although the essential unity of both kinds of fields cannot, by any means, be required a priori, it would undoubtedly be a great advance in the theory if this dualism could be over- come. Until a short while ago the sole attempt in this direction has been Weyl's theory[2].

[2] Weyl 1918a and 1918c, 1919b; for a summary of Weyl's approach see, e.g., Hermann Weyl to Einstein, March 1, 1918. (*The Collected Papers of Albert Einstein*, Vol. 8, Doc. 472).

Considerable misgivings about it exist, however. It does not do justice to the independence of the measuring rods and clocks, or atoms, from their prehistories. Furthermore, it does not remove this dualism to the extent that its Hamiltonian function is composed additively of two parts, an electromagnetic one and a gravitational one, which are not independent of each other. Furthermore, this theory leads to differential equations of fourth order while we have no indication that equations of second order would work out.

A short while ago a draft of a theory was presented to the Academy of Science in Berlin by Mr. Th[eodor] Kaluza that avoids all these troubles and is formally of astonishing simplicity[4].

[4] Theodor Kaluza (1885–1954) was Privatdozent in Mathematics at the University of Königsberg.

Let us first sketch Mr. Kaluza's thoughts and then move on to the question we wish to examine.

A five-dimensional manifold whose field variable does not depend on the fifth variable is (with a suitable choice of coordinates) equivalent to a four-dimensional continuum. It therefore does not signify any special physical hypothesis if we interpret the four-dimensional space-time manifold of physical experience as such a five-dimensional manifold, which we will call "cylindrical" with reference to x5. This is what Kaluza does. He furthermore assumes that physical reality in this continuum is characterized by a quadratic line element

$$ds^2 = g_{\mu\nu} dx_\mu dx_\nu \qquad (1)$$

whose coefficients ($g_{\mu\nu} = g_{\nu\mu}$)

g11 g12 g13 g14 g15
g21 g22 g23 g24 g25
— — — — —
— — — — —
g51 g52 g53 g54 g55

should not, according to what has been said, depend on x5. The components g_{11} ... g_{44} should describe the gravitational field g_{15}, g_{25}, g_{35}, g_{45} would be the electric potentials, g_{55} a field value that still awaits interpretation and may perhaps be related to the Poincaré pressure that played a kind of awkward stand-in role in the theory of the electron.

Kaluza's essential hypothesis, now, consists of the assumption that the laws of nature should be generally covariant in this five-dimensional world. Thus, the ways and means by which the electromagnetic potentials occur in the laws of nature are necessarily connected with the ways and means by which the gravitational potentials occur, which signifies a trenchant limitation of the possibilities. Thus, the possibility arises for us to construct the physical worldview on a *uniform* Hamiltonian function that does not contain heterogeneous terms superficially welded together by a plus sign. Mr. Kaluza introduces another tensor for the material current besides the magnitudes $g_{\mu\nu}$, however. But it is clear that the introduction of such a tensor only serves to give a preliminary, merely phenomenological description of matter, whereas the ultimate goal we envision today is a pure field theory in which the field variables represent the field of "empty space" as well as the electric elementary particles that make up "matter."

Nonetheless, the fundamental weak points of Kaluza's idea must not be left unmentioned. In the general theory of relativity, which operates with the four- dimensional continuum,

$$ds^2 = g_{\mu\nu}dx_\mu dx_\nu$$

means a directly measurable magnitude for a local inertial system using measuring rods and clocks, whereas the ds^2 of the five-dimensional manifold in Kaluza's extension initially stands for a pure abstraction that seems not to deserve direct metrical significance. Therefore, from the physical point of view, the requirement of general covariance of all equations in the five-dimensional continuum appears completely unfounded[7].

[7] See Einstein to Kaluza, 5 May 1919. (*The Collected Papers of Albert Einstein*, Vol. 9, Doc. 35).

Moreover, it is a questionable asymmetry that the requirement of the cylinder property distinguish one dimension above the others and yet with reference to the structure of the equations all five dimensions should be equivalent.

If one asks whether the $g_{\mu\nu}$'s alone suffice for a description of the total field, one can set, with a selection of coordinates in which the determinant $|g_{\mu\nu}| = g$ is assigned the value 1, the Hamiltonian function H

$$H = g^{\mu\nu}\Gamma^{\alpha}_{\mu\nu}\Gamma^{\beta}_{\nu\alpha}. \qquad (2)$$

Thus to first approximation, i.e., provided the deviations of the $g_{\mu\nu}$'s from the constants are slight, the field equations take the form

$$\partial\Gamma^{\sigma}_{\mu\nu}/\partial x_\sigma = 0. \qquad (3)$$

Mr. Kaluza already saw that in this way the laws of gravity and Maxwell's field equations in vacuo are correctly obtained to first approximation.

In this situation it is of interest to know whether the stringent equations (in three dimensions) corresponding to the Hamiltonian function (2) have centrally symmetric static solutions that are everywhere singularity-free [i.e., nowhere may the $g_{\mu\nu}$'s become infinite or their determinants vanish] and suitable for describing the ele- mentary electric charges[9].

[9] Einstein already in 1909 had considered the possibility of reducing particles to solutions of field equations in electromagnetic theory (see Einstein to Hendrik A. Lorentz, 23 May 1909. *The Collected Papers of Albert Einstein*, Vol. 5, Doc. 163). In Einstein (1919) [Do Gravitational Fields Play an Essential Role in the Structure of the Elementary Particles? *The Collected Papers of Albert Einstein*, Vol. 7, Doc. 17], he then tried to reduce particles

to configurations of (nonunified) electromagnetic and gravitational fields. The present document gives Einstein's first definition of the necessary and sufficient conditions that a solution to the field equations has to fulfill to be interpretable as representing a material particle. At the same time, the document gives the first formulation of the criterion that such particle solutions must exist in a satisfactory physical theory based on the field concept. This criterion is a recurring theme in Einstein's later work on unified field theories.
…

The following approach corresponds to a centrally symmetric solution:

1. For the three spatial indices, $g_{\alpha\beta} = \lambda\delta_{\alpha\beta} + \mu x_\alpha x_\beta$ ($\delta_{\alpha\beta} = 1$ or 0, for $\alpha = \beta$ or $\alpha \neq \beta$, respectively) must be valid.
2. $g_{14}, g_{24}, g_{34}, g_{15}, g_{25}, g_{35}$ should vanish throughout.
3. $g_{44}, g_{45}, g_{55}, \lambda, \mu$ should be functions of r $\{= \sqrt{((x_1^2 + x_2^2 + x_3^2))}\}$ alone.

If one now also sets the abbreviation

$$\gamma = g_{44}\, g_{55} - g^2_{45},$$

then for the Hamiltonian function from (2) one obtains:

$$r^2 H = \ldots . \qquad (4)$$

The variation of the action integral

$$\int H r^2\, dr$$

according to g_{44}, yields the equation

$$\ldots . \qquad (5)$$

The equations for g_{45} and g_{55} read analogously.

Through suitable combination of two of each of these equations, three similarly structured equations follow, of the type

$$\ldots . \qquad (6)$$

From this, by integration,

$$\ldots = \text{const.} \qquad (7)$$

The constant on the right-hand side has to vanish because the left-hand side for $r = 0$ vanishes. That is why, after another integration,

$$g_{45}/g_{44} = \text{const.} \qquad (8)$$

follows from (7). Likewise follows

$$g_{45}/g_{55} = \text{const.} \qquad (8a)$$

Within infinite space-time the manifold must be Euclidean and the electrostatic potential must vanish there. Hence the equations (8) and (8a) require that and vanish throughout. Therefore, no spatially variable electric potential exists and hence no electric field, either.

Thus, it is proven that Kaluza's theory possesses no centrally symmetric solution dependent on the $g_{\mu\nu}$'s alone that could be interpreted as a (singularity-free) electron.

Weyl, H. (April, 1929). Gravitation and the electron.

PNAS, 15, 4, 323–34, https://doi.org/10.1073/pnas.15.4.323; also in Weyl, H. (May, 1929). Elektron und Gravitation. (Electron and gravity.) *Zeit. Phys.*, 56, 330–352; https://doi.org/10.1007/ BF01339504.

Communicated March 7, 1929. Translated by H. P. Robertson.

Palmer Physical Laboratory, Princeton University.

Attempt to incorporate Dirac theory into the scheme of *general relativity*, introduces *gauge invariance* of *theory of coupled electromagnetic potentials* and Dirac *matter waves*, explains why "anti-symmetric" Pauli-Fermi statistics for electrons lead to "symmetric" Bose-Einstein statistics for photons, *barrier which hems progress of quantum theory is quantization of field equations.*

> [Dirac, P. A. M. (February, 1928). The Quantum Theory of the Electron. *Roy. Soc. Proc., A*, 117, 778, 610–24; https://doi.org/10.1098/rspa.1928.0023. See Underwood, T. G. *Quantum Electrodynamics – annotated sources.* Volume I, pp. 534-46.
>
> The new quantum mechanics applied to the problem of the *structure of the atom with point-charge electrons* results in discrepancies consisting of "duplexity" phenomena, observed number of stationary states for an electron in an atom twice the number given by the theory, Goudsmit and Uhlenbeck introduced the idea of an electron with a *spin*, previous *relativity* treatments by Gordon and Klein obtain the operator of the wave equation by the same procedure as in the *non-relativity* theory, substitution of classical *quantum differential operators* for the *momentum vector* in the amended *relativistic Hamiltonian equation* and application of resulting differential operator to the *wave function* to obtain the *Klein-Gordon equation*, gives rise to two difficulties, the *first difficulty* is in the physical interpretation of solutions of ψ as the *charge* and the *current*, satisfactory for emission and absorption of radiation, provides probability of any dynamical variable at any specific time having a value between specified limits if they refer to the position of the electron, but, unlike the *non-relativity* theory, *not if they refer to its momentum or any other dynamical variable*, the *second difficulty* is that the conjugate imaginary of the wave equation is the same as that for an electron with charge – e and negative energy, *this paper is concerned only with the removal of the first of difficulties*, the resulting theory is only an approximation but appears sufficient to address duplexity problems without further assumptions, applies the method of *q-numbers* and using non-commutative algebra exhibits the properties of a free electron and of an electron in a central field of electric force, shows that simplest Hamiltonian for a *point charge electron satisfying requirements of both relativity*

and the general transformation theory of quantum mechanics leads to explanation of all duplexity phenomena of number of stationary states being twice the observed value without further assumption about spin, in contrast to the Schrödinger equation which described wave functions of only one complex value Dirac introduces *vectors of four complex numbers* (known as bispinors), results in a *relativistic equation of motion* for the *wave function of the electron* $\{p_0 + \rho_1 (\boldsymbol{\sigma}, \mathbf{p}) + \rho_3 mc\} \psi = 0$, referred to as the *Dirac equation*, where **p** is the *momentum* vector, and $\boldsymbol{\sigma}$ denotes the vector (σ_1, σ_2, σ_3), includes term equal to spin correction given by Darwin and Pauli, describes all spin-½ particles with mass, does not address second class of solutions of the wave equation in which *charge of the electron is positive* and *energy of a free electron is negative*.

[This work led Dirac to predict the existence of the positron, the electron's antiparticle, which he interpreted in terms of what came to be called the *Dirac sea*. The positron was observed by Carl Anderson in 1932.]

Dirac, P. A. M. (March, 1928). The quantum theory of the Electron. Part II. *Roy. Soc. Proc., A*, 118, 779, 351-61; https://doi.org/10.1098/rspa.1928.0056. See Underwood, T. G. *Quantum Electrodynamics – annotated sources.* Volume I, pp. 547-57.

Application of the *Dirac equation* to the conservation theorem, the selection principle, the relative intensities of the lines of a multiplet, and to the Zeeman effect.]

The Problem. - The translation of Dirac's theory of the electron into *general relativity* is not only of formal significance, for, as we know, the Dirac equations applied to an electron in a spherically symmetric *electrostatic field* yield *in addition to the correct energy levels those - or rather the negative of those - of an "electron" with opposite charge but the same mass*. In order to do away with these superfluous terms the *wave function* ψ must be robbed of one of its pairs ψ_1^+, ψ_2^+; ψ_1^-, ψ_2^- of components. These two pairs occur unmixed in the *action principle* except for the term

$$m\,(\psi_1^+\psi^*{}_1^- + \psi_2^+\psi^*{}_2^- + \psi_1^-\psi^*{}_2^+ + \psi_2^-\psi^*{}_2^+) \qquad (1)$$

which contains the *mass* m of the electron as a factor. *But mass is a gravitational effect: it is the flux of the gravitational field through a surface enclosing the particle in the same sense that charge is the flux of the electric field.* In a satisfactory theory it must therefore be as impossible to introduce a non-vanishing *mass* without the *gravitational field* as it is to introduce *charge* without *electromagnetic field*. *It is therefore certain that the term (1) can at most be right in the large scale, but must really be replaced by one which includes gravitation*; this may at the same time remove the defects of the present theory.

The direction in which such a modification is to be sought is clear: the *field equations* arising from an *action principle* - which shall give the true laws of interaction between electrons, protons and photons only after quantization - contain at present only the Schrodinger-Dirac quantity ψ, *which describes the wave field of the electron*, in addition to the four *potentials* φ_p of the *electromagnetic field. It is unconditionally necessary to introduce the wave field of the proton before quantizing*. But since the ψ of the electron can only involve two components, ψ_1^+, ψ_2^+should be ascribed to the electron and ψ_1^-, ψ_2^- to the proton. *Obviously, the present expression,* $-e\,\psi\hat{}\,\psi$ *for charge-density*[#],

[#] The circumflex indicates transition to the conjugate of the transposed matrix (Hermitean conjugate). The four components of ψ are considered as the elements of a matrix with four rows and one column.

being necessarily negative, runs counter to this, and something must consequently be changed in this respect. Instead of one law for the conservation of charge we must have two, expressing the conservation of the number of electrons and protons separately.

If one introduces the quantities $e\varphi_p/ch$ instead of φ_p (and calls them φ_p), the field equations contain only the following combinations of atomistic constants: the pure number $\alpha = e^2/ch$ and h/mc, the "wave-length" of the electron[#].

[#] $h/2\pi$ is Planck's constant.

Hence the equations certainly do not alone suffice to explain the atomistic behavior of matter with the definite values of e, m and h. But *the subsequent quantization introduces the quantum of action h, and this together with the wave-length h/mc will be* sufficient, *since the velocity of light c is determined as an absolute measure of velocity by the theory of relativity.*

The introduction of the atomic constants by the quantum theory - or at least that of the *wavelength* - into the *field equations* has removed the support from under my *principle of gauge-invariance*, by means of which I had hoped to unify *electricity* and *gravitation*. But as I have remarked, *it possesses an equivalent in the field equations of quantum theory* which is its perfect counterpart in formal respects: the laws are invariant under the simultaneous substitution of $e^{i\lambda}\psi$ for ψ and $\varphi_p - \partial\lambda/\partial x_p$ for φ_p, where λ is an arbitrary function of position in space and time. The connection of this invariance with the *conservation law of electricity* remains exactly as before: the fact that the *action integral* is unaltered by the infinitesimal variation

$$\delta\psi = i\,\lambda\psi, \qquad \delta\varphi_p = -\,\partial\lambda/\partial x_p$$

(λ an arbitrary infinitesimal function) *signifies the identical fulfilment of a dependence between the material and the electromagnetic laws* which arise from the *action integral* by variations of the ψ and φ, respectively; it means that the *conservation of electricity* is a double consequence of them, that *it follows from the laws of matter as well as electricity.* This new *principle of gauge invariance*, which may go by the same name, has the character of *general relativity* since it contains an arbitrary function λ, and can certainly only be understood with reference to it.

It was such considerations as these, and not the desire for formal generalizations, which led me to attempt the incorporation of the Dirac theory into the scheme of general relativity. We establish the metric in a world point P by a "Cartesian" system of axes (instead of the g_{pq}) consisting of four vectors $e(\alpha)$ $\{\alpha = 0, 1, 2, 3\}$ of which e(l), e(2), e(3) are real space-like vectors while e(0)/i is a real time-like vector of which we expressly demand that it be directed toward the future. A rotation of these axes is an orthogonal or Lorentz transformation which leaves these conditions of reality and sign unaltered. The laws shall remain invariant when the axes in the various points P are subjected to arbitrary and independent rotations. In addition to these we need four (real) coordinates x_p (p = 0, 1, 2, 3) for the purpose of analytic expression. The components of $e(\alpha)$ in this coordinate system are designated by $e^p(\alpha)$. We need such local cartesian axes $e(\alpha)$ in each point P in order to be able to describe the quantity ψ by means of its components ψ_1^+, ψ_2^+; ψ_1^-, ψ_2^- for the law of transformation of the components ψ can only be given for orthogonal transformations as it corresponds to a representation of the orthogonal group which cannot be extended to the group of all linear transformations. *The tensor calculus is consequently an unusable instrument for considerations involving the* ψ[#].

[#] Attempts to employ only the tensor calculus have been made by Tetrode [Tetrode, H. (May, 1928). Allgemein-relativistische Quantentheorie des Elektrons. (General-relativistic quantum theory of the electron.) *Zeit. Phys.*, 50, 336-46; https://doi.org/10.1007/BF01347512; translation by D. H. Delphenich; https://neo-classical-physics.info/uploads/3/4/3/6/34363841/tetrode_-_impulse-energy_theorem.pdf)]; Whittaker, J. M. (1928). *Proc. Camb. Phil. Soc.*, 25, 501, and others; I consider them misleading.

In formal aspects our theory resembles the more recent attempts of Einstein to unify electricity and gravitation[$].

[$] Einstein, A. (1928). Riemanngeometrie mit Aufrechterhaltung des Begriffes des Fern-Parallelismus. (Riemannian Geometry with Preservation of the Concept of Distant Parallelism.) *Sitzungsber. Berl. Akad.*, 217-21; (1929). Neue Möglichkeit für eine einheitliche Feldtheorie von Gravitation und Elektrizität. (New Possibility for a Unified Field Theory of Gravity and Electricity.) *Ibidem*, 224-7.

But here there is no talk of "distant parallelism"; there is no indication that Nature has availed herself of such an artificial geometry. *I am convinced that if there is a physical content in Einstein's latest formal developments it must come to light in the present connection.* It seems to me that it is now hopeless to seek a unification of *gravitation* and *electricity* without taking *material waves* into account.

...

It should be noted that our *field equations* contain neither the theory of a single electron nor that of a single proton. *One might rather consider them as the laws governing a hydrogen atom consisting of an electron and a proton*; but here again, the problem of interaction between the two may first require quantization. *What we have obtained is solely a field scheme which can only be applied to and compared-with experience after the quantization has been accomplished.* We know from the Pauli *exclusion principle*[§] what commutation rules are to be applied in the quantization of ψ^+; those for ψ^- must be the same in our theory.

> [§] Jordan, P. & Wigner, E. (September, 1928). über das Paulische Äquivalenzverbot. (On the Paulian prohibition of equivalence.) *Ibidem*, 47, 631; https://doi.org/10.1007/BF01331938.

The commutation relations between ψ^+ and ψ^- are as yet entirely unknown. Those of the *electromagnetic field* (photons) are almost completely known. In this respect we know nothing concerning the *gravitational field.* The commutation rules for F are here almost completely fixed by those for ψ, by the condition that these latter be unaltered when ψ is given the increment $\delta\psi = i\ F(\alpha)\psi$ [where F_p are the components of *electromagnetic potential*]. *That the rules thus obtained are in agreement with experience is indeed a support for our theory; i.e., it tells us why the "anti-symmetric," Pauli-Fermi statistics for electrons leads to the "symmetric" Bose-Einstein statistics for photons.* A definite decision can, however, first be reached *when the barrier which hems the progress of quantum theory is overcome: the quantization of the field equations.*

Einstein, A., Podolsky, B. & Rosen, N. (May, 1935). Can Quantum-Mechanical Description of Physical Reality Be Considered Complete?

Phys. Rev., 47, 10, 777-80; https://doi.org/10.1103/PhysRev.47.777.

Received 25 March 1935.

Institute for Advanced Study, Princeton, New Jersey.

The description of reality as given by a wave function in quantum mechanics is not complete.

Abstract

In a complete theory there is an element corresponding to each element of reality. A sufficient condition for the reality of a physical quantity is the possibility of predicting it with certainty, without disturbing the system. In quantum mechanics in the case of two physical quantities described by non-commuting operators, the knowledge of one precludes the knowledge of the other. Then either (1) the description of reality given by the wave function in quantum mechanics is not complete or (2) these two quantities cannot have simultaneous reality. Consideration of the problem of making predictions concerning a system on the basis of measurements made on another system that had previously interacted with it leads to the result that if (1) is false then (2) is also false. One is thus led to conclude that the description of reality as given by a wave function is not complete.

Application of quantum theory to gravity - quantum entanglement.

Wikipedia: "*Quantum entanglement*" is the phenomenon that occurs when a duet of particles are generated, interact, or share spatial proximity *in such a way that the quantum state of each particle of the group cannot be described independently of the state of the others, including when the particles are separated by a large distance*. The topic of quantum entanglement is at the heart of the disparity between classical and quantum physics: entanglement is a primary feature of quantum mechanics not present in classical mechanics.

Measurements of physical properties such as position, momentum, spin, and polarization performed on entangled particles can, in some cases, be found to be perfectly correlated. For example, if a pair of entangled particles is generated such that their total spin is known to be zero, and one particle is found to have clockwise spin on a first axis, then the spin of the other particle, measured on the same axis, is found to be anticlockwise. However, this behavior gives rise to seemingly paradoxical effects: any measurement of a particle's properties results in an apparent and irreversible wave function collapse of that particle and changes the original quantum state. With entangled particles, such measurements affect the entangled system as a whole.

Such phenomena were the subject of a 1935 paper by Albert Einstein, Boris Podolsky, and Nathan Rosen,[1]

> [1] Einstein, A., Podolsky, B., Rosen, N. (1935). Can Quantum-Mechanical Description of Physical Reality Be Considered Complete? *Phys. Rev.*, 47, 10: 777–80. doi:10.1103/PhysRev.47.777.

and several papers by Erwin Schrödinger shortly thereafter,[2, 3]

> [2] Schrödinger, E. (1935). Discussion of probability relations between separated systems. *Mathematical Proceedings of the Cambridge Philosophical Society*. 31, 4: 555–63. doi:10.1017/S0305004100013554.
>
> [3] Schrödinger, E. (1936). Probability relations between separated systems. *Mathematical Proceedings of the Cambridge Philosophical Society*. 32, 3: 446–52. doi:10.1017/S0305004100019137.

describing what came to be known as the EPR paradox. Einstein and others considered such behavior impossible, as it violated the local realism view of causality (Einstein referring to it as "spooky action at a distance") and argued that the accepted formulation of quantum mechanics must therefore be incomplete.

Later, however, the counterintuitive predictions of quantum mechanics were verified in tests where polarization or spin of entangled particles were measured at separate locations, statistically violating Bell's inequality. In earlier tests, it could not be ruled out that the result at one point could have been subtly transmitted to the remote point, affecting the outcome at the second location. However, so-called "loophole-free" Bell tests have since been performed where the locations were sufficiently separated that communications at the speed of light would have taken longer—in one case, 10,000 times longer—than the interval between the measurements.

According to *some* interpretations of quantum mechanics, the effect of one measurement occurs instantly. Other interpretations which do not recognize wavefunction collapse dispute that there is any "effect" at all. However, all interpretations agree that entanglement produces correlation between the measurements, and that the mutual information between the entangled particles can be exploited, but that any transmission of information at faster-than-light speeds is impossible.[4]

[4] Penrose, R. (2004). *The Road to Reality: A Complete Guide to the Laws of the Universe,* London, p. 603.

Quantum entanglement has been demonstrated experimentally with photons, electrons, and even small diamonds. Despite popular thought to the contrary, quantum entanglement cannot be used for faster-than-light communication.

Einstein, Podolsky and Rosen's thought experiment attempted to show that "the quantum-mechanical description of physical reality given by wave functions is not complete." However, the three scientists did not coin the word entanglement, nor did they generalize the special properties of the quantum state they considered. Following the EPR paper, Erwin Schrödinger wrote a letter to Einstein in German in which he used the word Verschränkung (translated by himself as entanglement) "to describe the correlations between two particles that interact and then separate, as in the EPR experiment."
However, Schrödinger had discussed the phenomenon as early as 1932.[5]

[5] Christandl, M. (2006). The Structure of Bipartite Quantum States – Insights from Group Theory and Cryptography; arXiv:quant-ph/0604183.

Schrödinger shortly thereafter published his seminal paper defining and discussing the notion of "entanglement."[2]

[2] Schrödinger, E. (1935). Discussion of probability relations between separated systems. *Mathematical Proceedings of the Cambridge Philosophical Society*. 31, 4, 555–63. doi:10.1017/S0305004100013554.

In the paper, he recognized the importance of the concept, and stated: "I would not call [entanglement] one but rather the characteristic trait of quantum mechanics, the one that enforces its entire departure from classical lines of thought."

Like Einstein, Schrödinger was dissatisfied with the concept of entanglement, because it seemed to violate the speed limit on the transmission of information implicit in the theory of relativity. Einstein later famously derided entanglement as "spukhafte Fernwirkung" or "spooky action at a distance."

The EPR paper generated significant interest among physicists, which inspired much discussion about the foundations of quantum mechanics and Bohm's interpretation in particular, but produced relatively little other published work. Despite the interest, the weak point in EPR's argument was not discovered until 1964, when John Stewart Bell proved that one of their key assumptions, the principle of locality, as applied to the kind of hidden variables interpretation hoped for by EPR, was mathematically inconsistent with the predictions of quantum theory.

Specifically, Bell demonstrated an upper limit, seen in Bell's inequality, regarding the strength of correlations that can be produced in any theory obeying local realism, and showed that quantum theory predicts violations of this limit for certain entangled systems.[6]

[6] Bell, J. S. (1964). On the Einstein-Poldolsky-Rosen paradox. *Physics Physique Физика.* 1, 3, 195–200. doi:10.1103/PhysicsPhysiqueFizika.1.195.

His inequality is experimentally testable, and there have been numerous relevant experiments, starting with the pioneering work of Stuart Freedman and John Clauser in 1972 and Alain Aspect's experiments in 1982.

An early experimental breakthrough was due to Carl Kocher, who already in 1967 presented an apparatus in which two photons successively emitted from a calcium atom were shown to be entangled – the first case of *entangled visible light*. The two photons passed diametrically positioned parallel polarizers with higher probability than classically predicted but with correlations in quantitative agreement with quantum mechanical calculations. He also showed that the correlation varied as the squared cosine of the angle between the polarizer settings and decreased exponentially with time lag between emitted photons. Kocher's apparatus, equipped with better polarizers, was used by Freedman and Clauser who could confirm the cosine-squared dependence and use it to demonstrate a violation of Bell's inequality for a set of fixed angles. All these experiments have shown agreement with quantum mechanics rather than the principle of local realism.

For decades, each had left open at least one loophole by which it was possible to question the validity of the results. However, in 2015 an experiment was performed that simultaneously closed both the detection and locality loopholes, and was heralded as "loophole-free"; this experiment ruled out a large class of local realism theories with certainty. Aspect writes that "... no experiment ... can be said to be totally loophole-free," but he says the experiments "remove the last doubts that we should renounce" local hidden variables, and refers to examples of remaining loopholes as being "far-fetched" and "foreign to the usual way of reasoning in physics".

Bell's work raised the possibility of using these super-strong correlations as a resource for communication. It led to the 1984 discovery of quantum key distribution protocols, most famously BB84 by Charles H. Bennett and Gilles Brassard and E91 by Artur Ekert. Although BB84 does not use entanglement, Ekert's protocol uses the violation of a Bell's inequality as a proof of security.

In 2022, the Nobel Prize in Physics was awarded to Aspect, Clauser, and Anton Zeilinger "for experiments with entangled photons, establishing the violation of Bell inequalities and pioneering quantum information science".

Meaning of entanglement

An entangled system is defined to be one whose quantum state cannot be factored as a product of states of its local constituents; that is to say, they are not individual particles but are an inseparable whole. In entanglement, one constituent cannot be fully described without considering the other(s). The state of a composite system is always expressible as a sum, or superposition, of products of states of local constituents; it is entangled if this sum cannot be written as a single product term.

Quantum systems can become entangled through various types of interactions. Entanglement is broken when the entangled particles decohere through interaction with the environment; for example, when a measurement is made.

As an example of entanglement: a subatomic particle decays into an entangled pair of other particles. The decay events obey the various conservation laws, and as a result, the measurement outcomes of one daughter particle must be highly correlated with the measurement outcomes of the other daughter particle (so that the total momenta, angular momenta, energy, and so forth remains roughly the same before and after this process). For instance, a spin-zero particle could decay into a pair of spin-1/2 particles. Since the total spin before and after this decay must be zero (conservation of angular momentum), whenever the first particle is measured to be spin up on some axis, the other, when measured on the same axis, is always found to be spin down. (This is called the spin anti-correlated case; and if the prior probabilities for measuring each spin are equal, the pair is said to be in the singlet state.)

The above result may or may not be perceived as surprising. *A classical system would display the same property*, and a hidden variable theory would certainly be required to do so, based on conservation of angular momentum in classical and quantum mechanics alike. *The difference is that a classical system has definite values for all the observables all along, while the quantum system does not.* In a sense to be discussed below, the quantum system considered here seems to acquire a probability distribution for the outcome of a measurement of the spin along any axis of the other particle upon measurement of the first particle. This probability distribution is in general different from what it would be without measurement of the first particle. This may certainly be perceived as surprising in the case of spatially separated entangled particles.

Paradox

The paradox is that a measurement made on either of the particles apparently collapses the state of the entire entangled system—and does so instantaneously, before any information about the measurement result could have been communicated to the other particle (assuming that information cannot travel faster than light) and hence assured the "proper" outcome of the measurement of the other part of the entangled pair. In the Copenhagen interpretation, the result of a spin measurement on one of the particles is a collapse (of wave function) into a state in which each particle has a definite spin (either up or down) along the axis of measurement. The outcome is taken to be random, with each possibility having a probability of 50%. However, if both spins are measured along the same axis, they are found to be anti-correlated. This means that the random outcome of the measurement made on one particle seems to have been transmitted to the other, so that it can make the "right choice" when it too is measured.[7]

[7] Anderson, R. W. (March 28, 2015). *The Cosmic Compendium: Interstellar Travel* (First ed.). The Cosmic Compendium, p. 100.

The distance and timing of the measurements can be chosen so as to make the interval between the two measurements spacelike, hence, any causal effect connecting the events would have to travel faster than light. According to the principles of *special relativity*, it is not possible for any information to travel between two such measuring events. It is not even possible to say which of the measurements came first. For two spacelike separated events x1 and x2 there are inertial frames in which x1 is first and others in which x2 is first. Therefore, the correlation between the two measurements cannot be explained as one measurement determining the other: different observers would disagree about the role of cause and effect.

(In fact similar paradoxes can arise even without entanglement: the position of a single particle is spread out over space, and two widely separated detectors attempting to detect the particle in two different places must instantaneously attain appropriate correlation, so that they do not both detect the particle.)

Hidden variables theory

A possible resolution to the paradox is to assume that quantum theory is incomplete, and the result of measurements depends on predetermined "*hidden variables*". The state of the particles being measured contains some *hidden variables*, whose values effectively determine, right from the moment of separation, what the outcomes of the spin measurements are going to be. This would mean that each particle carries all the required information with it, and nothing needs to be transmitted from one particle to the other at the time of measurement. *Einstein and others originally believed this was the only way out of the paradox, and the accepted quantum mechanical description (with a random measurement outcome) must be incomplete*.

Violations of Bell's inequality

Local hidden variable theories fail, however, when measurements of the spin of entangled particles along different axes are considered. If a large number of pairs of such measurements are made (on a large number of pairs of entangled particles), then statistically, if the local realist or hidden variables view were correct, the results would always satisfy Bell's inequality. A number of experiments have shown in practice that Bell's inequality is not satisfied. However, prior to 2015, all of these had loophole problems that were considered the most important by the community of physicists. When measurements of the entangled particles are made in moving *relativisti*c reference frames, in which each measurement (in its own relativistic time frame) occurs before the other, the measurement results remain correlated.

The fundamental issue about measuring spin along different axes is that these measurements cannot have definite values at the same time—they are incompatible in the sense that these measurements' maximum simultaneous precision is constrained by the uncertainty principle. This is contrary to what is found in classical physics, where any number of properties can be measured simultaneously with arbitrary accuracy. It has been proven mathematically that compatible measurements cannot show Bell-inequality-violating correlations, and thus entanglement is a fundamentally non-classical phenomenon.

Emergence of time from quantum entanglement

There is a fundamental conflict, referred to as the problem of time, between the way the concept of time is used in quantum mechanics, and the role it plays in general relativity. In standard quantum theories time acts as an independent background through which states evolve, with the Hamiltonian operator acting as the generator of infinitesimal translations of quantum states through time. In contrast, general relativity treats time as a dynamical variable which relates directly with matter and moreover requires the Hamiltonian constraint to vanish. …

Emergent gravity

Based on AdS/CFT correspondence, Mark Van Raamsdonk suggested that spacetime arises as an emergent phenomenon of the quantum degrees of freedom that are entangled and live in the boundary of the space-time. *Induced gravity can emerge from the entanglement first law.*[8]

> [8] Lee, J-W, Kim, H-C, Lee, J. (2013). Gravity from quantum information. *Journal of the Korean Physical Society*. 63, 5, 1094–8. arXiv:1001.5445. doi:10.3938/jkps.63.1094.

…

Naturally entangled systems

The electron shells of multi-electron atoms always consist of entangled electrons. The correct ionization energy can be calculated only by consideration of electron entanglement.[9]

> [9] Jensen, F. (2007). *Introduction to Computational Chemistry*. Wiley.

…

Big bang theory.

Wikipedia: The Big Bang event is a physical theory that describes how the universe expanded from an initial state of high density and temperature. It was first proposed in 1927 by Roman Catholic priest and physicist Georges Lemaître. Various cosmological models of the Big Bang explain the evolution of the observable universe from the earliest known periods through its subsequent large-scale form. These models offer a comprehensive explanation for a broad range of observed phenomena, including the abundance of light elements, the *cosmic microwave background* (CMB) radiation, and large-scale structure. The overall uniformity of the Universe, known as the flatness problem, is explained through *cosmic inflation*: a sudden and very rapid expansion of space during the earliest moments. However, physics currently lacks a widely accepted *theory of quantum gravity* that can successfully model the earliest conditions of the Big Bang.

Crucially, these models are compatible with the Hubble–Lemaître law—the observation that the farther away a galaxy is, the faster it is moving away from Earth. Extrapolating this cosmic expansion backwards in time using the known laws of physics, the models describe an increasingly concentrated cosmos preceded by a singularity in which space and time lose meaning (typically named "the Big Bang singularity"). In 1964 the CMB was discovered, which convinced many cosmologists that the competing steady-state model of cosmic evolution was falsified, since the Big Bang models predict a uniform background radiation caused by high temperatures and densities in the distant past. A wide range of empirical evidence strongly favors the Big Bang event, which is now essentially universally accepted. Detailed measurements of the expansion rate of the universe place the Big Bang singularity at an estimated 13.787±0.020 billion years ago, which is considered the age of the universe.

There remain aspects of the observed universe that are not yet adequately explained by the Big Bang models. *After its initial expansion, the universe cooled sufficiently to allow the formation of subatomic particles, and later atoms*. The unequal abundances of matter and antimatter *that allowed this to occur* is an unexplained effect known as *baryon asymmetry*. These primordial elements—mostly hydrogen, with some helium and lithium—later coalesced through *gravity*, forming early stars and galaxies. *Astronomers observe the gravitational effects of an unknown dark matter surrounding galaxies*. Most of the gravitational potential in the universe seems to be in this form, and the Big Bang models and various observations indicate that *this excess gravitational potential is not created by baryonic matter, such as normal atoms*. Measurements of the redshifts of supernovae indicate that *the expansion of the universe is accelerating*, an observation attributed to an unexplained phenomenon known as *dark energy*.[1]

[1] Peebles, P. J. E., Ratra, B. (April 22, 2003). The cosmological constant and dark energy. *Reviews of Modern Physics*. 75, 2, 559–606; arXiv:astro-ph/0207347.

Features of the models

The Big Bang models offer a comprehensive explanation for a broad range of observed phenomena, including the abundances of the light elements, the CMB, large-scale structure, and Hubble's law. The models depend on two major assumptions: the universality of physical laws and the cosmological principle. The universality of physical laws is one of the underlying principles of the theory of relativity. The cosmological principle states that on large scales the universe is homogeneous and isotropic—appearing the same in all directions regardless of location.

These ideas were initially taken as postulates, but later efforts were made to test each of them. For example, the first assumption has been tested by observations showing that the largest possible deviation of the fine-structure constant over much of the age of the universe is of order 10^{-5}. ...

Horizons

An important feature of the Big Bang spacetime is the presence of particle horizons. Since the universe has a finite age, and light travels at a finite speed, there may be events in the past whose light has not yet had time to reach earth. This places a limit or a *past horizon* on the most distant objects that can be observed. Conversely, because space is expanding, and more distant objects are receding ever more quickly, light emitted by us today may never "catch up" to very distant objects. This defines a *future horizon*, which limits the events in the future that we will be able to influence. The presence of either type of horizon depends on the details of the FLRW model that describes our universe.

Our understanding of the universe back to very early times suggests that there is a past horizon, though in practice our view is also limited by the opacity of the universe at early times. So our view cannot extend further backward in time, though the horizon recedes in space. If the expansion of the universe continues to accelerate, there is a future horizon as well.

Thermalization

Some processes in the early universe occurred too slowly, compared to the expansion rate of the universe, to reach approximate thermodynamic equilibrium. Others were fast enough to reach thermalization. The parameter usually used to find out whether a process in the very early universe has reached thermal equilibrium is the ratio between the rate of the process (usually rate of collisions between particles) and the Hubble parameter. The larger the ratio, the more time particles had to thermalize before they were too far away from each other.

Timeline

According to the Big Bang models, the universe at the beginning was very hot and very compact, and since then it has been expanding and cooling.

Singularity

Extrapolation of the expansion of the universe backwards in time using *general relativity* yields an infinite density and temperature at a finite time in the past.[2]

> [2] Hawking, S. W., Ellis, G. F. R. (1973). *The Large-Scale Structure of Space-Time*. Cambridge University Press, Cambridge, UK.

This irregular behavior, known as the gravitational singularity, indicates that *general relativity is not an adequate description of the laws of physics in this regime. Models based on general relativity alone cannot fully extrapolate toward the singularity.* In some proposals, such as the emergent Universe models, the singularity is replaced by another cosmological epoch. A different approach identifies the initial singularity as a singularity predicted by some models of the Big Bang theory to have existed before the Big Bang.

Based on measurements of the expansion using Type Ia supernovae and measurements of temperature fluctuations in the cosmic microwave background, the time that has passed since that event—known as the "age of the universe"—is 13.8 billion years.[3]

> [3] Planck Collaboration (October 2016). Planck 2015 results. XIII. Cosmological parameters. *Astronomy & Astrophysics*. 594, A13; arXiv:1502.01589.

Despite being extremely dense at this time—far denser than is usually required to form a black hole—the universe did not re-collapse into a singularity. Commonly used calculations and limits for explaining gravitational collapse are usually based upon objects of relatively constant size, such as stars, and do not apply to rapidly expanding space such as the Big Bang. Since the early universe did not immediately collapse into a multitude of black holes, matter at that time must have been very evenly distributed with a negligible density gradient.

Inflation and baryogenesis

The *earliest phases of the Big Bang* are subject to much speculation, since astronomical data about them are not available. In the most common models the universe was filled homogeneously and isotropically with a very high energy density and huge temperatures and pressures, and *was very rapidly expanding and cooling*. The period up to 10^{-43} seconds into the expansion (*Planck time*), the Planck epoch, was *a phase in which the four fundamental forces—the electromagnetic force, the strong nuclear force, the weak nuclear force, and the gravitational force, were unified as one*.[4]

[4] Unruh, W. G., Semenoff, G. W., eds. (1988). *The Early Universe*. Reidel.

In this stage, *the characteristic scale length of the universe was the Planck length, 1.6×10^{-35} m*, and consequently had a temperature of approximately 10^{32} degrees Celsius. Even the very concept of a particle breaks down in these conditions. *A proper understanding of this period awaits the development of a theory of quantum gravity*. The Planck epoch was succeeded by the grand unification epoch beginning at 10^{-43} seconds, *where gravitation separated from the other forces as the universe's temperature fell*.[4]

At approximately 10^{-37} seconds into the expansion, *a phase transition caused a cosmic inflation, during which the universe grew exponentially, unconstrained by the light speed invariance, and temperatures dropped by a factor of 100,000*. This concept is motivated by the *flatness problem*, where the density of matter and energy is very close to the critical density needed to produce a flat universe. *That is, the shape of the universe has no overall geometric curvature due to gravitational influence*. Microscopic quantum fluctuations that occurred because of Heisenberg's uncertainty principle were "frozen in" by inflation, becoming amplified into the seeds that would later form the large-scale structure of the universe.[5]

[5] Guth, A. H. (1998). [Originally published 1997.] *The Inflationary Universe: Quest for a New Theory of Cosmic Origins*. Foreword by Alan Lightman. Vintage Books, London.

At a time around 10^{-36} seconds, the electroweak epoch begins when the strong nuclear force separates from the other forces, with only the electromagnetic force and weak nuclear force remaining unified.

Inflation stopped locally at around 10^{-33} to 10^{-32} seconds, with the observable universe's volume having increased by a factor of at least 10^{78}. Reheating occurred until the universe obtained the temperatures required for the production of a quark–gluon plasma as well as all other elementary particles. Temperatures were so high that the random motions of particles were at relativistic speeds, and particle–antiparticle pairs of all kinds were being continuously created and destroyed in collisions. *At some point, an unknown reaction called baryogenesis violated the conservation of baryon number, leading to a very small excess of quarks and leptons over antiquarks and antileptons—of the order of one part in 30 million. This resulted in the predominance of matter over antimatter in the present universe.*

Cooling

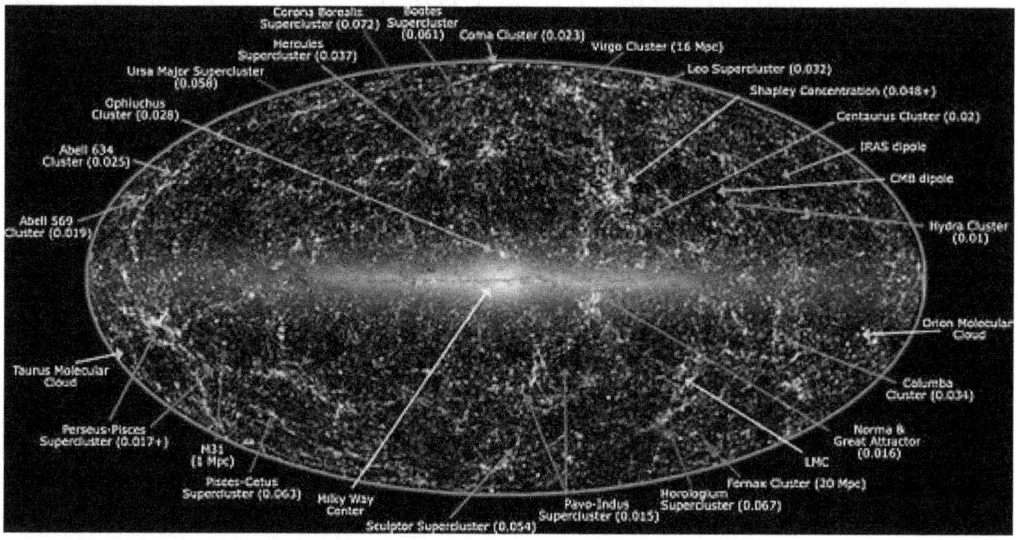

Panoramic view of the entire near-infrared sky reveals the distribution of galaxies beyond the Milky Way. Galaxies are color-coded by redshift.

The universe continued to decrease in density and fall in temperature, hence the typical energy of each particle was decreasing. Symmetry-breaking phase transitions put the fundamental forces of physics and the parameters of elementary particles into their present form, with *the electromagnetic force and weak nuclear force separating at about 10^{-12} seconds.*

After about 10^{-11} seconds, the picture becomes less speculative, since particle energies drop to values that can be attained in particle accelerators. *At about 10^{-6} seconds, quarks and gluons combined to form baryons such as protons and neutrons.* The small excess of quarks over antiquarks led to a small excess of baryons over antibaryons. The temperature was no longer high enough to create either new proton–antiproton or neutron–antineutron pairs. A *mass annihilation* immediately followed, leaving just one in 108 of the original matter particles and *none of their antiparticles*. A similar process happened at about 1 second for *electrons and positrons*. After these annihilations, the remaining protons, neutrons and electrons were no longer moving relativistically and *the energy density of the universe was dominated by photons* (with a minor contribution from neutrinos).

A few minutes into the expansion, when the temperature was about a billion kelvin and the density of matter in the universe was comparable to the current density of Earth's atmosphere, neutrons combined with protons to form the universe's deuterium and helium nuclei in a process called Big Bang nucleosynthesis (BBN). Most protons remained uncombined as hydrogen nuclei.

As the universe cooled, the *rest energy density of matter came to gravitationally dominate that of the photon radiation. After about 379,000 years, the electrons and nuclei combined into atoms (mostly hydrogen), which were able to emit radiation.* This relic radiation, which continued through space largely unimpeded, is known as the *cosmic microwave background.*

Structure formation

Over a long period of time, the slightly denser regions of the uniformly distributed matter gravitationally attracted nearby matter and thus grew even denser, forming gas clouds, stars, galaxies, and the other astronomical structures observable today. The details of this process depend on the amount and type of matter in the universe. The four possible types of matter are known as cold dark matter (CDM), warm dark matter, hot dark matter, and baryonic matter. The best measurements available, from the Wilkinson Microwave Anisotropy Probe (WMAP), show that the data is well-fit by a Lambda-CDM model in which dark matter is assumed to be cold. (Warm dark matter is ruled out by early reionization.) This CDM is estimated to make up about 23% of the matter/energy of the universe, while baryonic matter makes up about 4.6%.[6]

[6] Jarosik, N., Bennett, C. L., Dunkley, J., et al. (February 2011). Seven-Year Wilkinson Microwave Anisotropy Probe (WMAP) Observations: Sky Maps, Systematic Errors, and Basic Results. *The Astrophysical Journal Supplement Series*. 192, 2, 14; arXiv:1001.4744; doi:10.1088/0067-0049/192/2/14.

In an "extended model" which includes hot dark matter in the form of neutrinos, then the "physical baryon density" $\Omega_b h^2$ is estimated at 0.023. (This is different from the 'baryon density' Ω_b expressed as a fraction of the total matter/energy density, which is about 0.046.) The corresponding cold dark matter density $\Omega_c h^2$ is about 0.11, and the corresponding neutrino density $\Omega_v h^2$ is estimated to be less than 0.0062.

Cosmic acceleration

Independent lines of evidence from Type Ia supernovae and the CMB imply that the universe today is dominated by a mysterious form of energy known as *dark energy*, which appears to homogeneously permeate all of space. Observations suggest that 73% of the total energy density of the present-day universe is in this form. When the universe was very young it was likely infused with *dark energy, but with everything closer together, gravity predominated,* braking the expansion. Eventually, after billions of years of expansion, *the declining density of matter relative to the density of dark energy allowed the expansion of the universe to begin to accelerate.*[1]

[1] *Dark energy* in its simplest formulation is modeled by a cosmological constant term in Einstein field equations of general relativity, but its composition and mechanism are unknown. More generally, the details of its equation of state and relationship with the Standard Model of particle physics continue to be investigated both through observation and theory.

All of this cosmic evolution after the inflationary epoch can be rigorously described and modeled by the lambda-CDM model of cosmology, which uses the independent frameworks of quantum mechanics and *general relativity.* There are no easily testable models that would describe the situation prior to approximately 10^{-15} seconds. Understanding this earliest of eras in the history of the universe is one of the greatest unsolved problems in physics.

Concept history

Etymology

English astronomer Fred Hoyle is credited with coining the term "Big Bang" during a talk for a March 1949 BBC Radio broadcast, saying: "These theories were based on the hypothesis that all the matter in the universe was created in one big bang at a particular time in the remote past." However, it did not catch on until the 1970s.

Development

The Big Bang models developed from observations of the structure of the universe and from theoretical considerations. In 1912, Vesto Slipher measured the first Doppler shift of a "spiral nebula" (spiral nebula is the obsolete term for spiral galaxies), and soon discovered that almost all such nebulae were receding from Earth. He did not grasp the cosmological implications of this fact, and indeed at the time it was highly controversial whether or not these nebulae were "island universes" outside our Milky Way. Ten years later, Alexander Friedmann, a Russian cosmologist and mathematician, derived the Friedmann equations from the *Einstein field equations*, showing that the universe might be expanding in contrast to the static universe model advocated by Albert Einstein at that time.[7]

[7] Friedman, A. (December 1922). Über die Krümmung des Raumes. *Zeitschrift für Physik* (in German). 10, 1, 377–86; doi:10.1007/BF01332580; translated in: Friedmann,

Alexander (December 1999). On the Curvature of Space. *General Relativity and Gravitation*. 31,12, 1991–2000; doi:10.1023/A:1026751225741.

In 1924, American astronomer Edwin Hubble's measurement of the great distance to the nearest spiral nebulae showed that these systems were indeed other galaxies. Starting that same year, Hubble painstakingly developed a series of distance indicators, the forerunner of the cosmic distance ladder, using the 100-inch (2.5 m) Hooker telescope at Mount Wilson Observatory. This allowed him to estimate distances to galaxies whose redshifts had already been measured, mostly by Slipher. In 1929, *Hubble discovered a correlation between distance and recessional velocity—now known as Hubble's law*.[8]

[8] Hubble, E. (March 15, 1929). A Relation Between Distance and Radial Velocity Among Extra-Galactic Nebulae. *Proceedings of the National Academy of Sciences*. 15, 3, 168–73; doi:10.1073/pnas.15.3.168.

[*Measurement of extragalactic distances.*

The *cosmic distance ladder* is the succession of methods by which astronomers determine the distances to celestial objects. A direct distance measurement of an astronomical object is possible only for those objects that are "close enough" (within about a thousand parsecs) to Earth. The techniques for determining distances to more distant objects are all based on various measured correlations between methods that work at close distances and methods that work at larger distances. Several methods rely on a standard candle, which is an astronomical object that has a known luminosity.

The ladder analogy arises because no single technique can measure distances at all ranges encountered in astronomy. Instead, one method can be used to measure nearby distances, a second can be used to measure nearby to intermediate distances, and so on. Each rung of the ladder provides information that can be used to determine the distances at the next higher rung.

The extragalactic distance scale.

The extragalactic distance scale is a series of techniques used today by astronomers to determine the distance of cosmological bodies beyond our own galaxy, which are not easily obtained with traditional methods. Some procedures use properties of these objects, such as stars, globular clusters, nebulae, and galaxies as a whole. Other methods are based more on the statistics and probabilities of things such as entire galaxy clusters.

Wilson–Bappu effect

Discovered in 1956 by Olin Wilson and M.K. Vainu Bappu, the Wilson–Bappu effect uses the effect known as spectroscopic parallax. Many stars have features in their spectra, such as the calcium K-line, that indicate their absolute magnitude. The distance to the star can then be calculated from its apparent magnitude using the distance modulus.

There are major limitations to this method for finding stellar distances. The calibration of the spectral line strengths has limited accuracy and it requires a correction for interstellar extinction. Though in theory this method has the ability to provide reliable distance calculations to stars up to 7 megaparsecs (Mpc), it is generally only used for stars at hundreds of kiloparsecs (kpc).

Classical Cepheids

Beyond the reach of the Wilson–Bappu effect, the next method relies on the period-luminosity relation of classical Cepheid variable stars to calculate the distance to Galactic and extragalactic classical Cepheids. Several problems complicate the use of Cepheids as standard candles and are actively debated, chief among them are: the nature and linearity of the period-luminosity relation in various passbands and the impact of metallicity on both the zero-point and slope of those relations, and the effects of photometric contamination (blending) and a changing (typically unknown) extinction law on Cepheid distances.

These unresolved matters have resulted in cited values for the Hubble constant ranging between 60 km/s/Mpc and 80 km/s/Mpc. Resolving this discrepancy is one of the foremost problems in astronomy since some cosmological parameters of the Universe may be constrained significantly better by supplying a precise value of the Hubble constant.

Cepheid variable stars were the key instrument in Edwin Hubble's 1923 conclusion that M31 (Andromeda) was an external galaxy, as opposed to a smaller nebula within the Milky Way. He was able to calculate the distance of M31 to 285 kpc, today's value being 770 kpc.

As detected thus far, NGC 3370, a spiral galaxy in the constellation Leo, contains the farthest Cepheids yet found at a distance of 29 Mpc. Cepheid variable stars are in no way perfect distance markers: at nearby galaxies they have an error of about 7% and up to a 15% error for the most distant.

Supernovae

There are several different methods for which supernovae can be used to measure extragalactic distances.

Measuring a supernova's photosphere

We can assume that a supernova expands in a spherically symmetric manner. If the supernova is close enough such that we can measure the angular extent, $\theta(t)$, of its photosphere, we can use the equation

$$\omega = \Delta\theta/\Delta t,$$

where ω is angular velocity, θ is angular extent. In order to get an accurate measurement, it is necessary to make two observations separated by time Δt. Subsequently, we can use

$$d = V_{ej}/\omega,$$

where d is the distance to the supernova, V_{ej} is the supernova's ejecta's radial velocity (it can be assumed that V_{ej} equals V_θ if spherically symmetric).

This method works only if the supernova is close enough to be able to measure accurately the photosphere. Similarly, the expanding shell of gas is in fact not perfectly spherical nor a perfect blackbody. Also, interstellar extinction can hinder the accurate measurements of the photosphere. This problem is further exacerbated by core-collapse supernova. All of these factors contribute to the distance error of up to 25%.

Type Ia light curves

Type Ia supernovae are some of the best ways to determine extragalactic distances. Ia's occur when a binary white dwarf star begins to accrete matter from its companion star. As the white dwarf gains matter, eventually it reaches its Chandrasekhar limit of 1.4 $M_\odot$.

Once reached, the star becomes unstable and undergoes a runaway nuclear fusion reaction. Because all Type Ia supernovae explode at about the same mass, their absolute magnitudes are all the same. This makes them very useful as standard candles. All Type Ia supernovae have a standard blue and visual magnitude of $MB \approx MV \approx -19.3 \pm 0.3$.

Therefore, when observing a Type Ia supernova, if it is possible to determine what its peak magnitude was, then its distance can be calculated. It is not intrinsically necessary to capture the supernova directly at its peak magnitude; using the multicolor light curve shape method (MLCS), the shape of the light curve (taken at any reasonable time after the initial explosion) is compared to a family of parameterized curves that will determine the absolute magnitude at the maximum brightness. This method also takes into effect interstellar extinction/dimming from dust and gas.

Similarly, the stretch method fits the particular supernovae magnitude light curves to a template light curve. This template, as opposed to being several light curves at different wavelengths (MLCS) is just a single light curve that has been stretched (or compressed) in time. By using this Stretch Factor, the peak magnitude can be determined.

Using Type Ia supernovae is one of the most accurate methods, particularly since supernova explosions can be visible at great distances (their luminosities rival that of the galaxy in which they are situated), much farther than Cepheid Variables (500 times farther). Much time has been devoted to the refining of this method. The current uncertainty approaches a mere 5%, corresponding to an uncertainty of just 0.1 magnitudes.

Novae in distance determinations

Novae can be used in much the same way as supernovae to derive extragalactic distances. There is a direct relation between a nova's max magnitude and the time for its visible light to decline by two magnitudes.

After novae fade, they are about as bright as the most luminous Cepheid variable stars, therefore both these techniques have about the same max distance: ~ 20 Mpc. The error in this method produces an uncertainty in magnitude of about ± 0.4.

Globular cluster luminosity function

Based on the method of comparing the luminosities of globular clusters (located in galactic halos) from distant galaxies to that of the Virgo Cluster, the globular cluster luminosity function carries an uncertainty of distance of about 20% (or 0.4 magnitudes).

US astronomer William Alvin Baum first attempted to use globular clusters to measure distant elliptical galaxies. He compared the brightest globular clusters in Virgo A galaxy with those in Andromeda, assuming the luminosities of the clusters were the same in both. Knowing the distance to Andromeda, Baum has assumed a direct correlation and estimated Virgo A's distance.

Baum used just a single globular cluster, but individual formations are often poor standard candles. Canadian astronomer René Racine assumed the use of the globular cluster luminosity function (GCLF) would lead to a better approximation.

It is assumed that globular clusters all have roughly the same luminosities within the universe. There is no universal globular cluster luminosity function that applies to all galaxies.

Planetary nebula luminosity function

Like the GCLF method, a similar numerical analysis can be used for planetary nebulae within far off galaxies. The planetary nebula luminosity function (PNLF) was first proposed in the late 1970s by Holland Cole and David Jenner. They suggested that all planetary nebulae might all have similar maximum intrinsic brightness, now calculated to be $M = -4.53$. This would therefore make them potential standard candles for determining extragalactic distances.

Surface brightness fluctuation method

The following method deals with the overall inherent properties of galaxies. These methods, though with varying error percentages, have the ability to make distance estimates beyond 100 Mpc, though it is usually applied more locally.

The surface brightness fluctuation (SBF) method takes advantage of the use of CCD cameras on telescopes. Because of spatial fluctuations in a galaxy's surface brightness, some pixels on these cameras will pick up more stars than others. However, as distance increases the picture will become increasingly smoother. Analysis of this describes a magnitude of the pixel-to-pixel variation, which is directly related to a galaxy's distance.

Sigma-D relation

The Sigma-D relation (or Σ-D relation), used in elliptical galaxies, relates the angular diameter (D) of the galaxy to its velocity dispersion. It is important to describe exactly what D represents, in order to understand this method. It is, more precisely, the galaxy's angular diameter out to the surface brightness level of 20.75 B-mag $arcsec^{-2}$. This surface brightness is independent of the galaxy's actual distance from us. Instead, D is inversely proportional to the galaxy's distance, represented as d. Thus, this relation does not employ standard candles. Rather, D provides a standard ruler.

This method has the potential to become one of the strongest methods of galactic distance calculators, perhaps exceeding the range of even the Tully–Fisher method. As of today, however, elliptical galaxies are not bright enough to provide a calibration for this method through the use of techniques such as Cepheids. Instead, calibration is done using more crude methods.

Overlap and scaling

A succession of distance indicators, which is the distance ladder, is needed for determining distances to other galaxies. The reason is that objects bright enough to be recognized and measured at such distances are so rare that few or none are present nearby, so there are too few examples close enough with reliable trigonometric parallax to calibrate the indicator. For example, Cepheid variables, one of the best indicators for nearby spiral galaxies, cannot yet be satisfactorily calibrated by parallax alone, though the Gaia space mission can now weigh in on that specific problem. The situation is further complicated by the fact that different stellar populations generally do not have all types of stars in them. Cepheids in particular are massive stars, with short lifetimes, so they will only be found in places where stars have very recently been formed. Consequently, because elliptical galaxies usually have long ceased to have large-scale star formation, they will not have Cepheids. Instead, distance indicators whose origins are in an older stellar population (like novae and RR Lyrae variables) must be used. However, RR Lyrae variables are less luminous than Cepheids, and novae are unpredictable and an intensive monitoring program—and luck during that program—is needed to gather enough novae in the target galaxy for a good distance estimate.

Because the more distant steps of the cosmic distance ladder depend upon the nearer ones, the more distant steps include the effects of errors in the nearer steps, both systematic and statistical ones. The result of these propagating errors means that distances in astronomy are rarely known to the same level of precision as measurements in the other sciences, and that the precision necessarily is poorer for more distant types of object.

Another concern, especially for the very brightest standard candles, is their "standardness": how homogeneous the objects are in their true absolute magnitude. For some of these different standard candles, the homogeneity is based on theories about the formation and evolution of stars and galaxies, and is thus also subject to uncertainties in those aspects. For the most luminous of distance indicators, the Type Ia supernovae, this homogeneity is known to be poor; however, no other class of object is bright enough to be detected at such large distances, so the class is useful simply because there is no real alternative.

The observational result of Hubble's Law, the proportional relationship between distance and the speed with which a galaxy is moving away from us (usually referred to as redshift) is a product of the cosmic distance ladder. Edwin Hubble observed that fainter galaxies are more redshifted. Finding the value of the Hubble constant was the result of decades of work by many astronomers, both in amassing the measurements of galaxy redshifts and in calibrating the steps of the distance ladder. Hubble's Law is the primary means we have for estimating the distances of quasars and distant galaxies in which individual distance indicators cannot be seen.]

Independently deriving Friedmann's equations in 1927, Georges Lemaître, a Belgian physicist and Roman Catholic priest, proposed that the recession of the nebulae was due to the expansion of the universe. He inferred the relation that Hubble would later observe, given the cosmological principle. In 1931, Lemaître went further and suggested that the evident expansion of the universe, if projected back in time, meant that the further in the past the smaller the universe was, until at some finite time in the past all the mass of the universe was concentrated into a single point, a "primeval atom" where and when the fabric of time and space came into existence.

After World War II, two distinct possibilities emerged. One was Fred Hoyle's steady-state model, whereby new matter would be created as the universe seemed to expand. In this model the universe is roughly the same at any point in time.[9]

[9] Hoyle, F. (October 1948). A New Model for the Expanding Universe. *Monthly Notices of the Royal Astronomical Society*, 108, 5, 372–82; doi:10.1093/mnras/108.5.372.

The other was Lemaître's Big Bang theory, advocated and developed by George Gamow, who introduced BBN[10]

[10] Alpher, R. A., Bethe, H., Gamow, G. (April 1, 1948). The Origin of Chemical Elements. *Physical Review*, 73, 7, 803–4; doi:10.1103/PhysRev.73.803.

and whose associates, Ralph Alpher and Robert Herman, predicted the CMB.[11]

[11] Alpher, R. A., Herman, R. (November 13, 1948). Evolution of the Universe. *Nature*, 162, 4124, 774–5; doi:10.1038/162774b0.

For a while, support was split between these two theories. Eventually, the observational evidence, most notably from radio source counts, began to favor Big Bang over steady state. The discovery and confirmation of the CMB in 1964 secured the Big Bang as the best theory of the origin and evolution of the universe.

In 1968 and 1970, Roger Penrose, Stephen Hawking, and George F. R. Ellis published papers where they showed that mathematical singularities were an inevitable initial condition of *relativistic* models of the Big Bang.[12]

[12] Hawking, S. W., Penrose, R. (January 27, 1970). The Singularities of Gravitational Collapse and Cosmology. *Proceedings of the Royal Society A: Mathematical, Physical and Engineering Sciences*, 314, 1519, 529–48; doi:10.1098/rspa.1970.0021.

Then, from the 1970s to the 1990s, cosmologists worked on characterizing the features of the Big Bang universe and resolving outstanding problems. In 1981, Alan Guth made a breakthrough in theoretical work on resolving certain outstanding theoretical problems in the Big Bang models with the introduction of an epoch of rapid expansion in the early universe he called "inflation".[13]

[13] Guth, A. (January 15, 1981). Inflationary universe: A possible solution to the horizon and flatness problems. *Physical Review, D.* 23, 2, 347–56; doi:10.1103/PhysRevD.23.347.

Meanwhile, during these decades, two questions in observational cosmology that generated much discussion and disagreement were over the precise values of the Hubble Constant[14]

[14] Huchra, J. P. (2008). The Hubble Constant. *Science*, 256, 5055, 321–5; doi:10.1126/science.256.5055.321.

and the matter-density of the universe (before the discovery of dark energy, thought to be the key predictor for the eventual fate of the universe).

In the mid-1990s, observations of certain globular clusters appeared to indicate that they were about 15 billion years old, which conflicted with most then-current estimates of the age of the universe (and indeed with the age measured today). This issue was later resolved when new computer simulations, which included the effects of mass loss due to stellar winds, indicated a much younger age for globular clusters.

Significant progress in Big Bang cosmology has been made since the late 1990s as a result of advances in telescope technology as well as the analysis of data from satellites such as the Cosmic Background Explorer (COBE), the Hubble Space Telescope and WMAP.

Cosmologists now have fairly precise and accurate measurements of many of the parameters of the Big Bang model, and have made the unexpected discovery that the *expansion of the universe appears to be accelerating.*[15, 16]

[15] Reiss, A. G., Filippenko, A. V., Challis, P., Clocchiatti, A., Diercks, A., Garnavich, P. M., et al. (1998). Observational Evidence from Supernovae for an Accelerating Universe and a Cosmological Constant. *The Astronomical Journal*, 116, 3, 1009–38; arXiv:astro-ph/9805201; doi:10.1086/300499.

[16] Perlmutter, S., Aldering, G., Goldhaber, G., Knop, R.A., Nugent, P., Castro, P.G., et al. (1999). Measurements of Omega and Lambda from 42 High-Redshift Supernovae. *The Astrophysical Journal*, 517, 2, 565–86; arXiv:astro-ph/9812133; doi:10.1086/307221.

Observational evidence

"[The] big bang picture is too firmly grounded in data from every area to be proved invalid in its general features."
— Lawrence Krauss[85]

[85] Krauss, L. M. (2012*). A Universe From Nothing: Why there is Something Rather than Nothing*. Afterword by Richard Dawkins (1st Free Press hardcover ed.). Free Press, New York.

The earliest and most direct observational evidence of the validity of the theory are the expansion of the universe according to Hubble's law (as indicated by the redshifts of galaxies), discovery and measurement of the cosmic microwave background and the relative abundances of light elements produced by Big Bang nucleosynthesis (BBN). More recent evidence includes observations of galaxy formation and evolution, and the distribution of large-scale cosmic structures, These are sometimes called the "four pillars" of the Big Bang models.

Precise modern models of the Big Bang appeal to various exotic physical phenomena that have not been observed in terrestrial laboratory experiments or incorporated into the Standard Model of particle physics. Of these features, *dark matter* is currently the subject of most active laboratory investigations. Remaining issues include the cuspy halo problem and the dwarf galaxy problem of *cold dark matter*. *Dark energy* is also an area of intense interest for scientists, but it is not clear whether direct detection of dark energy will be possible. *Inflation and baryogenesis remain more speculative features of current Big*

Bang models. Viable, quantitative explanations for such phenomena are still being sought. *These are unsolved problems in physics*.

Hubble's law and the expansion of the universe

Observations of distant galaxies and quasars show that *these objects are redshifted*: the light emitted from them has been shifted to longer wavelengths. This can be seen by taking a frequency spectrum of an object and matching the spectroscopic pattern of emission or absorption lines corresponding to atoms of the chemical elements interacting with the light. These redshifts are uniformly isotropic, distributed evenly among the observed objects in all directions. *If the redshift is interpreted as a Doppler shift, the recessional velocity of the object can be calculated.* For some galaxies, it is possible to estimate distances via the *cosmic distance ladder*.

When the recessional velocities are plotted against these distances, a linear relationship known as *Hubble's law* is observed:[8]

> [8] Hubble, E. (March 15, 1929). A Relation Between Distance and Radial Velocity Among Extra-Galactic Nebulae. *Proceedings of the National Academy of Sciences*, 15, 3, 168–73; doi:10.1073/pnas.15.3.168.

$$v = H_0 D$$

where

v is the recessional velocity of the galaxy or other distant object,
D is the proper distance to the object, and
H_0 is Hubble's constant, measured to be $70.4^{+1.3}_{-1.4}$ km/s/Mpc by the WMAP.

Hubble's law implies that the universe is uniformly expanding everywhere. This cosmic expansion was predicted from *general relativity* by Friedmann in 1922[7] and Lemaître in 1927, well before Hubble made his 1929 analysis and observations, and it remains the cornerstone of the Big Bang model as developed by Friedmann, Lemaître, Robertson, and Walker.

The theory requires the relation $v = H_0 D$ to hold at all times, where D is the proper distance, v is the recessional velocity, and v, H, and D vary as the universe expands (hence we write H_0 to denote the present-day Hubble "constant"). For distances much smaller than the size of the observable universe, the Hubble redshift can be thought of as the Doppler shift corresponding to the recession velocity v. For distances comparable to the size of the observable universe, the attribution of the cosmological redshift becomes more ambiguous, although its interpretation as a kinematic Doppler shift remains the most natural one.

An unexplained discrepancy with the determination of the Hubble constant is known as Hubble tension. Techniques based on observation of the CMB suggest a lower value of this constant compared to the quantity derived from measurements based on the cosmic distance ladder.[17]

[17] Di Valentino, Eleonora; Mena, Olga; Pan, Supriya; Visinelli, Luca; Yang, Weiqiang; Melchiorri, Alessandro; Mota, David F.; Riess, Adam G.; Silk, Joseph. (2021). In the realm of the Hubble tension—a review of solutions. *Classical and Quantum Gravity*. 38, 15, 153001. arXiv:2103.01183. doi:10.1088/1361-6382/ac086d.

Cosmic microwave background radiation

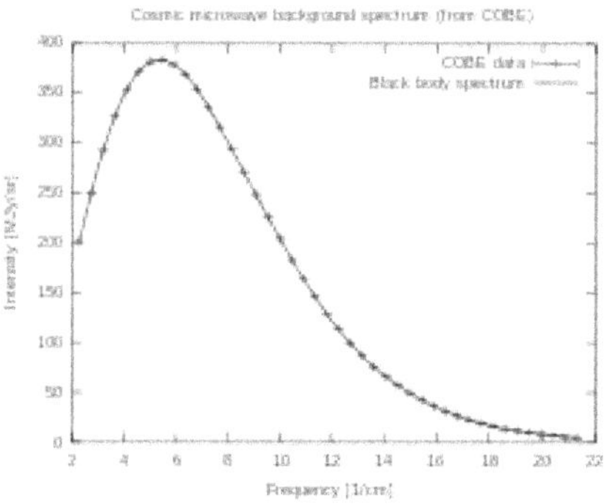

The cosmic microwave background spectrum measured by the FIRAS instrument on the COBE satellite is the most-precisely measured blackbody spectrum in nature. The data points and error bars on this graph are obscured by the theoretical curve.

In 1964, Arno Penzias and Robert Wilson serendipitously discovered the cosmic background radiation, an omnidirectional signal in the microwave band.

[18] Penzias, A. A., Wilson, R. W. (July 1965). A Measurement of Excess Antenna Temperature at 4080 Mc/s. *The Astrophysical Journal*, 142, 419–21; doi:10.1086/148307.

Their discovery provided substantial confirmation of the big-bang predictions by Alpher, Herman and Gamow around 1950. Through the 1970s, the radiation was found to be approximately consistent with a blackbody spectrum in all directions; *this spectrum has been redshifted by the expansion of the universe, and today corresponds to approximately 2.725 K.* This tipped the balance of evidence in favor of the Big Bang model, and Penzias and Wilson were awarded the 1978 Nobel Prize in Physics.

The *surface of last scattering* corresponding to emission of the CMB occurs shortly after *recombination*, the epoch when neutral hydrogen becomes stable. Prior to this, the universe comprised a hot dense photon-baryon plasma sea where photons were quickly scattered from free charged particles. Peaking at around 372±14 kyr, *the mean free path for a photon becomes long enough to reach the present day and the universe becomes transparent.*

In 1989, NASA launched COBE, which made two major advances: in 1990, high-precision spectrum measurements showed that the CMB frequency spectrum is an almost perfect blackbody with no deviations at a level of 1 part in 10^4, and measured a residual temperature of 2.726 K (more recent measurements have revised this figure down slightly to 2.7255 K); then in 1992, further COBE measurements discovered tiny fluctuations (anisotropies) in the CMB temperature across the sky, at a level of about one part in 10^5. John C. Mather and George Smoot were awarded the 2006 Nobel Prize in Physics for their leadership in these results.

During the following decade, CMB anisotropies were further investigated by a large number of ground-based and balloon experiments. In 2000–2001, several experiments, most notably BOOMERanG, *found the shape of the universe to be spatially almost flat* by measuring the typical angular size (the size on the sky) of the anisotropies.

In early 2003, the first results of the Wilkinson Microwave Anisotropy Probe were released, yielding what were at the time the most accurate values for some of the cosmological parameters. The results disproved several specific cosmic inflation models, but are consistent with the inflation theory in general.[19]

[19] Spergel, D. N.; Bean, R., Doré, O., et al. (June 2007). Three-Year Wilkinson Microwave Anisotropy Probe (WMAP) Observations: Implications for Cosmology. *The Astrophysical Journal Supplement Series*. 170, 2, 377–408; arXiv:astro-ph/0603449; doi:10.1086/513700.

The *Planck* space probe was launched in May 2009. Other ground and balloon-based cosmic microwave background experiments are ongoing.

Abundance of primordial elements

Using Big Bang models, it is possible to calculate the expected concentration of the isotopes helium-4 (^4He), helium-3 (^3He), deuterium (^2H), and lithium-7 (^7Li) in the universe as ratios to the amount of ordinary hydrogen. *The relative abundances depend on a single parameter, the ratio of photons to baryons.* This value can be calculated independently from the detailed structure of CMB fluctuations. The ratios predicted (by mass, not by abundance) are about 0.25 for $^4He{:}H$, about 10^{-3} for $^2H{:}H$, about 10^{-4} for $^3He{:}H$, and about 10^{-9} for $^7Li{:}H$.

The measured abundances all agree at least roughly with those predicted from a single value of the baryon-to-photon ratio. The agreement is excellent for deuterium, close but formally discrepant for ^4He, and off by a factor of two for ^7Li (this anomaly is known as the *cosmological lithium problem*); in the latter two cases, there are substantial systematic uncertainties. Nonetheless, the general consistency with abundances predicted by BBN is strong evidence for the Big Bang, as the theory is the only known explanation for the relative abundances of light elements, and it is virtually impossible to "tune" the Big Bang to produce much more or less than 20–30% helium. Indeed, there is no obvious reason outside of the Big Bang that, for example, the young universe before star formation, as determined by studying matter supposedly free of stellar nucleosynthesis products, should have more helium than deuterium or more deuterium than ^3He, and in constant ratios, too.

Galactic evolution and distribution

Detailed observations of the morphology and distribution of galaxies and quasars are in agreement with the current Big Bang models. A combination of observations and theory suggest that the first quasars and galaxies formed within a billion years after the Big Bang, and since then, larger structures have been forming, such as galaxy clusters and superclusters.

Populations of stars have been aging and evolving, so that distant galaxies (which are observed as they were in the early universe) appear very different from nearby galaxies (observed in a more recent state). Moreover, galaxies that formed relatively recently, appear markedly different from galaxies formed at similar distances but shortly after the Big Bang. These observations are strong arguments against the steady-state model. Observations of star formation, galaxy and quasar distributions and larger structures, agree well with Big Bang simulations of the formation of structure in the universe, and are helping to complete details of the theory.

> [A *galaxy* is a system of stars, stellar remnants, interstellar gas, dust, and dark matter bound together by gravity. The word is derived from the Greek galaxias (γαλαξίας), literally 'milky', a reference to the Milky Way galaxy that contains the Solar System. Galaxies, averaging an estimated 100 billion stars, range in size from dwarfs with less than a hundred million stars, to the largest galaxies known – supergiants with one hundred trillion stars, each orbiting its galaxy's center of mass. Most of the mass in a typical galaxy is in the form of dark matter, with only a few percent of that mass visible in the form of stars and nebulae. Supermassive black holes are a common feature at the centers of galaxies.
>
> Galaxies are categorized according to their visual morphology as elliptical, spiral, or irregular. Many are thought to have supermassive black holes at their centers. The Milky Way's central black hole, known as Sagittarius A*, has a mass four million times greater than the Sun.

It is estimated that there are between 200 billion (2×10^{11}) to 2 trillion galaxies in the observable universe. Most galaxies are 1,000 to 100,000 parsecs in diameter (approximately 3,000 to 300,000 light years) and are separated by distances on the order of millions of parsecs (or megaparsecs). For comparison, the Milky Way has a diameter of at least 26,800 parsecs (87,400 light years) and is separated from the Andromeda Galaxy (with diameter of about 152,000 light years), its nearest large neighbor, by 780,000 parsecs (2.5 million light years.)

The space between galaxies is filled with a tenuous gas (the *intergalactic medium*) *with an average density of less than one atom per cubic meter*. Most galaxies are gravitationally organized into groups, clusters and superclusters. The Milky Way is part of the Local Group, which it dominates along with the Andromeda Galaxy. The group is part of the Virgo Supercluster. At the largest scale, these associations are generally arranged into sheets and filaments surrounded by immense voids. Both the Local Group and the Virgo Supercluster are contained in a much larger cosmic structure named Laniakea.

Formation

Galactic formation and evolution is an active area of research in astrophysics.

Current models of the formation of galaxies in the early universe are based on the ΛCDM model. About 300,000 years after the big bang, atoms of hydrogen and helium began to form, in an event called recombination. Nearly all the hydrogen was neutral (non-ionized) and readily absorbed light, and no stars had yet formed. As a result, this period has been called the "dark ages". It was from density fluctuations (or anisotropic irregularities) in this primordial matter that larger structures began to appear. As a result, masses of baryonic matter started to condense within cold dark matter halos. These primordial structures eventually became the galaxies we see today.

Early galaxy formation

Evidence for the appearance of galaxies very early in the Universe's history was found in 2006, when it was discovered that the galaxy IOK-1 has an unusually high redshift of 6.96, corresponding to just 750 million years after the Big Bang and making it the most distant and earliest-to-form galaxy seen at that time. While some scientists have claimed other objects (such as Abell 1835 IR1916) have higher redshifts (and therefore are seen in an earlier stage of the universe's evolution), IOK-1's age and composition have been more reliably established. In December 2012, astronomers reported that UDFj-39546284 is the most distant object known and has a redshift value of 11.9. The object, estimated to have existed around 380 million years after the Big Bang (which was about 13.8 billion years ago), is about 13.42 billion light travel distance years away. The existence of galaxies so soon after the Big Bang suggests that protogalaxies must have grown in the so-called "dark ages". As of May 5, 2015, the galaxy EGS-zs8-1 is the most distant and earliest galaxy measured, forming 670 million years after the Big Bang. The light from EGS-zs8-1 has taken 13 billion years to reach Earth, and is now 30 billion light-years away, because of the expansion of the universe during 13 billion years. On 17 August 2022, NASA released a large mosaic image of 690 individual frames taken by the Near Infrared Camera (NIRCam) on the James Webb Space Telescope (JWST) of numerous very early galaxies.

In May 2023, a study in the journal *Nature* identified an ultra-faint galaxy named JD1. Galaxy JD1 was observed by the JWST using the near-infrared spectrograph instrument NIRSpec and was found to have a distance value of redshift z=9.79. This means that JD1 was observed at 480 million years after the Big Bang when the universe was only about 4% of its present age. Observations of this ultra-faint galaxy were aided by the effect of a gravitational lens in the galaxy cluster Abell 2744 which helped make the image of JD1 larger and 13 times brighter than it otherwise would be. This effect and the use of the JWST's NIRCam showed JD1's structure to be three star forming clumps of dust and gas.

The detailed process by which the earliest galaxies formed is an open question in astrophysics. Theories can be divided into two categories: top-down and bottom-up. In top-down correlations (such as the Eggen–Lynden-Bell–Sandage [ELS] model), protogalaxies form in a large-scale simultaneous collapse lasting about one hundred million years. In bottom-up theories (such as the Searle-Zinn [SZ] model), small structures such as globular clusters form first, and then a number of such bodies accrete to form a larger galaxy. *Once protogalaxies began to form and contract, the first halo stars (called Population III stars) appeared within them.* These were composed almost entirely of hydrogen and helium and may have been more massive than 100 times the Sun's mass. If so, these huge stars would have quickly consumed their supply of fuel and became supernovae, releasing heavy elements into the interstellar medium. This first generation of stars re-ionized the surrounding neutral hydrogen, creating expanding bubbles of space through which light could readily travel.

In June 2015, astronomers reported evidence for Population III stars in the Cosmos Redshift 7 galaxy at z = 6.60. Such stars are likely to have existed in the very early universe (i.e., at high redshift), and may have started the production of chemical elements heavier than hydrogen that are needed for the later formation of planets and life as we know it.

In late 2023, scientists reported results suggesting that newborn galaxies in the very early universe were "banana"-shaped, much to the surprise of researchers.

Evolution

Within a billion years of a galaxy's formation, key structures begin to appear. Globular clusters, the central supermassive black hole, and a galactic bulge of metal-poor Population II stars form. The creation of a supermassive black hole appears to play a key role in actively regulating the growth of galaxies by limiting the total amount of additional matter added. During this early epoch, galaxies undergo a major burst of star formation.

During the following two billion years, the accumulated matter settles into a galactic disc. A galaxy will continue to absorb infalling material from high-velocity clouds and dwarf galaxies throughout its life. This matter is mostly hydrogen and helium. The cycle of stellar birth and death slowly increases the abundance of heavy elements, eventually allowing the formation of planets.

The evolution of galaxies can be significantly affected by interactions and collisions. Mergers of galaxies were common during the early epoch, and the majority of galaxies were peculiar in morphology. Given the distances between the stars, the great majority of stellar systems in colliding galaxies will be unaffected. However, gravitational stripping of the interstellar gas and dust that makes up the spiral arms produces a long train of stars known as tidal tails. Examples of these formations can be seen in NGC 4676 or the Antennae Galaxies.

The Milky Way galaxy and the nearby Andromeda Galaxy are moving toward each other at about 130 km/s, and—depending upon the lateral movements—the two might collide in about five to six billion years. Although the Milky Way has never collided with a galaxy as large as Andromeda before, evidence of past collisions of the Milky Way with smaller dwarf galaxies is increasing.

Such large-scale interactions are rare. As time passes, mergers of two systems of equal size become less common. Most bright galaxies have remained fundamentally unchanged for the last few billion years, and the net rate of star formation probably also peaked about ten billion years ago.

Future trends

Spiral galaxies, like the Milky Way, produce new generations of stars as long as they have dense molecular clouds of interstellar hydrogen in their spiral arms. Elliptical galaxies are largely devoid of this gas, and so form few new stars. The supply of star-forming material is finite; once stars have converted the available supply of hydrogen into heavier elements, new star formation will come to an end.

The current era of star formation is expected to continue for up to one hundred billion years, and then the "stellar age" will wind down after about ten trillion to one hundred trillion years (10^{13}–10^{14} years), as the smallest, longest-lived stars in the visible universe, tiny red dwarfs, begin to fade. At the end of the stellar age, galaxies will be composed of compact objects: brown dwarfs, white dwarfs that are cooling or cold ("black dwarfs"), neutron stars, and black holes. Eventually, *as a result of gravitational relaxation,* all stars will either fall into central supermassive black holes or be flung into intergalactic space as a result of collisions.

Deep-sky surveys show that galaxies are often found in groups and clusters. Solitary galaxies that have not significantly interacted with other galaxies of comparable mass in the past billion years are relatively scarce. Only about 5% of the galaxies surveyed are truly isolated; however, they may have interacted and even merged with other galaxies in the past, and may still be orbited by smaller satellite galaxies. Isolated galaxies can produce stars at a higher rate than normal, *as their gas is not being stripped by other nearby galaxies.*

On the largest scale, the universe is continually expanding, resulting in an average increase in the separation between individual galaxies (see Hubble's law). *Associations of galaxies can overcome this expansion on a local scale through their mutual gravitational attraction.* These associations formed early, as clumps of *dark matter* pulled their respective galaxies together. Nearby groups later merged to form larger-scale clusters. This ongoing merging process (as well as an influx of infalling gas) heats the intergalactic gas in a cluster to very high temperatures of 30–100 megakelvins. About 70–80% of a cluster's mass is in the form of *dark matter*, with 10–30% consisting of this heated gas and the remaining few percent in the form of galaxies.

Most galaxies are gravitationally bound to a number of other galaxies. These form a fractal-like hierarchical distribution of clustered structures, with the smallest such associations being termed groups. A group of galaxies is the most common type of galactic cluster; these formations contain the majority of galaxies (as well as most of the baryonic mass) in the universe. *To remain gravitationally bound to such a group, each member galaxy must have a sufficiently low velocity to prevent it from escaping.* If there is insufficient kinetic energy, however, the group may evolve into a smaller number of galaxies through mergers.

Clusters of galaxies consist of hundreds to thousands of galaxies bound together by gravity. Clusters of galaxies are often dominated by a single giant elliptical galaxy, known as the brightest cluster galaxy, which, over time, tidally destroys its satellite galaxies and adds their mass to its own.

Superclusters contain tens of thousands of galaxies, which are found in clusters, groups and sometimes individually. At the supercluster scale, galaxies are arranged into sheets and filaments surrounding vast empty voids. Above this scale, the universe appears to be the same in all directions (isotropic and homogeneous), though this notion has been challenged in recent years by numerous findings of large-scale structures that appear to be exceeding this scale. The Hercules–Corona Borealis Great Wall, currently the largest structure in the universe found so far, is 10 billion light-years (three gigaparsecs) in length.

Seven Main Stages of a Star. Stars come in a variety of masses and the mass determines how radiantly the star will shine and how it dies. Massive stars transform into supernovae, neutron stars and black holes while average stars like the sun, end life as a white dwarf surrounded by a disappearing planetary nebula. All stars, irrespective of their size, follow the same 7 stage cycle, they start as a gas cloud and end as a star remnant.

1. *Giant Gas Cloud*. Stars are thought to form inside giant clouds of cold molecular hydrogen—giant molecular clouds of roughly 300,000 M⊙ and 65 light-years (20 pc) in diameter. The temperature in the cloud is low enough for the synthesis of molecules. The Orion cloud complex in the Orion system is an example of a star in this stage of life.

2. *Protostar. When the gas particles in the molecular cloud run into each other, heat energy is produced.* This results in the formation of a warm clump of molecules referred to as the *Protostar*. Several Protostars can be formed in one cloud, depending on the size of the molecular cloud. Over millions of years, giant molecular clouds are prone to collapse and fragmentation. These fragments then form small, dense cores, which in turn collapse into stars. The cores range in mass from a fraction to several times that of the Sun and are called protostellar (protosolar) nebulae. They possess diameters of 2,000 – 20,000 astronomical units

(0.01–0.1 pc) and a *relatively low particle number density of roughly 10^{10} to $10^{11}/m^3$*. (The particle number density of the air at sea level is $2.8 \times 10^{25}/m^3$.) *Every nebula begins with a certain amount of angular momentum.* Gas in the central part of the nebula, with relatively low angular momentum, *undergoes fast compression* and forms a hot hydrostatic (non-contracting) core containing a small fraction of the mass of the original nebula. This core forms the seed of what will become a star.

3. *T-Tauri Phase.* A T-Tauri star begins when materials stop falling into the Protostar and release tremendous amounts of energy. As the collapse continues, conservation of angular momentum dictates that the rotation of the infalling envelope accelerates, which eventually forms a disk. (Protostars do not possess accretion disks.) As the infall of material from the disk continues, the envelope eventually becomes thin and transparent and the young stellar object (YSO) becomes observable, initially in far-infrared light and later in the visible. *Around this time the protostar begins to fuse deuterium. If the protostar is sufficiently massive (above 80 MJ), hydrogen fusion follows.* Otherwise, if its mass is too low, the object becomes a brown dwarf. This birth of a new star occurs approximately 100,000 years after the collapse begins. Objects at this stage are known as Class I protostars, which are also called young T Tauri stars, evolved protostars, or young stellar objects. *By this time, the forming star has already accreted much of its mass*; the total mass of the disk and remaining envelope does not exceed 10–20% of the mass of the central YSO. At the next stage, the envelope completely disappears, having been gathered up by the disk, and the protostar becomes a classical T Tauri star. The latter have accretion disks and continue to accrete hot gas, which manifests itself by strong emission lines in their spectrum. The mean temperature of the Tauri star is not enough to support nuclear fusion at its core. The T-Tauri star lasts for about 100 million years, following which it enters the most extended phase of development – the Main sequence phase.

4. *Main Sequence.* The main sequence phase is the stage in development where the core temperature reaches the point for the fusion to commence. In this process, the protons of hydrogen are converted into atoms of helium. This reaction is exothermic; it gives off more heat than it requires and so the core of a main-sequence star releases a tremendous amount of energy. Classical T Tauri stars evolve into weakly lined T Tauri stars. This happens after about 1 million years. The mass of the disk around a classical T Tauri star is about 1–3% of the stellar mass, and it is accreted at a rate of 10^{-7} to 10^{-9} M⊙ per year. A pair of bipolar jets is usually present as well. The accretion explains all of the peculiar properties of

classical T Tauri stars: strong flux in the emission lines (up to 100% of the intrinsic luminosity of the star), magnetic activity, photometric variability and jets. The emission lines actually form as the accreted gas hits the "surface" of the star, which happens around its magnetic poles. The jets are byproducts of accretion: they carry away excessive angular momentum. The classical T Tauri stage lasts about 10 million years. The disk eventually disappears due to accretion onto the central star, planet formation, ejection by jets, and photoevaporation by ultraviolet radiation from the central star and nearby stars. As a result, the young star becomes a weakly lined T Tauri star, which, over hundreds of millions of years, evolves into an ordinary Sun-like star, dependent on its initial mass.

5. *Red Giant*. A star converts hydrogen atoms into helium over its course of life at its core. When the star exhausts the hydrogen fuel in its core, nuclear reactions can no longer continue at the core and so the core begins to contract *due to the diminishing force of the fusion, which used to push against gravity*, and results in the core heating up. The increased temperature of the core causes hydrogen in a shell around the core to be burned and the star to expand. As it expands, the star first becomes a subgiant star and then a *red giant*. *Red giants* have cooler surfaces than the main-sequence star, and because of this, they appear red than yellow.

6. *The Fusion of Heavier Elements*. *Helium molecules fuse at the core, as the star expands*. The energy of this reaction prevents the core from collapsing. The core shrinks and begins *fusing carbon*, once the helium fusion ends. This process repeats until *iron appears at the core*. The iron fusion reaction absorbs energy, which causes the core to collapse. This implosion transforms massive stars into a *supernova* while smaller stars like the sun contract into *white dwarfs*.

7. *Supernovae and Planetary Nebulae*. Most of the star material is blasted away into space, but the core implodes into a neutron star or a singularity known as the black hole. Less massive stars do not explode, their cores contract instead into a tiny, hot star known as the *white dwarf* while the outer material drifts away. Stars smaller than the sun, do not have enough mass to burn with anything but a red glow during their main sequence. These *red dwarves* are difficult to spot, but they may be the most common stars that can burn for trillions of years.]

Primordial gas clouds

In 2011, astronomers found what they believe to be pristine clouds of primordial gas by analyzing absorption lines in the spectra of distant quasars. Before this discovery, all other astronomical objects have been observed to contain heavy elements that are formed in stars. Despite being sensitive to carbon, oxygen, and silicon, these three elements were not detected in these two clouds. Since the clouds of gas have no detectable levels of heavy elements, they likely formed in the first few minutes after the Big Bang, during BBN.

Other lines of evidence

The age of the universe as estimated from the Hubble expansion and the CMB is now in agreement with other estimates using the ages of the oldest stars, both as measured by applying the theory of stellar evolution to globular clusters and through radiometric dating of individual Population II stars. It is also in agreement with age estimates based on measurements of the expansion using Type Ia supernovae and measurements of temperature fluctuations in the cosmic microwave background. *The agreement of independent measurements of this age supports the Lambda-CDM (ΛCDM) model*, since the model is used to relate some of the measurements to an age estimate, and all estimates in turn agree. Still, *some observations of objects from the relatively early universe (in particular quasar APM 08279+5255) raise concern as to whether these objects had enough time to form so early in the ΛCDM model.*

The prediction that the CMB temperature was higher in the past has been experimentally supported by observations of very low temperature absorption lines in gas clouds at high redshift. This prediction also implies that the amplitude of the Sunyaev–Zel'dovich effect in clusters of galaxies does not depend directly on redshift. Observations have found this to be roughly true, but this effect depends on cluster properties that do change with cosmic time, making precise measurements difficult.

Future observations

Future gravitational-wave observatories *might be able to detect primordial gravitational waves*, relics of the early universe, up to less than a second after the Big Bang.

Problems and related issues in physics

As with any theory, a number of mysteries and problems have arisen as a result of the development of the Big Bang models. Some of these mysteries and problems have been resolved while others are still outstanding. Proposed solutions to some of the problems in the Big Bang model have revealed new mysteries of their own. For example, the *horizon problem*, the *magnetic monopole problem*, and the *flatness problem* are most commonly resolved with inflation theory, *but the details of the inflationary universe are still left unresolved* and many, including some founders of the theory, say it has been disproven. What follows are a list of the mysterious aspects of the Big Bang concept still under intense investigation by cosmologists and astrophysicists.

Baryon asymmetry

It is not yet understood why the universe has more matter than antimatter. It is generally assumed that when the universe was young and very hot it was in statistical equilibrium and contained equal numbers of baryons and antibaryons. However, *observations suggest that the universe, including its most distant parts, is made almost entirely of normal matter, rather than antimatter*. A process called baryogenesis was hypothesized to account for the asymmetry. For baryogenesis to occur, the Sakharov conditions must be satisfied. These require that baryon number is not conserved, that C-symmetry and CP-symmetry are violated and that the universe depart from thermodynamic equilibrium. All these conditions occur in the Standard Model, but *the effects are not strong enough to explain the present baryon asymmetry*.

Dark energy

Measurements of the redshift–magnitude relation for type Ia supernovae indicate that *the expansion of the universe has been accelerating since the universe was about half its present age*. To explain this acceleration, *general relativity* requires that much of the energy in the universe consists of a component with large negative pressure, dubbed "*dark energy*".[10]

> [10] Peebles, P. J. E.; Ratra, B. (April 22, 2003). The cosmological constant and dark energy. *Reviews of Modern Physics*, 75, 2, 559–606; arXiv:astro-ph/0207347; doi:10.1103/RevModPhys.75.559.

Dark energy, though speculative, solves numerous problems. Measurements of the cosmic microwave background indicate that the universe is very nearly spatially flat, and therefore according to *general relativity* the universe must have almost exactly the critical density of mass/energy. But the mass density of the universe can be measured from its gravitational clustering, and is found to have only about 30% of the critical density. Since theory suggests that *dark energy* does not cluster in the usual way it is the best explanation for the "missing" energy density. *Dark energy* also helps to explain two geometrical measures of the overall curvature of the universe, one using the frequency of gravitational lenses, and the other using the characteristic pattern of the large-scale structure as a cosmic ruler.

Negative pressure is believed to be a property of vacuum energy, but *the exact nature and existence of dark energy remains one of the great mysteries of the Big Bang. Results from the WMAP team in 2008 are in accordance with a universe that consists of 73% dark energy, 23% dark matter, 4.6% regular matter and less than 1% neutrinos*. According to theory, the energy density in matter decreases with the expansion of the universe, but the *dark energy density* remains constant (or nearly so) as the universe expands. Therefore, matter made up a larger fraction of the total energy of the universe in the past than it does today, but its fractional contribution will fall in the far future as *dark energy* becomes even more dominant.

The *dark energy* component of the universe has been explained by theorists using a variety of competing theories including Einstein's cosmological constant but also extending to more exotic forms of quintessence or other modified gravity schemes.[20]

[20] Tanabashi, M.; et al. (Particle Data Group) (2018). Review of Particle Physics. *Physical Review* D. ,8, 3, 1–708, pp. 406–413; doi:10.1103/PhysRevD.98.030001.

A *cosmological constant problem*, sometimes called the "most embarrassing problem in physics", *results from the apparent discrepancy between the measured energy density of dark energy, and the one naively predicted from Planck units.*[21]

[21] Rugh, S. E., Zinkernagel, H. (December 2002). The quantum vacuum and the cosmological constant problem". *Studies in History and Philosophy of Science Part B*, 33, 4, 663–705; arXiv:hep-th/0012253; doi:10.1016/S1355-2198(02)00033-3.

Dark matter

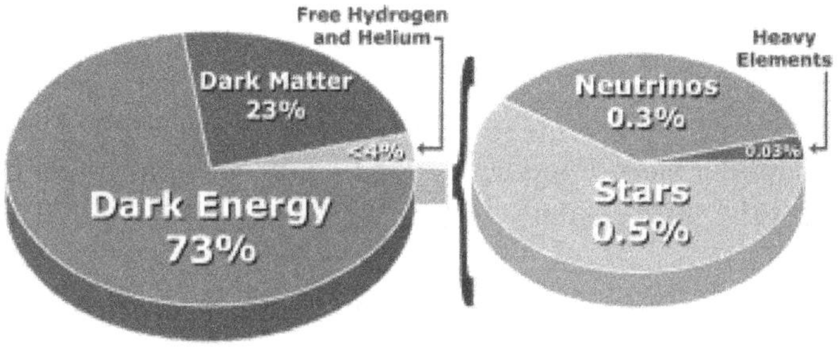

Chart shows the proportion of different components of the universe – about 95% is *dark matter* and *dark energy*.

During the 1970s and the 1980s, various observations showed that *there is not sufficient visible matter in the universe to account for the apparent strength of gravitational forces within and between galaxies*. This led to the idea that up to 90% of the matter in the universe is *dark matter* that does not emit light or interact with normal baryonic matter. In addition, *the assumption that the universe is mostly normal matter led to predictions that were strongly inconsistent with observations*. In particular, the universe today is far more lumpy and contains far less deuterium than can be accounted for without *dark matter*. While *dark matter* has always been controversial, it is inferred by various observations: the anisotropies in the CMB, galaxy cluster velocity dispersions, large-scale structure distributions, gravitational lensing studies, and X-ray measurements of galaxy clusters.

Indirect evidence for dark matter comes from its gravitational influence on other matter, as no dark matter particles have been observed in laboratories. Many particle physics candidates for dark matter have been proposed, and several projects to detect them directly are underway.[20]

[20] Tanabashi, M. (2018), pp. 396–405.

Additionally, there are outstanding problems associated with the currently favored *cold dark matter model* which include the *dwarf galaxy problem* and the *cuspy halo problem. Alternative theories have been proposed that do not require a large amount of undetected matter, but instead modify the laws of gravity established by Newton and Einstein*; yet no alternative theory has been as successful as the cold dark matter proposal in explaining all extant observations.

Horizon problem

The *horizon problem* results from the premise that information cannot travel faster than light. In a universe of finite age this sets a limit—the particle horizon—on the separation of any two regions of space that are in causal contact. The observed isotropy of the CMB is problematic in this regard: if the universe had been dominated by radiation or matter at all times up to the epoch of last scattering, the particle horizon at that time would correspond to about 2 degrees on the sky. There would then be no mechanism to cause wider regions to have the same temperature.

A resolution to this apparent inconsistency is offered by *inflation theory* in which a homogeneous and isotropic scalar energy field dominates the universe at some very early period (before baryogenesis). During inflation, the universe undergoes exponential expansion, and the particle horizon expands much more rapidly than previously assumed, so that regions presently on opposite sides of the observable universe are well inside each other's particle horizon. The observed isotropy of the CMB then follows from the fact that this larger region was in causal contact before the beginning of inflation.

Heisenberg's uncertainty principle predicts that during the inflationary phase there would be quantum thermal fluctuations, which would be magnified to a cosmic scale. These fluctuations served as the seeds for all the current structures in the universe. Inflation predicts that the primordial fluctuations are nearly scale invariant and Gaussian, which has been confirmed by measurements of the CMB.

A related issue to the classic horizon problem arises because in most standard cosmological inflation models, inflation ceases well before electroweak symmetry breaking occurs, so inflation should not be able to prevent large-scale discontinuities in the electroweak vacuum since distant parts of the observable universe were causally separate when the electroweak epoch ended.[21]

[21] Penrose, R. (2007) [Originally published: London: Jonathan Cape, 2004*]. The Road to Reality,* (1st Vintage Books ed.). Vintage Books, New York.

Magnetic monopoles

The magnetic monopole objection was raised in the late 1970s. Grand unified theories (GUTs) predicted topological defects in space that would manifest as magnetic monopoles. These objects would be produced efficiently in the hot early universe, resulting in a density much higher than is consistent with observations, given that no monopoles have been found. This problem is resolved by cosmic inflation, which removes all point defects from the observable universe, in the same way that it drives the geometry to flatness.

Flatness problem

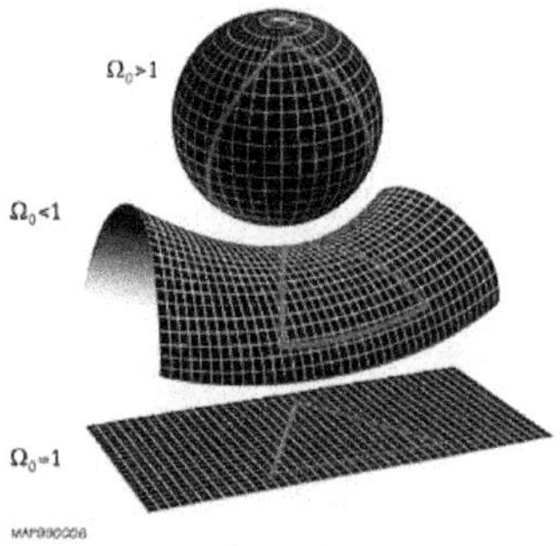

The overall geometry of the universe is determined by whether the Omega cosmological parameter is less than, equal to or greater than 1. Shown from top to bottom are a closed universe with positive curvature, a hyperbolic universe with negative curvature and a flat universe with zero curvature.

The *flatness problem* (also known as the oldness problem) is an observational problem associated with a FLRW. The universe may have positive, negative, or zero spatial curvature depending on its total energy density. Curvature is negative if its density is less than the critical density; positive if greater; and zero at the critical density, in which case space is said to be *flat*. Observations indicate the universe is consistent with being flat.

The problem is that any small departure from the critical density grows with time, and yet the universe today remains very close to flat. Given that a natural timescale for departure from flatness might be the Planck time, 10^{-43} seconds, *the fact that the universe has reached neither a heat death nor a Big Crunch after billions of years requires an explanation*. For instance, even at the relatively late age of a few minutes (the time of nucleosynthesis), the density of the universe must have been within one part in 10^{14} of its critical value, or it would not exist as it does today.

Misconceptions

One of the common misconceptions about the Big Bang model is that it fully explains the origin of the universe. However, the Big Bang model does not describe how energy, time, and space were caused, but rather it describes the emergence of the present universe from an ultra-dense and high-temperature initial state. It is misleading to visualize the Big Bang by comparing its size to everyday objects. When the size of the universe at Big Bang is described, it refers to the size of the observable universe, and not the entire universe.

Another common misconception is that the Big Bang must be understood as the expansion of space and not in terms of the contents of space exploding apart. In fact, *either description can be accurate*. The expansion of space (implied by the FLRW metric) is only a mathematical convention, corresponding to a choice of coordinates on spacetime. There is no generally covariant sense in which space expands.

The recession speeds associated with Hubble's law are not velocities in a relativistic sense (for example, they are not related to the spatial components of 4-velocities). Therefore, it is not remarkable that according to Hubble's law, galaxies farther than the Hubble distance recede faster than the speed of light. Such recession speeds do not correspond to faster-than-light travel.

Many popular accounts attribute the cosmological redshift to the expansion of space. This can be misleading because the expansion of space is only a coordinate choice. The most natural interpretation of the cosmological redshift is that it is a Doppler shift.

Implications

Given current understanding, scientific extrapolations about the future of the universe are only possible for finite durations, albeit for much longer periods than the current age of the universe. Anything beyond that becomes increasingly speculative. Likewise, at present, a proper understanding of the origin of the universe can only be subject to conjecture.

Pre–Big Bang cosmology

The Big Bang explains the evolution of the universe from a starting density and temperature that is well beyond humanity's capability to replicate, so extrapolations to the most extreme conditions and earliest times are necessarily more speculative. Lemaître called this initial state the "*primeval atom*" while Gamow called the material "*ylem*". How the initial state of the universe originated is still an open question, but the Big Bang model does constrain some of its characteristics. For example, specific laws of nature most likely came to existence in a random way, but as inflation models show, some combinations of these are far more probable.[22]

> [22] Hawking, S. W. (1988). *A Brief History of Time: From the Big Bang to Black Holes*. Introduction by Carl Sagan; illustrations by Ron Miller. Bantam Dell Publishing Group, New York.

A flat universe implies a balance between gravitational potential energy and other energy forms, requiring no additional energy to be created.

The Big Bang theory, built upon the equations of classical *general relativity*, indicates a singularity at the origin of cosmic time, and such an infinite energy density may be a physical impossibility. However, *the physical theories of general relativity and quantum mechanics as currently realized are not applicable before the Planck epoch*, and *correcting this will require the development of a correct treatment of quantum gravity*.

Certain quantum gravity treatments, such as the Wheeler–DeWitt equation, imply that time itself could be an emergent property. As such, physics may conclude that time did not exist before the Big Bang.

While it is not known what could have preceded the hot dense state of the early universe or how and why it originated, or even whether such questions are sensible, speculation abounds on the subject of "cosmogony".

Some speculative proposals in this regard, each of which entails untested hypotheses, are:

- The simplest models, in which the Big Bang was caused by quantum fluctuations. That scenario had very little chance of happening, but, according to the totalitarian principle, even the most improbable event will eventually happen. It took place instantly, in our perspective, due to the absence of perceived time before the Big Bang.
- Emergent Universe models, which feature a low-activity past-eternal era before the Big Bang, resembling ancient ideas of a cosmic egg and birth of the world out of primordial chaos.
- Models in which the whole of spacetime is finite, including the Hartle–Hawking no-boundary condition. For these cases, the Big Bang does represent the limit of time but without a singularity. In such a case, the universe is self-sufficient.
- Brane cosmology models, in which inflation is due to the movement of branes in string theory; the pre-Big Bang model; the ekpyrotic model, in which the Big Bang is the result of a collision between branes; and the cyclic model, a variant of the ekpyrotic model in which collisions occur periodically. In the latter

model the Big Bang was preceded by a Big Crunch and the universe cycles from one process to the other.

- Eternal inflation, in which universal inflation ends locally here and there in a random fashion, each end-point leading to a *bubble universe*, expanding from its own big bang.

Proposals in the last two categories see the Big Bang as an event in either a much larger and older universe or in a multiverse.

Ultimate fate of the universe

Before observations of dark energy, cosmologists considered two scenarios for the future of the universe. If the mass density of the universe were greater than the critical density, then the universe would reach a maximum size and then begin to collapse. It would become denser and hotter again, ending with a state similar to that in which it started—a Big Crunch.

Alternatively, if the density in the universe were equal to or below the critical density, the expansion would slow down but never stop. Star formation would cease with the consumption of interstellar gas in each galaxy; stars would burn out, leaving white dwarfs, neutron stars, and black holes. Collisions between these would result in mass accumulating into larger and larger black holes. The average temperature of the universe would very gradually asymptotically approach absolute zero—a Big Freeze. Moreover, if protons are unstable, then baryonic matter would disappear, leaving only radiation and black holes. Eventually, black holes would evaporate by emitting Hawking radiation. The entropy of the universe would increase to the point where no organized form of energy could be extracted from it, a scenario known as heat death.

Modern observations of accelerating expansion imply that more and more of the currently visible universe will pass beyond our event horizon and out of contact with us. The eventual result is not known. The ΛCDM model of the universe contains dark energy in the form of a cosmological constant. This theory suggests that only gravitationally bound systems, such as galaxies, will remain together, and they too will be subject to heat death as the universe expands and cools. Other explanations of dark energy, called phantom energy theories, suggest that ultimately galaxy clusters, stars, planets, atoms, nuclei, and matter itself will be torn apart by the ever-increasing expansion in a so-called Big Rip.

Accretion of planets

It is claimed that self-accretion of *cosmic dust* accelerates the growth of particles into boulder-sized planetesimals. The more massive planetesimals accrete some smaller ones, while others shatter in collisions. Accretion disks are common around smaller stars, stellar remnants in a close binary, or black holes surrounded by material (such as those at the centers of galaxies). Some dynamics in the disk, such as dynamical friction, are necessary to allow orbiting gas to lose angular momentum and fall onto the central massive object. Occasionally, this can result in stellar surface fusion.

In the formation of terrestrial planets or planetary cores, several stages can be considered. First, when gas and dust grains collide, *they agglomerate by microphysical processes like van der Waals forces and electromagnetic forces*, forming micrometer-sized particles; during this stage, accumulation mechanisms *are largely non-gravitational in nature*.

However, *planetesimal formation in the centimeter-to-meter range is not well understood, and no convincing explanation is offered as to why such grains would accumulate rather than simply rebound.* In particular, it is still *not clear how these objects grow to become 0.1–1 km (0.06–0.6 mi) sized planetesimals*; this problem is known as the "meter size barrier".

According to this theory, as dust particles grow by coagulation, they acquire increasingly large relative velocities with respect to other particles in their vicinity, as well as a systematic inward drift velocity, that leads to destructive collisions, and thereby limit the growth of the aggregates to some maximum size. *Ward (1996) suggests that when slow moving grains collide, the very low, yet non-zero, gravity of colliding grains impedes their escape.* It is also thought that grain fragmentation plays an important role replenishing small grains and keeping the disk thick, but also in maintaining a relatively high abundance of solids of all sizes.

A number of mechanisms have been proposed for crossing the 'meter-sized' barrier. Local concentrations of pebbles may form, *which then gravitationally collapse into planetesimals the size of large asteroids*. These concentrations can occur passively due to the structure of the gas disk, for example, between eddies, at pressure bumps, at the edge of a gap created by a giant planet, or at the boundaries of turbulent regions of the disk. Or, the particles may take an active role in their concentration via a feedback mechanism referred to as a *streaming instability*. In a streaming instability the interaction between the solids and the gas in the protoplanetary disk results in the growth of local concentrations, as new particles accumulate in the wake of small concentrations, causing them to grow into massive filaments.

Alternatively, if the grains that form due to the agglomeration of dust are highly porous *their growth may continue until they become large enough to collapse due to their own gravity*. The low density of these objects allows them to remain strongly coupled with the gas, thereby avoiding high velocity collisions which could result in their erosion or fragmentation.

Grains eventually stick together to form mountain-size (or larger) bodies called *planetesimals*. Collisions and *gravitational interactions between planetesimals* combine to produce Moon-size planetary embryos (protoplanets) over roughly 0.1–1 million years. Finally, the planetary embryos collide to form planets over 10–100 million years.

The planetesimals are massive enough that mutual gravitational interactions are significant enough to be taken into account when computing their evolution. Growth is aided by orbital decay of smaller bodies due to gas drag, which prevents them from being stranded between orbits of the embryos. Further collisions and accumulation lead to terrestrial planets or the core of giant planets.

If the planetesimals *formed via the gravitational collapse of local concentrations of pebbles*, their growth into planetary embryos and the cores of giant planets is dominated by the further accretions of pebbles. Pebble accretion is aided by the gas drag felt by objects as they accelerate toward a massive body. Gas drag slows the pebbles below the escape velocity of the massive body causing them to spiral toward and to be accreted by it. Pebble accretion may accelerate the formation of planets by a factor of 1000 compared to the accretion of planetesimals, allowing giant planets to form before the dissipation of the gas disk. *Yet, core growth via pebble accretion appears incompatible with the final masses and compositions of Uranus and Neptune.*

The formation of *terrestrial planets* differs from that of *giant gas planets*, also called *Jovian planets*. The particles that make up the terrestrial planets are made from metal and rock that condensed in the inner Solar System. However, *Jovian planets began as large, icy planetesimals, which then captured hydrogen and helium gas from the solar nebula.* Differentiation between these two classes of planetesimals arise due to the frost line of the solar nebula.

Cosmic dust

Cosmic dust also called extraterrestrial dust, space dust, or star dust – is dust that occurs in outer space or has fallen onto Earth. Most cosmic dust particles measure between a few molecules and 0.1 mm (100 μm), such as micrometeoroids. Larger particles are called meteoroids. Cosmic dust can be further distinguished by its astronomical location: *intergalactic dust*, *interstellar dust*, *interplanetary dust* (as in the zodiacal cloud), and *circumplanetary dust* (as in a planetary ring). There are several methods to obtain space dust measurement.

In the Solar System, *interplanetary dust* causes the zodiacal light. Solar System dust includes *comet dust*, *planetary dust* (like from Mars), *asteroidal dust*, *dust from the Kuiper belt*, and *interstellar dust passing through the Solar System*. Thousands of tons of cosmic dust are estimated to reach Earth's surface every year, with most grains having a mass between 10^{-16} kg (0.1 pg) and 10^{-4} kg (0.1 g). The density of the dust cloud through which the Earth is traveling is approximately 10^{-6} dust grains/m^3.

Cosmic dust contains some complex organic compounds (amorphous organic solids with a mixed aromatic–aliphatic structure) that could be created naturally, and rapidly, by stars. A smaller fraction of dust in space is "*stardust*" consisting of larger refractory minerals that condensed as matter left by stars.

The large grains in interstellar space are probably complex, with refractory cores that condensed within stellar outflows topped by layers acquired during incursions into cold dense interstellar clouds. That cyclic process of growth and destruction outside of the clouds has been modeled to demonstrate that the cores live much longer than the average lifetime of dust mass. Those cores mostly start with silicate particles condensing in the atmospheres of cool, oxygen-rich red-giants and carbon grains condensing in the atmospheres of cool carbon stars. Red giants have evolved or altered off the main sequence and have entered the giant phase of their evolution and are the major source of refractory dust grain cores in galaxies. Those refractory cores are also called *stardust*, which is a scientific term for the small fraction of cosmic dust that condensed thermally within stellar gases as they were ejected from the stars. Several percent of refractory grain cores have condensed within expanding interiors of supernovae, a type of cosmic decompression chamber. Meteoriticists who study *refractory stardust* (extracted from meteorites) often call it presolar grains but that within meteorites is only a small fraction of all presolar dust.

Stardust condenses within the stars via considerably different condensation chemistry than that of the bulk of *cosmic dust*, which accretes cold onto preexisting dust in dark molecular clouds of the galaxy. Those molecular clouds are very cold, typically less than 50K, so that ices of many kinds may accrete onto grains, in cases only to be destroyed or split apart by radiation and sublimation into a gas component. Finally, as the *Solar System* formed many interstellar dust grains were further modified by coalescence and chemical reactions in the planetary accretion disk. *The history of the various types of grains in the early Solar System is complicated and only partially understood.*

Astronomers know that the dust is formed in the envelopes of late-evolved stars from specific observational signatures. In infrared light, emission at 9.7 micrometers is a signature of silicate dust in cool evolved oxygen-rich giant stars. Emission at 11.5 micrometers indicates the presence of silicon carbide dust in cool evolved carbon-rich giant stars. These help provide evidence that the small silicate particles in space came from the ejected outer envelopes of these stars.

Conditions in *interstellar space* are generally not suitable for the formation of silicate cores. *This would take excessive time to accomplish*, even if it might be possible. The arguments are that: given an observed typical grain diameter a, the time for a grain to attain a, and given the temperature of interstellar gas, *it would take considerably longer than the age of the Universe for interstellar grains to form*. On the other hand, grains are seen to have recently formed in the vicinity of nearby stars, in nova and supernova ejecta, and in R Coronae Borealis variable stars which seem to eject discrete clouds containing both gas and dust. So mass loss from stars is unquestionably where the refractory cores of grains formed.

Peebles, P. J. E.* & Ratra, B.** (April 2003). The Cosmological Constant and Dark Energy.

Reviews of Modern Physics, 75, 2, 559–606; arXiv:astro-ph/0207347; https://arxiv.org/pdf/astro-ph/0207347.pdf.

* Joseph Henry Laboratories, Princeton University, Princeton, NJ 08544;
** Department of Physics, Kansas State University, Manhattan, KS 66506.

Physics invites the idea that *space contains energy whose gravitational effect approximates that of Einstein's cosmological constant, Λ*; nowadays the concept is termed *dark energy* or quintessence. Physics also suggests the dark energy could be dynamical, allowing the arguably appealing picture that the *dark energy density is evolving to its natural value, zero*, and is small now because the expanding universe is old. This alleviates the classical problem of the curious energy scale of order a millielectronvolt associated with a constant Λ. Dark energy may have been detected by recent advances in the cosmological tests. The tests establish a good scientific case for the context, in the *relativistic* Friedmann-Lemaˆıtre model, including the gravitational inverse square law applied to the scales of cosmology. We have well-checked evidence that the *mean mass density* is not much more than one quarter of the critical Einstein-de Sitter value. The case for detection of dark energy is serious but not yet as convincing; we await more checks that may come out of work in progress. Planned observations might be capable of detecting evolution of the dark energy density; a positive result would be a considerable stimulus to attempts to understand the microphysics of dark energy. *This review presents the basic physics and astronomy of the subject*, reviews the history of ideas, assesses the state of the observational evidence, and comments on recent developments in the search for a fundamental theory.

Contents

VI. ***Appendix: Recent dark energy scalar field research***

References

I. INTRODUCTION

There is significant observational evidence for the detection of *Einstein's cosmological constant*, Λ, or a component of the material content of the universe that varies only slowly with time and space and so acts like Λ. *We will use the term dark energy for Λ or a component that acts like it*. Detection of dark energy would be a new clue to an old puzzle, *the gravitational effect of the zero-point energies of particles and fields*. The total with other energies that are close to homogeneous and nearly independent of time act as dark energy. *The puzzle has been that the value of the dark energy density has to be tiny compared to what is suggested by dimensional analysis*; the startling new evidence is that it may be different from the only other natural value, zero.

The main question to consider now has to be *whether to accept the evidence for detection of dark energy*. We outline the nature of the case in this section. After reviewing the basic concepts of the relativistic world model, in Sec. II, and in Sec. III reviewing the history of ideas, we present in Sec. IV a more detailed assessment of the state of the cosmological tests and the evidence for detection of Λ or its analog in dark energy.

There is little new to report on the big issue for physics — *why the dark energy density is so small* — since Weinberg's (1989) review in this Journal.[1]

[1] Weinberg, S. (1989). *Rev. Mod. Phys.*, 61, 1.

But there have been analyses of a simpler idea: *can we imagine the present dark energy density is evolving, perhaps approaching zero*? Models are introduced in Secs. II.C and III.E, and recent work is summarized in more detail in the Appendix. Feasible advances in the cosmological tests could detect evolution of the dark energy density, and maybe its gravitational response to the large-scale fluctuations in the mass distribution. That would really drive the search for a more fundamental physics model for dark energy.

A. *The issues for observational cosmology*

We have to make two points. First, cosmology has a substantial observational and experimental basis that shows many aspects of the standard model almost certainly are good approximations to reality. Second, *the empirical basis is not nearly as strong as it is for the standard model for particle physics*: in cosmology it is not yet a matter of measuring the parameters in a well-established theory.

To explain the second point, we must remind those more accustomed to experiments in the laboratory than observations in astronomy of the astronomers' tantalus principle: one can look at distant objects but never touch them. For example, the observations of supernovae in distant galaxies offer evidence for the detection of dark energy, under the assumption that distant and nearby supernovae are drawn from the same statistical sample (that is, that they are statistically similar enough for the purpose of this test). There is no direct way to check this, and it is easy to imagine differences between distant and nearby supernovae of the same nominal type. More distant supernovae are seen in younger galaxies, because of the light travel time, and these younger galaxies tend to have more massive rapidly evolving stars with lower heavy element abundances. How do we know the properties of the supernovae are not also different? We recommend Leibundgut's (2001, Sec. 4) discussion of the astrophysical hazards. Astronomers have checks for this and other issues of interpretation of the observations used in the cosmological tests. But it takes nothing away from this careful and elegant work to note that the checks seldom can be convincing, because the astronomy is complicated and what can be observed is sparse. What is more, we don't know ahead of time that the physics that is well tested on scales ranging from the laboratory to the Solar System survives the enormous extrapolation to cosmology.

...

The case for detection of Λ or *dark energy* commences with the *Friedmann-Lemaıtre cosmological model*. In this model the expansion history of the universe is determined by a set of dimensionless [density] parameters whose sum is normalized to unity,

$$\Omega_{M0} + \Omega_{R0} + \Omega_{\Lambda 0} + \Omega_{K0} = 1. \qquad (1)$$

The first, Ω_{M0}, is a measure of the *present mean mass density in nonrelativistic matter*, mainly baryons and nonbaryonic dark matter. The second, $\Omega_{R0} \sim 1\times10^{-4}$, is a measure of the *present mass in the relativistic 3 K thermal cosmic microwave background radiation* that almost homogeneously fills space, and the accompanying low mass neutrinos. The third [$\Omega_{\Lambda 0}$] is a measure of Λ or *the present value of the dark energy equivalent*. The fourth, Ω_{K0}, is an effect of *the curvature of space*.

> [This is hardly a model. It is the sum of descriptions of the presumed different phases of the evolution of the universe, with the term "dark energy" (or *Einstein's cosmological constant* Λ) used to signify the unexplained expansion, and "dark matter" to signify unexplained attraction.]
>
> [The *density parameter* Ω is defined as the ratio of the actual (or observed) density ρ to the critical density ρ_c of the Friedmann universe. The relation between the *actual density* and the *critical density* determines the overall geometry of the

universe; when they are equal, the geometry of the universe is flat (Euclidean). *In earlier models, which did not include a cosmological constant term, critical density was initially defined as the watershed point between an expanding and a contracting Universe.*

To date, the *critical density* is estimated to be approximately five atoms (of monatomic hydrogen) per cubic meter, whereas the average density of ordinary matter in the Universe is believed to be 0.2–0.25 atoms per cubic meter.

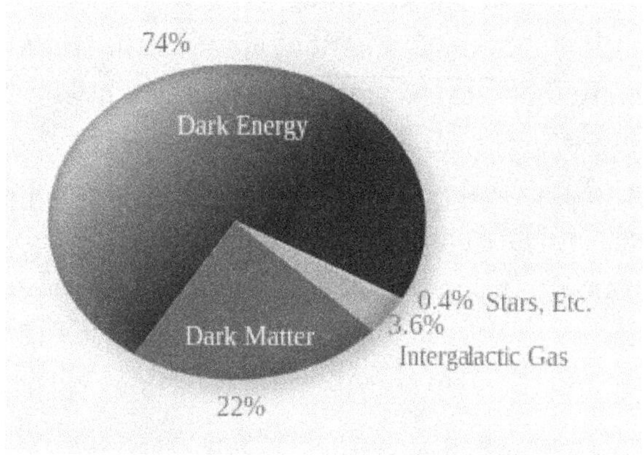

Estimated relative distribution for components of the *energy density* of the universe. *Dark energy* dominates the total *energy* (74%) while *dark matter* (22%) constitutes most of the *mass*. Of the remaining baryonic matter (4%), only one tenth is compact. In February 2015, the European-led research team behind the Planck cosmology probe released new data refining these values to 4.9% ordinary matter, 25.9% *dark matter* and 69.1% *dark energy*.

A much greater *density* comes from the unidentified *dark matter*; *both ordinary and dark matter contribute in favor of contraction of the universe*. However, *the largest part comes from so-called dark energy*, which accounts for the *cosmological constant term*. Although the total *density* is equal to the *critical density* (exactly, up to measurement error), the *dark energy* does not lead to contraction of the universe but rather *may accelerate its expansion*.

An expression for the *critical density* is found by assuming Λ to be zero (as it is for all basic Friedmann universes) and setting the normalized spatial curvature, k, equal to zero. When the substitutions are applied to the first of the *Friedmann equations* we find:

$$\rho_c = 3H^2/8\pi G = 1.8788 \times 10^{-26}\, h^2 kg\ m^{-3} = 2.7754 \times 10^{11}\, h^2\, M_O\, Mpc^{-3},$$

(where $h = H_0/(100\ km/s/Mpc)$. For $H_o = 67.4$ km/s/Mpc, i.e. $h = 0.674$,

$\rho_c = 8.5\times10^{-27}$ kg/m^3).

The *density parameter* Ω (useful for comparing different cosmological models) is then defined as:

$$\Omega = \rho/\rho_c$$

This term originally was used as a means to determine the spatial geometry of the universe, where ρ_c is *the critical density for which the spatial geometry is flat* (or Euclidean). Assuming a *zero vacuum energy density*, if Ω is larger than unity, the space sections of the universe are closed; the universe will eventually stop expanding, then collapse. If Ω is less than unity, they are open; and the universe expands forever. However, one can also subsume the spatial curvature and vacuum energy terms into a more general expression for Ω in which case this density parameter equals exactly unity. Then it is a matter of measuring the different components, usually designated by subscripts. *According to the ΛCDM model, there are important components of Ω due to baryons, cold dark matter and dark energy.* The spatial geometry of the universe has been measured by the WMAP spacecraft to be nearly flat. This means that the universe can be well approximated by a model where the spatial curvature parameter k is zero; however, this does not necessarily imply that the universe is infinite: it might merely be that the universe is much larger than the part we see.

The first *Friedmann equation* is often seen in terms of the present values of the density parameters, that is

$$H^2/H_0^2 = \Omega_{M0}(1 + z)^3 + \Omega_{R0}(1 + z)^4 + \Omega_{\Lambda 0} + \Omega_{K0}(1 + z)^2\}.$$

Here Ω_{R0} is the radiation density today (when at the present epoch $z = 0$), Ω_{M0} is the matter (dark plus baryonic) density today, $\Omega_{K0} = 1 - \Omega_0$ is the "spatial curvature density" today, and $\Omega_{\Lambda 0}$ is the cosmological constant or vacuum density today.]

We review some details of these parameters in the next section, and of their measurements in Sec. IV.

The most direct evidence for detection of *dark energy* comes from observations of supernovae of a type whose intrinsic luminosities are close to uniform (after subtle astronomical corrections, a few details of which are discussed in Sec. IV.B.4). The observed brightness as a function of the wavelength shift of the radiation probes the geometry of spacetime, in what has come to be called the redshift-magnitude relation.[2]

[2] The apparent magnitude is m = −2.5 $\log_{10}$ f plus a constant, where f is the detected *energy flux density* in a chosen wavelength band. The standard measure of the wavelength shift, due to the expansion of the universe, is the redshift z defined in Eq. (7) below.

The measurements agree with the *relativistic* cosmological model with $\Omega_{K0} = 0$, meaning *no space curvature*, and $\Omega_{\Lambda 0}$ [*the present value of the dark energy equivalent*] ~ 0.7, meaning nonzero Λ. A model with $\Omega_{\Lambda 0} = 0$ is two or three standard deviations off the best fit, depending on the data set and analysis technique. This is an important indication, but 2 to 3σ is not convincing, even when we can be sure the systematic errors are under reasonable control. And we have to consider that there may be a significant systematic error from differences between distant, high redshift, and nearby, low redshift, supernovae.

There is a check, based on the *Cold Dark Matter (CDM) model* for *structure formation.* The fit of the model to the observations reviewed in Sec. IV.B yields two key constraints. First, the angular power spectrum of fluctuations in the temperature of the 3 K thermal cosmic microwave background radiation across the sky indicates Ω_{K0} [the effect of *the curvature of space*] is small. Second, the power spectrum of the spatial distribution of the galaxies requires Ω_{M0} [*the present mean mass density in nonrelativistic matter*] ~ 0.25. Similar estimates of Ω_{M0} follow from independent lines of observational evidence. The rate of gravitational lensing prefers a somewhat larger value (if Ω_{K0} is small), and some dynamical analyses of systems of galaxies prefer lower Ω_{M0}. But the differences could all be in the measurement uncertainties. Since Ω_{R0} [*the present mass in the relativistic 3 K thermal cosmic microwave background radiation*] in Eq. (1) is small, the conclusion is $\Omega_{\Lambda 0}$ [*the present value of the dark energy equivalent*] is large, in excellent agreement with what the supernovae say.

> [The *Cold Dark Matter (CDM) model* for *structure formation* confirms the obvious. The effect of the expansion of the universe is large! It also confirms little evidence for *general relativity*, the curvature of space.]

Caution is in order, however, *because this check depends on the CDM model for structure formation.* We can't see the *dark matter*, so we naturally assign it the simplest properties we can get away with. Maybe it is significant that *the model has observational problems with galaxy formation,* as discussed in Sec. IV.A.2, or maybe these problems are only apparent, from the complications of the astronomy. We are going to have to determine which it is before we can have a lot of confidence in the role of the CDM model in the cosmological tests. We will get a strong hint from the precision measurements in progress of the angular distribution of the 3 K thermal cosmic microwave background radiation. If the results match in all detail the prediction of the *relativistic* model for *cosmology* and the CDM model for *structure formation*, with parameter choices that agree with the constraints from all the other cosmological tests, it will be strong evidence that we are approaching a good approximation to reality [???], and the completion of the great program of cosmological tests that commenced in the 1930s. But all that is to come.

We emphasize that the advances in the empirical basis for cosmology already are very real and substantial. How firm is the conclusion depends on the issue, of course. Every competent cosmologist we know accepts as established beyond reasonable doubt that the universe is expanding and cooling in a near homogeneous and isotropic way from a hotter denser state: how else could space, which is transparent now, have been filled with radiation that has relaxed to a thermal spectrum? *The debate is about when the expansion commenced or became a meaningful concept.* Some whose opinions and research we respect *question the extrapolation of the gravitational inverse square law*, in its use in estimates of masses in galaxies and systems of galaxies, and of Ω_{M0} [*the present mean mass density in nonrelativistic matter*].

> [The inverse square law is founded on a universe in a three-dimensional space.]

We agree that this law is one of the hypotheses to be tested. Our conclusion from the cosmological tests in Sec. IV is that the law passes significant though not yet complete tests, and that we already have a strong scientific case, resting on the abundance of cross-checks, that the matter density parameter Ω_{M0} [*the present mean mass density in nonrelativistic matter*] is about one quarter. The case for detection of $\Omega_{\Lambda 0}$ [*the present value of the dark energy equivalent*] is significant too, but not yet as compelling.

For the most part the results of the cosmological tests agree wonderfully well with accepted theory. But *the observational challenges to the tests are substantial: we are drawing profound conclusions from very limited information.* We have to be liberal in considering ideas about what the universe is like, conservative in accepting ideas into the established canon.

...

C. **Some explanations**

We have to explain our choice of nomenclature. Basic concepts of physics say *space contains homogeneous zero-point energy*, and maybe also energy that is homogeneous or nearly so in other forms, real or effective (*as from counter terms in the gravity physics*, which make the net energy density cosmologically acceptable).

> [*Zero-point energy* (ZPE) is the lowest possible energy that a quantum mechanical system may have. Unlike in classical mechanics, quantum systems constantly fluctuate in their lowest energy state as described by the Heisenberg uncertainty

principle. Therefore, even at absolute zero, atoms and molecules retain some vibrational motion. Apart from atoms and molecules, the empty space of the vacuum also has these properties. According to *quantum field theory*, the universe can be thought of not as isolated particles but continuous fluctuating fields: *matter fields*, whose quanta are fermions (i.e., leptons and quarks), and *force fields*, whose quanta are bosons (e.g., photons and gluons). *All these fields have zero-point energy*. These fluctuating zero-point fields lead to a kind of reintroduction of an ether in physics since *some systems can detect the existence of this energy*. However, this ether cannot be thought of as a physical medium if it is to be Lorentz invariant such that there is no contradiction with *Einstein's theory of special relativity*.

The notion of a *zero-point energy* is also important for *cosmology*, and *physics currently lacks a full theoretical model for understanding zero-point energy in this context*; in particular, the discrepancy between theorized and observed vacuum energy in the universe is a source of major contention. Physicists Richard Feynman and John Wheeler calculated the *zero-point radiation of the vacuum* to be *an order of magnitude greater than nuclear energy*, with a single light bulb containing enough energy to boil all the world's oceans. Yet according to Einstein's theory of *general relativity*, *any such energy would gravitate*, and the experimental evidence from the *expansion of the universe*, *dark energy* and the *Casimir effect* shows *any such energy to be exceptionally weak*. A popular proposal that attempts to address this issue is to say that the *fermion field* has a *negative zero-point energy*, while the *boson field* has *positive zero-point energy* and thus these energies somehow cancel each other out. This idea would be true if supersymmetry were an exact symmetry of nature; however, the LHC at CERN has so far found no evidence to support it. Moreover, it is known that if supersymmetry is valid at all, it is at most a broken symmetry, only true at very high energies, and *no one has been able to show a theory where zero-point cancellations occur in the low-energy universe we observe today*. This discrepancy is known as *the cosmological constant problem* and it is one of the greatest unsolved mysteries in physics. Many physicists believe that "the vacuum holds the key to a full understanding of nature".]

In the literature this near homogeneous energy has been termed the *vacuum energy*, the *sum of vacuum energy and quintessence* (Caldwell, Dave, and Steinhardt, 1998), and the *dark energy* (Turner, 1999). We have adopted the last term, and we will refer to the *dark energy density* ρ_Λ that manifests itself as an effective version of Einstein's *cosmological constant*, but one that *may vary slowly with time and position*.

Our subject involves two quite different traditions, in physics and astronomy. Each has familiar notation, and familiar ideas that may be "in the air" but not in the recent literature. Our attempt to take account of these traditions commences with the summary in Sec. II of the basic notation with brief explanations. …

We offer in Sec. III our reading of the history of ideas on Λ [*Einstein's cosmological constant*] and its generalization to *dark energy*. This is a fascinating and we think edifying illustration of how science may advance in unexpected directions. It is relevant to an understanding of the present state of research in cosmology, because traditions inform opinions, and people have had mixed feelings about Λ [*Einstein's cosmological constant*] ever since Einstein (1917)[#] introduced it 85 years ago.

> [#] Einstein, A. (February, 1917). Kosmologische Betrachtungen zur allgemeinen Relativitätstheorie. (Cosmological Considerations in the General Theory of Relativity.) *Koniglich Preußische Akademie der Wissenschaften, Sitzungsberichte* (Berlin), 142–52; translation by W. Perrett & G. B. Jeffery in A. Engel (translator), E. Schuckling (consultant). (1997). *The Collected Papers of Albert Einstein*, Volume 6: The Berlin Years: Writings, 1914-1917, Princeton University Press, Princeton, Doc. 43, 421-32.

The concept never entirely dropped out of sight in cosmology because a series of observations hinted at its presence, and because to some cosmologists Λ fits the formalism too well to be ignored. The search for the physics of the vacuum, and its possible relation to Λ, has a long history too. Despite the common and strong suspicion that Λ must be negligibly small, because any other acceptable value is absurd, all this history has made the community well prepared for the recent observational developments that argue for the detection of Λ.

Our approach in Sec. IV to the discussion of the evidence for detection of Λ, from the cosmological tests, also requires explanation. One occasionally reads that the tests will show us how the world ends. That certainly seems interesting, but it is not the main point: why should we trust an extrapolation into the indefinite future of a theory we can at best show is a good approximation to reality? As we remarked in Sec. I.A, the purpose of the tests is to check the approximation to reality, *by checking the physics and astronomy of the standard relativistic cosmological model*, along with any viable alternatives that may be discovered. We take our task to be to identify the aspects of the standard theory that enter the interpretation of the measurements and thus are or may be empirically checked or measured.

II. BASIC CONCEPTS

A. **The Friedmann-Lemaıtre model**

The standard world model is close to homogeneous and isotropic on large scales, and lumpy on small scales — the effect of the mass concentrations in galaxies, stars, people, and all that. The length scale at the transition from nearly smooth to strongly clumpy is about 10 Mpc. We use here and throughout the standard astronomers' length unit,

$$1 \text{ Mpc} = 3.1 \times 1024 \text{ cm} = 3.3 \times 10^6 \text{ light years.} \quad (3)$$

...

The expansion of the universe means the *distance* $l(t)$ between two well-separated galaxies varies with world time, t, as

$$l(t) \propto a(t), \quad (4)$$

where the *expansion or scale factor*, $a(t)$, is independent of the choice of galaxies. It is an interesting exercise, for those who have not already thought about it, to check that Eq. (4) is required to preserve homogeneity and isotropy.

The rate of change of the distance in Eq. (4) is the speed

$$v = dl/dt = Hl, \qquad H = \dot{a}/a, \quad (5)$$

where the dot means the derivative with respect to world time t and H is the time-dependent Hubble parameter. *When v is small compared to the speed of light this is Hubble's law*. ($v = H_0 D$.) The *present value* of H is Hubble's constant, H_0. When needed we will use

$$H_0 = 100h \text{ km s}^{-1}\text{Mpc}^{-1} = 67 \pm 7 \text{ km s}^{-1}\text{Mpc}^{-1} = (15 \pm 2 \text{ Gyr})^{-1}, \quad (6)$$

at two standard deviations. The first equation defines the dimensionless parameter h.

Another measure of the expansion follows by considering the stretching of the wavelength of light received from a distant galaxy. The observed wavelength, λ_{obs}, of a feature in the spectrum that had wavelength λ_{em} at emission satisfies

$$1 + z = \lambda_{obs}/\lambda_{em} = a(t_{obs})/a(t_{em}), \quad (7)$$

where the *expansion factor a* is defined in Eq. (4) and z is the *redshift*. That is, *the wavelength of freely traveling radiation stretches in proportion to the factor by which the universe expands*. To understand this, imagine a large part of the universe is enclosed in a cavity with perfectly reflecting walls. The cavity expands with the general expansion, the widths proportional to a(t). Electromagnetic radiation is a sum of the normal modes that fit the cavity. At interesting wavelengths, the mode frequencies are much larger than the rate of expansion of the universe, so adiabaticity says a photon in a mode stays there, and its wavelength thus must vary as $\lambda \propto a(t)$, as stated in Eq. (7). The cavity disturbs the long wavelength part of the radiation, but the disturbance can be made exceedingly small by choosing a large cavity.

Equation (7) defines the *redshift* z. The redshift is a convenient label for epochs in the early universe, where z exceeds unity. A good exercise for the student is to check that when z is small Eq. (7) reduces to Hubble's law, where λz is the first-order Doppler shift in the wavelength λ, and Hubble's parameter H is given by Eq. (5). Thus, Hubble's law may be written as $cz = Hl$ (where we have put in the speed of light).

These results follow from the symmetry of the cosmological model and conventional local physics; *we do not need general relativity theory*. When z >~ 1 we need the relativistic theory to compute the relations among the redshift and other observables. An example is the relation between redshift and apparent magnitude used in the supernova test. Other cosmological tests check consistency among these relations, and this checks the world model.

In *general relativity* the second time derivative of the *expansion factor* satisfies

$$\ddot{a}/a = -\,4/3\ \pi G(\rho + 3p). \qquad (8)$$

The *gravitational constant* is G. Here and throughout, we choose units to set the velocity of light to unity. The *mean mass density, ρ(t)*, and the *pressure, p(t)*, counting all contributions including *dark energy*, satisfy the local energy conservation law,

$$\dot{\rho} = -3\ \dot{a}/a\ (\rho + p). \qquad (9)$$

The first term on the right-hand side represents the decrease of *mass density* due to the expansion that more broadly disperses the matter. The pdV work in the second term is a familiar local concept, and meaningful in *general relativity*. But one should note that *energy does not have a general global meaning in this theory*.

The first integral of Eqs. (8) and (9) is the *Friedmann equation*

$$\dot{a}^2 = 8/3\ \pi G\rho a^2 + \text{constant}. \qquad (10)$$

[The *Friedmann equations* are a set of equations in physical cosmology that govern the expansion of space in homogeneous and isotropic models of the universe within the context of *general relativity*. They were first derived by Alexander Friedmann in 1922 from Einstein's field equations of gravitation for the *Friedmann–Lemaître–Robertson–Walker metric* and a *perfect fluid with a given mass density ρ and pressure p*. The equations for negative spatial curvature were given by Friedmann in 1924.

The Friedmann equations start with the simplifying assumption that *the universe is spatially homogeneous and isotropic*, that is, the cosmological principle; empirically, this is justified on scales larger than the order of 100 Mpc. The cosmological principle implies that the metric of the universe must be of the form

$$-\,ds^2 = a(t)^2\, ds_3{}^2 - c^2\, dt^2$$

where $ds_3{}^2$ is a three-dimensional metric that must be one of (a) flat space, (b) a sphere of constant positive curvature or (c) a hyperbolic space with constant negative curvature. This metric is called *Friedmann–Lemaître–Robertson–Walker (FLRW) metric*. The parameter k takes the value 0, 1, −1, or the Gaussian curvature, in these three cases respectively. It is this fact that allows us to sensibly speak of a "scale factor" $a(t)$.

Einstein's equations now relate the evolution of this scale factor to the pressure and energy of the matter in the universe. From FLRW metric we compute Christoffel symbols, then the Ricci tensor. With the *stress–energy tensor for a perfect fluid*, we substitute them into Einstein's field equations and the resulting equations are described below.

There are two independent *Friedmann equations* for modelling a homogeneous, isotropic universe. The first is:

$$(a^{\cdot 2} + kc^2)/a^2 = (8\pi G\rho - \Lambda c^2)/3,$$

$$[(a^{\cdot}/a)^2 + kc^2/a^2 - \Lambda c^2/3 = 8\pi G/3\ \rho \text{ below.}]$$

which is derived from the 00 component of the *Einstein field equations*. The second is:

$$a^{\cdot\cdot}/a = -\,4\pi G/3\ (\rho + 3p/c^2) + \Lambda c^2/3.$$

$$[2a^{\cdot\cdot}/a + (a^{\cdot}/a)^2 + kc^2/a^2 - \Lambda c^2 = -\,8\pi G/c^2\ p \text{ below.}]$$

which is derived from the first together with the trace of Einstein's field equations (the dimension of the two equations is time^{-2}).

The pair of equations given above is equivalent to the previous pair of equations

$$(a^{\cdot 2} + kc^2)/a^2 = (8\pi G\rho - \Lambda c^2)/3,$$
$$2a^{\cdot\cdot}/a + (a^{\cdot}/a)^2 + kc^2/a^2 - \Lambda c^2 = - 8\pi G/c^2\ p$$

with k, the spatial curvature index, serving as a constant of integration for the first equation.

The first equation can be derived also from thermodynamical considerations and is equivalent to the first law of thermodynamics, assuming the expansion of the universe is an adiabatic process (which is implicitly assumed in the derivation of the Friedmann–Lemaître–Robertson–Walker metric).

The second equation states that both the *energy density* and the *pressure* cause the expansion rate of the universe $a^{\cdot}$ to decrease, i.e., *both cause a deceleration in the expansion of the universe. This is a consequence of gravitation, with pressure playing a similar role to that of energy (or mass) density, according to the principles of general relativity.* The *cosmological constant*, on the other hand, *causes an acceleration in the expansion of the universe*.

The *cosmological constant* term can be omitted if we make the following replacements

$$\rho \rightarrow \rho - \Lambda c^2/8\pi G$$
$$p \rightarrow p - \Lambda c^4/8\pi G$$

Therefore, the *cosmological constant* can be interpreted as arising from *a form of energy which has negative pressure, equal in magnitude to its (positive) energy density*:

$$p = - \rho c^2$$

which is an *equation of state of vacuum with dark energy*. In order to get a term which causes an *acceleration of the universe expansion*, it is enough to have a scalar field which satisfies

$$p < - \rho c^2/3.$$

Such a field is sometimes called *quintessence*.

The *Friedmann–Lemaître–Robertson–Walker (FLRW) metric* is a metric *based on the exact solution of the Einstein field equations of general relativity*. The metric describes a homogeneous, isotropic, expanding (or otherwise, contracting) universe that is path-connected, but not necessarily simply connected. The general form of the metric follows from the geometric properties of homogeneity and isotropy; Einstein's field equations are only needed to derive the scale factor of the universe as a function of time.

The FLRW metric starts with the assumption of homogeneity and isotropy of space. It also assumes that the spatial component of the metric can be time-dependent. The generic metric which meets these conditions is

$$-c^2\,d\tau^2 = -c^2\,dt^2 + a(t)^2\,d\Sigma^2$$

where Σ ranges over a 3-dimensional space of uniform curvature, that is, elliptical space, Euclidean space, or hyperbolic space. It is normally written as a function of three spatial coordinates, but there are several conventions for doing so, detailed below. dΣ does not depend on *t* – all of the time dependence is in the function $a(t)$, known as the "*scale factor*".

Einstein's field equations are not used in deriving the general form for the metric: *it follows from the geometric properties of homogeneity and isotropy*. However, *determining the time evolution of a(t) does require Einstein's field equations* together with a way of calculating the density, ρ(t), such as a *cosmological equation of state*.

This metric has an analytic solution to *Einstein's field equations*

$$G_{\mu\nu} + \Lambda g_{\mu\nu} = 8\pi G/c^4\ T_{\mu\nu}$$

giving the *Friedmann equations* when the energy–momentum tensor is similarly assumed to be isotropic and homogeneous. The resulting equations are:

$$(a^{\cdot}/a)^2 + kc^2/a^2 - \Lambda c^2/3 = 8\pi G/3\ \rho$$
$$2a^{\cdot\cdot}/a + (a^{\cdot}/a)^2 + kc^2/a^2 - \Lambda c^2 = -\ 8\pi G/c^2\ p.$$

These equations are the basis of the standard *Big Bang cosmological model* including the current ΛCDM model. Because the FLRW model assumes homogeneity, some popular accounts mistakenly assert that the *Big Bang model* cannot account for the observed lumpiness of the universe. In a strictly FLRW model, there are no clusters of galaxies or stars, since these are objects much denser than a typical part of the universe. Nonetheless, the FLRW model is used as a first approximation for the evolution of the real, lumpy universe because it is simple to calculate, and models which calculate the lumpiness in the universe are added onto the FLRW models as extensions. Most cosmologists agree that the observable universe is well approximated by an *almost FLRW model*, i.e., a model which follows the FLRW metric *apart from primordial density fluctuations*. As of 2003, the theoretical implications of the various extensions to the FLRW model appear to be well understood, and the goal is to make these consistent with observations from COBE and WMAP.]

It is conventional to rewrite this as

$$(\dot{a}/a)^2 = H_0^2E(z)^2$$
$$= H_0^2\{\Omega_{M0}(1 + z)^3 + \Omega_{R0}(1 + z)^4 + \Omega_{\Lambda 0} + \Omega_{K0}(1 + z)^2\}. \quad (11)$$

The first equation defines the function E(z) that is introduced for later use. The second equation assumes constant Λ; the time-dependent *dark energy* case is reviewed in Secs. II.C and III.E. The first term in the last part of Eq. (11) represents *non-relativistic matter* with negligibly small *pressure*; one sees from Eqs. (7) and (9) that *the mass density in this form varies with the expansion of the universe* as $\rho M \propto a^{-3} \propto (1 + z)^3$. The second term represents *radiation and relativistic matter*, with pressure $p_R = \rho_R/3$, whence

$\rho_R \propto (1 + z)^4$. The third term is the effect of Einstein's *cosmological constant*, or a *constant dark energy density*. The last term, discussed in more detail below, is the constant of integration in Eq. (10). The four density parameters Ω_{i0} are the fractional contributions to the square of Hubble's constant, $H^2{}_0$, that is, $\Omega_{i0}(t) = 8\pi G\rho_{i0}/(3H^2{}_0)$. At the present epoch, z = 0, the present value of $\dot{a}/a$ is H_0, and the Ω_{i0} sum to unity (Eq. [1]).

In this notation, Eq. (8) is

$$\ddot{a}/a = -H_0^2 \{\Omega_{M0}(1 + z)^3/2 + \Omega_{R0}(1 + z)^4 - \Omega_{\Lambda 0}\}. \quad (12)$$

The constant of integration in Eqs. (10) and (11) is related to the geometry of spatial sections at constant world time. Recall that in general relativity events in spacetime are labeled by the four coordinates x^μ of time and space. Neighboring events 1 and 2 at separation dx^μ have invariant separation ds defined by the line element

$$ds^2 = g_{\mu\nu}\, dx^\mu dx^\nu. \quad (13)$$

The repeated indices are summed, and the metric tensor $g_{\mu\nu}$ is a function of position in spacetime. If ds^2 is positive then ds is the proper (physical) time measured by an observer who moves from event 1 to 2; if negative, |ds| is the proper distance between events 1 and 2 measured by an observer who is moving so the events are seen to be simultaneous.

...

It is useful for what follows to recall that the metric tensor in Eq. (15) satisfies *Einstein's field equation*, a differential equation we can write as

$$G_{\mu\nu} = 8\pi G T_{\mu\nu}. \quad (17)$$

The left side is a function of $g_{\mu\nu}$ and its first two derivatives and represents the geometry of spacetime. The stress-energy tensor $T_{\mu\nu}$ represents the material contents of the universe, including particles, radiation, fields, and zero-point energies. An observer in a homogeneous and isotropic universe, moving so the universe is observed to be isotropic, would measure the stress-energy tensor to be

$$T_{\mu\nu} = \begin{matrix} \rho & 0 & 0 & 0 \\ 0 & p & 0 & 0 \\ 0 & 0 & p & 0 \\ 0 & 0 & 0 & p. \end{matrix} \quad (18)$$

This diagonal form is a consequence of the symmetry; the diagonal components define the *pressure* and *energy density*. With Eq. (18), the differential equation (17) yields the expansion rate equations (11) and (12).

B. ***The cosmological constant***.

...

There are two measures of gravitational interactions with a fluid: the passive *gravitational mass density determines how the fluid streaming velocity is affected by an applied gravitational field*, and *the active gravitational mass density determines the gravitational field produced by the fluid*. When the fluid velocity is *nonrelativistic* the expression for the former in general relativity is $\rho + p$, as one sees by writing out the covariant divergence of $T^{\mu\nu}$. This vanishes when $p = -\rho$, consistent with the loss of meaning of the streaming velocity. The latter is $\rho + 3p$, as one sees from Eq. (8). Thus, a fluid with $p = -\rho/3$, if somehow kept homogeneous and static, would produce no gravitational field. In the model in Eqs. (19) and (21) the active gravitational mass density is negative when ρ_Λ is positive. When this positive ρ_Λ dominates the stress-energy tensor $\ddot{a}$ is positive: *the rate of expansion of the universe increases*. In the language of Eq. (20), *this cosmic repulsion is a gravitational effect of the negative active gravitational mass density*, not a new force law.

The homogeneous *active mass* represented by Λ changes the equation of relative motion of freely moving test particles in the *nonrelativistic* limit to

$$d^2r^{\rightarrow}/dt^2 = g^{\rightarrow} + \Omega_{\Lambda 0}H_0^2\, r^{\rightarrow}, \qquad (24)$$

where $g^{\rightarrow}$ is the *relative gravitational acceleration* produced by the distribution of ordinary matter.

…

C. ***Inflation and dark energy***

The negative *active gravitational mass density* associated with a positive *cosmological constant* is an early precursor of the inflation picture of the early universe; inflation in turn is one precursor of the idea that *Λ might generalize into evolving dark energy*. To begin, we review some aspects of causal relations between events in spacetime. *Neglecting space curvature*, a light ray moves proper distance $dl = a(t)dx = dt$ in time interval dt, so the integrated coordinate displacement is

$$x = \int dt/a(t). \qquad (26)$$

If $\Omega_{\Lambda 0} = 0$ this integral converges in the past — we see distant galaxies that at the time of observation cannot have seen us since the singular start of expansion at $a = 0$. This "particle horizon problem" is curious: how could distant galaxies in different directions in the sky know to look so similar? The *inflation* idea is that in the early universe the
expansion history approximates that of de Sitter's (1917) solution to *Einstein's field equation* for $\Lambda > 0$ and $T_{\mu\nu} = 0$ in Eq. (20).

[$G\mu\nu = 8\pi G(T_{\mu\nu} + \rho_\Lambda g_{\mu\nu})$. (20)

This is Einstein's (1917) revision of the field equation of general relativity, where ρ_Λ is proportional to his cosmological constant Λ.]

We can choose the coordinate labels in this *de Sitter spacetime* so space curvature vanishes. Then Eqs. (11) and (12) say the expansion parameter is

$$a \propto e^{H\Lambda t}, \qquad (27)$$

where H_Λ is a constant. As one sees by working the integral in Eq. (26), here everyone can have seen everyone else in the past. The details need not concern us; for the following discussion two concepts are important. First, the early universe acts like an approximation to de Sitter's solution because it is dominated by a large effective cosmological "constant", or *dark energy density*. Second, the *dark energy* is modeled as that of a near homogeneous field, Φ.

In this scalar field model, motivated by grand unified models of very high energy particle physics, the *action* of the real scalar field, Φ (in units chosen so Planck's constant ħ is unity) is

$$S = Z\, d4x \sqrt{(-g)}[½\, g^{\mu\nu}\, \partial_\mu \Phi \partial_\nu \Phi - V(\Phi)]. \qquad (28)$$

The potential energy density V is a function of the field Φ, and g is the determinant of the metric tensor. When the field is spatially homogeneous (in the line element of Eq. [15]), and *space curvature may be neglected*, the *field equation* is

$$\ddot{\Phi} + 3\dot{a}/a\, \dot{\Phi} + dV/d\Phi = 0. \qquad (29)$$

...

III. HISTORICAL REMARKS

...

A. Einstein's thoughts

Einstein disliked the idea of an island universe in asymptotically flat spacetime, because a particle could leave the island and move arbitrarily far from all the other matter in the universe, yet preserve all its inertial properties, which he considered a violation of Mach's idea of the relativity of inertia. Einstein's (1917) cosmological model accordingly assumes the universe is *homogeneous* and *isotropic, on average*, thus removing the possibility of arbitrarily isolated particles. Einstein had no empirical support for this assumption, yet it agrees with modern precision tests. There is no agreement on whether this is more than a lucky guess.

Motivated by the observed low velocities of the then known stars, Einstein assumed that the large-scale structure of the universe is static. *He introduced the cosmological constant to reconcile this picture with his general relativity theory.* In the notation of Eq. 12,

$$[\ddot{a}/a = -H_0^2\, \{\Omega_{M0}(1 + z)^3/2 + \Omega_{R0}(1 + z)^4 - \Omega_{\Lambda 0}\}. \qquad (12)]$$

one sees that a positive value of the density parameter $\Omega_{\Lambda 0}$ can balance the positive values of Ω_{M0} and Ω_{R0} for consistency with $\ddot{a} = 0$. The balance is unstable: a small perturbation to the *mean mass density* or the *mass distribution* causes expansion or contraction of the whole or parts of the universe. One sees this in Eq. (24):

$$[\qquad d^2\vec{r}/dt^2 = \vec{g} + \Omega_{\Lambda 0} H_0^2\, \vec{r}, \qquad (24)$$

where $\vec{g}$ is the *relative gravitational acceleration* produced by the distribution of ordinary matter] the mass distribution can be chosen so the two terms on the right-hand side cancel, but the balance can be upset by redistributing the mass.

Einstein did not consider the cosmological constant to be part of the stress-energy term: his form for the field equation (in the streamlined notation of Eq. 17 is

$$G_{\mu\nu} - 8\pi G\rho_\Lambda g_{\mu\nu} = 8\pi G T_{\mu\nu}. \qquad (34)$$

The left-hand side contains the metric tensor and its derivatives; a new constant of nature, Λ, appears in the addition to Einstein's original field equation. …

…

The record shows Einstein never liked the Λ term. His view of how general relativity might fit Mach's principle was disturbed by de Sitter's (1917) solution to Eq. (34) for empty space ($T_{\mu\nu} = 0$) with $\Lambda > 0$. Pais, A., 1982, *Subtle is the Lord ...* (Oxford University, New York, p. 288) points out that Einstein in a letter to Weyl in 1923 comments on the effect of Λ in Eq. (24): "According to De Sitter two material points that are sufficiently far apart, continue to be accelerated and move apart. If there is no quasistatic world, then away with the cosmological term." We do not know whether at this time Einstein was influenced by Slipher's redshifts or Friedmann's expanding world model.

The earliest published comments we have found on Einstein's opinion of Λ within the evolving world model (Einstein, A. (1931), *Sitzungsber. Preuss. Akad. Wiss.*, p. 235.; Einstein and de Sitter, 1932) make the point that, since not all the terms in the expansion rate Eq. (11) are logically required, and the matter term surely is present and likely dominates over radiation at low redshift, a reasonable working model drops Ω_{K0} and $\Omega_{\Lambda 0}$ and ignores Ω_{R0}. This simplifies the expansion rate equation to what has come to be called the Einstein-de Sitter model,

$$\dot{a}^2/a^2 = 8/3\ \pi G\rho_M, \qquad (35)$$

where ρ_M is the *mass density* in *non-relativistic matter*; here $\Omega_M = 8\pi G\rho_M/(3H^2)$ is unity. The left side is a measure of the kinetic energy of expansion per unit mass, and the right-hand side a measure of the negative of the gravitational potential energy. In effect, this model universe is expanding with escape velocity.

Einstein and de Sitter point out that Hubble's estimate of H_0 and de Sitter's estimate of the mean *mass density* in galaxies are not inconsistent with Eq. (35) (and since both quantities scale with distance in the same way, this result is not affected by the error in the distance scale that affected Hubble's initial measurement of H_0). But the evidence now is that the *mass density* is about one quarter of what is predicted by this equation, as we will discuss.

…

Further to this point, in the appendix of the second edition of his book, *The Meaning of Relativity*, Einstein (1945, p. 127) states that the "introduction of the 'cosmologic member' " — Einstein's terminology for Λ — "into the equations of gravity, though possible from the point of view of relativity, *is to be rejected from the point of view of logical economy*", and that if "Hubble's expansion had been discovered at the time of the creation of the general theory of relativity, the cosmologic member would never have been introduced. It seems now so much less justified to introduce such a member into the field equations, since its introduction loses its sole original justification, — that of leading to a natural solution of the cosmologic problem." Einstein knew that without the cosmological constant the expansion time derived from Hubble's estimate of H_0 is uncomfortably short compared to estimates of the ages of the stars, and opined that that might be a problem with the star ages. The big error, the value of H_0, was corrected by 1960 [Sandage, A. (1958). *Astrophys. J.* 127, 513; (1962). *Problems of Extragalactic Research*, edited by G. C. McVittie (McMillan, New York), p. 359].

Gamow, G. (1970). [*My World Line*, Viking, New York, p. 44] recalls that "when I was discussing cosmological problems with Einstein, he remarked that the introduction of the cosmological term was the biggest blunder he ever made in his life." This certainly is consistent with all of Einstein's written comments we have seen on the cosmological constant per se; we do not know whether Einstein was also referring to the missed chance to predict the evolution of the universe.

B. The development of ideas

…

C. Inflation

1. *The scenario*

The deep issue that *inflation* addresses is the origin of the large-scale homogeneity of the observable universe. In a *relativistic* model with positive pressure, we can see distant galaxies that have not been in causal contact with each other since the singular start of expansion (Sec. II.C, Eq. [26]); they are said to be outside each other's particle horizon. Why do apparently causally unconnected parts of space look so similar? Sato (1981a, 1981b), Kazanas (1980), and Guth (1981) make the key point: if the early universe were dominated by the *energy density* of a relatively *flat real [dark energy] scalar field* (inflation) *potential* V(Φ) that acts like Λ, the particle horizon could spread beyond the universe we can see. This would allow for the possibility that microphysics during inflation could smooth inhomogeneities sufficiently to provide an explanation of the observed large-scale homogeneity. (We are unaware of a definitive demonstration of this idea, however.)

In the *inflation* scenario the [*dark energy*] field Φ rolls down its potential until eventually V(Φ) steepens enough to terminate inflation. *Energy in the scalar field is supposed to decay to matter and radiation, heralding the usual Big Bang expansion of the universe. …*

A signal achievement of *inflation* is that it offers a theory for the origin of the departures from homogeneity. *Inflation* tremendously stretches length scales, so cosmologically significant lengths now correspond to extremely short lengths during *inflation*. On these tiny length scales quantum mechanics governs: the wavelengths of zero-point field fluctuations generated during inflation are stretched by the inflationary expansion, and these fluctuations are converted to classical density fluctuations in the late time universe. …

D. The cold dark matter model

[The *Lambda-CDM, Lambda cold dark matter* or *ΛCDM model* is a mathematical model of the Big Bang theory with three major components:

1. a cosmological constant denoted by lambda (Λ) associated with dark energy,
2. the postulated cold dark matter, and
3. ordinary matter.

It is frequently referred to as the standard model of Big Bang cosmology because it is the simplest model that provides a reasonably good account of:

- the existence and structure of the cosmic microwave background
- the large-scale structure in the distribution of galaxies
- the observed abundances of hydrogen (including deuterium), helium, and lithium
- the accelerating expansion of the universe observed in the light from distant galaxies and supernovae

The model assumes that general relativity is the correct theory of gravity on cosmological scales. The ΛCDM model can be extended by adding cosmological inflation, quintessence, and other areas of speculation and research in cosmology.

Some alternative models challenge the assumptions of the ΛCDM model. Examples of these are *modified Newtonian dynamics*, *entropic gravity*, *modified gravity*, *theories of large-scale variations in the matter density of the universe*, *bimetric gravity*, *scale invariance of empty space*, and *decaying dark matter* (DDM).]

Some of the present cosmological tests were understood in the 1930s; others are based on new ideas about structure formation. A decade ago, a half dozen models for structure formation were under discussion; now the known viable models have been winnowed to one class: cold dark matter (CDM) and variants. We comment on the present state of tests of the CDM model in Sec. IV.A.2, and in connection with the cosmological tests in Sec. IV.B.

The CDM model assumes the mass of the universe now is dominated by dark matter that is nonbaryonic and acts like a gas of massive, weakly interacting particles with negligibly small primeval velocity dispersion. Structure is supposed to have formed by the gravitational growth of primeval departures from homogeneity that are adiabatic, scale-invariant, and Gaussian. The early discussions also assume an Einstein-de Sitter universe. These features all are naturally implemented in simple models for *inflation*, and the CDM model may have been inspired in part by the developing ideas of *inflation*. But the motivation in writing down this model was to find a simple way to show that the observed present-day mass fluctuations can agree with the growing evidence that the anisotropy of the 3 K thermal cosmic microwave background radiation is very small (Peebles, 1982). The first steps toward turning this picture into a model for structure formation were taken by Blumenthal et al. (1984).

In the decade commencing about 1985 the standard cosmology for many active in research in this subject was the Einstein-de Sitter model, and for good reason: it eliminates the coincidences problem, it avoids the curiosity of nonzero dark energy, and it fits the condition from conventional inflation that space sections have zero curvature. But unease about the astronomical problems with the high mass density of the Einstein-de Sitter model led to occasional discussions of a low-density universe with or without a cosmological constant, and the CDM model played an important role in these considerations, as we now discuss.

When the CDM model was introduced, it was known that the observations disfavor the high mass density of the Einstein-de Sitter model, unless the mass is more smoothly distributed than the visible matter (Sec. III.C). …

…

E. Dark energy

The idea that the universe contains close to homogeneous *dark energy* that approximates a time-variable cosmological "constant" arose in particle physics, through the discussion of phase transitions in the early universe and through the search for a dynamical cancellation of the vacuum energy density; in cosmology, through the discussions of how to reconcile a cosmologically flat universe with the small mass density indicated by galaxy peculiar velocities; and on both sides by the thought that Λ might be very small now because *it has been rolling toward zero for a very long time.*[44]

> [44] This last idea is similar in spirit to Dirac's (1937, 1938) attempt to explain the large dimensionless numbers of physics. [Dirac, P. A. M., (1937). *Nature*, 139, 323; Dirac, P. A. M., (1938). *Proc. R. Soc. London, Ser. A*, 165, 199.] He noted that the gravitational force *between two protons* is much smaller than the electromagnetic force, and that that might be because the gravitational constant G is decreasing in inverse proportion to the world time. This is the earliest discussion we know of what has come to be called the hierarchy problem, that is, the search for a mechanism that might be responsible for the large ratio between a possibly more fundamental high energy scale, for example, that of grand unification or the Planck scale (where quantum gravitational effects become significant) and a lower possibly less fundamental energy scale, for example that of electroweak unification (see, for example, Georgi, Quinn, and Weinberg, 1974).

The idea that the dark energy is decaying by emission of matter or radiation is now strongly constrained by the condition that the decay energy must not significantly disturb the spectrum of the 3 K cosmic microwave background radiation. But the history of the idea is interesting, and decay to *dark matter* still a possibility, so we comment on both here. The picture of *dark energy* in the form of defects in cosmic fields has not received much attention in recent years, in part because the computations are difficult, but might yet prove to be productive. Much discussed nowadays is *dark energy* in a slowly varying scalar field.

…

…

1. The XCDM parametrization

…

2. *Decay by emission of matter or radiation*

Bronstein (1933) introduced the idea that the *dark energy density* $\rho_{\_}$ *is decaying by the emission of matter or radiation. …*

…

3. *Cosmic field defects*

…

4. *Dark energy scalar field*

At the time of writing the popular picture for dark energy is a classical scalar field with a *self-interaction potential* V(Φ) that is shallow enough that the *field energy density* decreases with the expansion of the universe more slowly than the *energy density in matter*.

…

…

IV. THE COSMOLOGICAL TESTS

…

A. The theories

1. *General relativity*

Some early discussions of the cosmological tests, as in Robertson (1955) and Bondi (1960), make the point that observationally important elements of a spatially homogeneous cosmology follow by symmetry, independent of general relativity. This means *some empirical successes of the cosmology are not tests of relativity. …*

…

One sobering detail is that *in the standard cosmology the two dominant contributions to the stress-energy tensor — dark energy and dark matter — are hypothetical, introduced to make the theories fit the observations* [Eq. (2)]. This need not mean there is anything wrong with *general relativity* — we have no reason to expect Nature to have made all matter readily observable other than by its gravity — but it is a cautionary example of the challenges. *Milgrom's (1983) modified Newtonian dynamics (MOND) replaces the dark matter hypothesis with a hypothetical modification of the gravitational force law*. MOND gives remarkably successful fits to observed motions within galaxies, *without dark matter* (de Blok et al., 2001). So why should we believe there really is cosmologically significant mass in nonbaryonic *dark matter*? …

… This indirect chain of evidence for *dark matter* is becoming tight. A new example — the prospect for *a test of the inverse square law for gravity* on the length scales of cosmology — is striking enough for special mention here.

Consider the equation of motion of a freely moving test particle with *nonrelativistic* peculiar velocity $\vec{v}$ in a universe with expansion factor a(t),

$$\partial \vec{v}/\partial t + \dot{a}/a\, \vec{v} = \vec{g} = -\, 1/a\, \nabla\phi. \qquad (51)$$

The particle always is moving toward receding observers, which produces the second term in the left-most expression. The peculiar *gravitational acceleration* $\vec{g}$ relative to the homogeneous background model is computed from the Poisson equation for the *gravitational potential* ϕ,

$$\nabla^2\phi = 4\pi G a^2[\rho(\vec{x}, t) - \langle\rho\rangle]. \quad (52)$$

The *mean mass density* $\langle\rho\rangle$ is subtracted because $\vec{g}$ is computed relative to the homogeneous model. The *equation of mass conservation* expressed in terms of the *density contrast* $\delta = \rho/\langle\rho\rangle - 1$ of the *mass distribution* modeled as a continuous pressureless fluid is

$$\partial\delta/\partial t + 1/a\, \nabla \cdot (1 + \delta)\, \vec{v} = 0. \quad (53)$$

In linear perturbation theory in $\vec{v}$ and δ these equations give

$$\partial^2\delta/\partial t^2 + 2\, \dot{a}/a\, \partial\delta/\partial t = 4\pi G \langle\rho\rangle\delta, \quad \delta(\vec{x}, t) = f(\vec{x})D(t). \quad (54)$$

Here D(t) is the growing solution to the first equation.[61]

> [61] The general solution is a sum of the growing and decaying solutions, but because the universe has expanded by a large factor since nongravitational forces were last important on large scales we can ignore the decaying part.

The velocity field belonging to the solution D(t) is the inhomogeneous solution to Eq. (53) in linear perturbation theory,

$$\vec{v} = f H_0 a/4\pi \int (\vec{y} - \vec{x})/|\vec{y} - \vec{x}|^3\, \delta(\vec{y})\, d^3y, \quad f \simeq \Omega^{0.6}_{M0}. \quad (55)$$

The factor f = d logD/d log a depends on the cosmological model; the second equation is a good approximation if $\Lambda = 0$ or space curvature vanishes. One sees from Eq. (55) that the peculiar velocity is proportional to the *gravitational acceleration*, as one would expect in linear theory.

The key point of Eq. (54) for the present purpose is that the evolution of the *density contrast* δ at a given position is not affected by the value of δ anywhere else. *This is a consequence of the inverse square law*. The mass fluctuation in a chosen volume element produces a peculiar *gravitational acceleration* $\delta\vec{g}$ that produces a peculiar velocity field $\delta\vec{v} \propto \vec{g}$ that has zero divergence and so the mass inside the volume element does not affect the exterior.

…

There does not seem to be a coherent pattern to the present list of challenges to the CDM model. The rotation curves of low surface brightness galaxies suggest we want to suppress the primeval density fluctuations on small scales, but the observations of what seem to be mature elliptical galaxies at high redshifts suggest we want to increase small-scale fluctuations, or maybe postulate non-Gaussian fluctuations that grow into the central engines for quasars at $z \sim 6$. We do not want these central engines to appear in low surface brightness galaxies, of course.

It would not be at all surprising if the confusion of challenges proved to be at least in part due to the difficulty of comparing necessarily schematic analytic and numerical model analyses to the limited and indirect empirical constraints. But it is also easy to imagine that the CDM model has to be refined because the physics of the dark sector of matter and energy is more complicated than ΛCDM, and maybe even more complicated than any of the alternatives now under discussion. Perhaps some of the structure formation ideas people were considering a decade ago, which invoke good physics, also will prove to be significant factors in relieving the problems with structure formation. And the important point for our purpose is that we do not know how the relief might affect the cosmological tests.

B. The tests

The literature on the cosmological tests is enormous compared to what it was just a decade ago, and growing. Our references to this literature are much sparser than in Sec. III, on the principle that no matter how complete the list it will be out of date by the time this review is published. For the same reason, we do not attempt to present the best values of the cosmological parameters based on their joint fit to the full suite of present measurements. The situation will continue to evolve as the measurements improve, and the state of the art is best followed on astro-ph. We do take it to be our assignment to consider what the tests are testing, and to assess the directions the results seem to be leading us. The latter causes us to return many times to two results that seem secure because they are so well checked by independent lines of evidence, as follows.

First, at the present state of the tests, optically selected galaxies are useful mass tracers. By that we mean the assumption that visible galaxies trace mass does not seriously degrade the accuracy of analyses of the observations. This will change as the measurements improve, of course, but the case is good enough now that we suspect the evidence will continue to be that optically selected galaxies are good indicators of where most of the mass is at the present epoch. Second, the *mass density* in matter is significantly less than the critical Einstein-de Sitter value. The case is compelling because it is supported by so many different lines of evidence (as summarized in Sec. IV.C). Each could be compromised by systematic error, to be sure, but it seems quite unlikely the evidence could be so consistent yet misleading. A judgement of the range of likely values of the *mass density* is more difficult. Our estimate, based on the measurements we most trust, is

$$0.15 <\sim \Omega_{M0} <\sim 0.4, \qquad (59)$$

and we would put the central value at $\Omega_{M0} \simeq 0.25$. The spread is meant in the sense of two standard deviations: we would be surprised to find Ω_{M0} is outside this range.

...

Our remarks are ordered by our estimates of the model dependence.

1. *The thermal cosmic microwave background radiation*

We are in a sea of radiation with spectrum very close to Planck at T = 2.73 K, and isotropic to one part in 10^5 (after correction for a dipole term that usually is interpreted as the result of our motion relative to the rest frame defined by the radiation). *The thermal spectrum indicates thermal relaxation*, for which the optical depth has to be large on the scale of the Hubble length H_0^{-1}. We know space now is close to transparent at the wavelengths of this radiation, because radio galaxies are observed at high redshift. Thus, the universe has to have expanded from a state quite different from now, when it was hotter, denser, and optically thick. This is strong evidence our universe is evolving. ...

The 3 K thermal cosmic background radiation is a centerpiece of modern cosmology, *but its existence does not test general relativity.*

2. Light element abundance

The best evidence that the expansion and cooling of the universe traces back to high redshift is the success of the standard model *for the origin of deuterium and isotopes of helium and lithium, by reactions among radiation, leptons, and atomic nuclei as the universe expands and cools through temperature $T \sim 1$ MeV at redshift $z \sim 10^{10}$*. The free parameter in the standard model is the present baryon number density. The model assumes the baryons are uniformly distributed at high redshift, so this parameter with the known present radiation temperature fixes the baryon number density as a function of temperature and the temperature as a function of time. The latter follows from the expansion rate Eq. (11), which at the epoch of light element formation may be written as

$$(\dot{a}/a)^2 = 8/3\ \pi G \rho_r, \qquad (61)$$

where the *mass density* ρ_r counts radiation, which is now at T = 2.73 K, the associated neutrinos, and $e^{\pm}$ pairs. *The curvature and Λ terms are unimportant*, unless the *dark energy mass density* varies quite rapidly.

. . .

How is *general relativity* probed? The only part of the computation that depends specifically on this theory is the *pressure* term in the *active gravitational mass density*, in the expansion rate equation (8).

$$[\ddot{a}/a = -4/3\ \pi G(\rho + 3p). \qquad (8)]$$

If we did not have general relativity, a simple Newtonian picture might have led us to write down $\ddot{a}/a = -4\pi G(\rho_r/3)$ instead of Eq. (8). With $\rho_r \sim 1/a^4$, as appropriate since most of the mass is fully *relativistic* at the redshifts of light element production, this would predict the expansion time $a/\dot{a}$ is $2^{1/2}$ times the standard expression (that from Eq. [61].

$$[(\dot{a}/a)^2 = 8/3\ \pi G\rho_r, \qquad (61)].$$

The larger expansion time would hold the neutron to proton number density ratio close to that at thermal equilibrium, $n/p = e^{-Q/kT}$, where Q is the difference between the neutron and proton masses, to lower temperature. It would also allow more time for free decay of the neutrons after thermal equilibrium is broken. Both effects decrease the final ^4He abundance. The factor $2^{1/2}$ increase in expansion time would reduce the helium abundance by mass to Y ~ 0.20. This is significantly less than what is observed in objects with the lowest heavy element abundances, and so *seems to be ruled out* (Steigman, 2002). *That is, we have positive evidence for the relativistic expression for the active gravitational mass density* at redshift $z \sim 10^{10}$, a striking result.

3. *Expansion times*

The predicted time of expansion from the very early universe to redshift z is

$$t(z) = \int da/\dot{a} = H_0^{-1} \int_z^{\infty} dz/\{(1 + z)E(z)\}, \qquad (64)$$

where E(z) is defined in Eq. (11).

$$[(\dot{a}/a)^2 = H^2_0E(z)^2$$
$$= H^2_0\ \{\Omega_{M0}(1 + z)^3 + \Omega_{R0}(1 + z)^4 + \Omega_{\Lambda 0} + \Omega_{K0}(1 + z)^2\}. \qquad (11)]$$

If Λ = 0 the present age is $t_0 < H_0^{-1}$. In the Einstein-de Sitter model the present age is $t_0 = 2/(3H_0)$. If the *dark energy density* is significant and evolving, we may write $\rho_\Lambda = \rho_{\Lambda 0}f(z)$, where the function of redshift is normalized to f(0) = 1. Then E(z) generalizes to

$$E(z) = \{\Omega_{M0}(1 + z)^3 + \Omega_{R0}(1 + z)^4 + \Omega_{K0}(1 + z)^2 + \Omega_{\Lambda 0}\, f(z)\}^{1/2}. \qquad (65)$$

...

The *relativistic* correction to the *active gravitational mass density* (Eq. [8]) *is not important* at the redshifts at which galaxies can be observed and the ages of their star populations estimated. At moderately high redshift, where the *nonrelativistic* matter term dominates, Eq. (64) is approximately

$$t(z) \simeq 2/(3H_0\Omega_{M0}^{1/2})\,(1 + z)^{-3/2}. \qquad (66)$$

That is, the ages of star populations at high redshift are an interesting probe of Ω_{M0} but they are *not very sensitive to space curvature or to a near constant dark energy density.*

Recent analyses of the ages of old stars indicate the expansion time is in the range

$$11\ \text{Gyr} \lesssim t_0 \lesssim 17\ \text{Gyr}, \qquad (67)$$

at 95% confidence, with central value $t_0 \simeq 13$ Gyr. ...

...

13. *The gravitational inverse square law*

The *inverse square law for gravity* determines the relation between the *mass distribution* and the *gravitationally driven peculiar velocities* that enter estimates of the *matter density* parameter Ω_{M0}. The peculiar velocities also figure in the evolution of the mass distribution, and hence the relation between the present mass fluctuation spectrum and the spectrum of cosmic microwave background temperature fluctuations imprinted at redshift $z \sim 1000$. We are starting to see demanding tests of both aspects of the inverse square law.

We have a reasonably well checked set of measurements of the apparent value of Ω_{M0} on scales ranging from 100 kpc to 10 Mpc (as reviewed under test [7]). Most agree with a constant value of the apparent Ω_{M0}, within a factor of three or so. *This is not the precision one would like*, but the subject has been under discussion for a long time, and, we believe, is now pretty reliably understood, within the factor of three or so. If galaxies were biased tracers of mass, one might have expected to have seen that Ω_{M0} increases with increasing length scale, as the increasing scale includes the outer parts of extended massive halos. *Maybe that is masked by a gravitational force law that decreases more rapidly than the inverse square law at large distance*. But the much more straightforward reading is that the slow variation of Ω_{M0} sampled over two orders of magnitude in length scale agrees with the evidence from tests (7) to (10) that galaxies are useful mass tracers, and that the inverse square law therefore is a useful approximation on these scales.

…

C. The state of the cosmological tests

Precision cosmology is not very interesting if it is based on faulty physics or astronomy. That is why we have emphasized the tests of the standard gravity physics and structure formation model, and the checks of consistency of measures based on different aspects of the astronomy.

There are now five main lines of evidence that significantly constrain the value of Ω_{M0} to the range of Eq. (59):

$$[0.15 <\sim \Omega_{M0} <\sim 0.4, \qquad (59)]$$

the *redshift-magnitude relation* (test [4]), *gravitational dynamics and weak lensing* (test [7]), the *baryon mass fraction in clusters of galaxies* (test [8]), the *abundance of clusters as a function of mass and redshift* (test [9]), and the *large-scale galaxy distribution* (test [12]). There are indications for larger values of Ω_{M0}, from analyses of the *rate of strong lensing of quasars by foreground galaxies* (test [6]) and some *analyses of large-scale flows* (test [7]), though we know of no well-developed line of evidence that points to the Einstein-de Sitter value $\Omega_{M0} = 1$. Each of these measures of Ω_{M0} may suffer from systematic errors: we must bear in mind the tantalus principle mentioned in Sec. I.A, and we have to remember that the interpretations could be corrupted by a failure of standard physics. But the general pattern of results from a considerable variety of independent approaches seems so close to consistent as to be persuasive. Thus, *we conclude that there is a well-checked scientific case for the proposition that the measures of the mean mass density of matter in forms capable of clustering are physically meaningful*, and that the *mass density parameter* almost certainly is in the range $0.15 <\sim \Omega_{M0} <\sim 0.4$.

In the standard cosmology the *masses* of the galaxies are dominated by *dark matter*, with *mass density* parameter $\Omega_{DM0} \sim 0.2$, that is not baryonic (or acts that way). We do not have the direct evidence of a laboratory detection; this is based on two indirect lines of argument. First, *the successful model for the origin of the light elements* (test [2]) requires baryon density $\Omega_{B0} \sim 0.05$. *It is difficult to see how to reconcile a mass density this small with the mass estimates from dynamics and lensing*; the hypothesis that Ω_{M0} is dominated by matter that is not baryonic allows us to account for the difference. Second, *the nonbaryonic matter allows us to reconcile the theory of the anisotropy of the cosmic microwave background radiation with the distributions of galaxies and groups and clusters of galaxies, and the presence of galaxies at $z \sim 3$* (tests [11] and [12]). This interpretation requires a value for Ω_{B0} that is in line with test (2). The consistency is impressive. But the case is not yet as convincing as the larger network of evidence that Ω_{M0} is well below unity.

…

V. CONCLUDING REMARKS

…

A decade ago, the high energy physics community had a well-defined challenge: show why the *dark energy density* vanishes. Now there seems to be a new challenge and clue: show why the *dark energy density* is exceedingly small but not zero. The present state of ideas can be compared to the state of research on structure formation a decade ago: in

both situations there are many lines of thought but not a clear picture of which is the best direction to take. The big difference is that a decade ago we could be reasonably sure that observations in progress would guide us to a better understanding of how structure formed. *Untangling the physics of dark matter and dark energy and their role in gravity physics is a much more subtle challenge*, but, we may hope, will also be guided by advances in the exploration of the phenomenology. *Perhaps in another ten years* that will include detection of *evolution of the dark energy*, and maybe detection of the *gravitational response of the dark energy distribution to the large-scale mass distribution*. There may be three unrelated phenomena to deal with: *dark energy*, *dark matter*, and a *vanishing sum of zero-point energies* and whatever goes with them. Or the phenomena may be related. Because *our only evidence of dark matter and dark energy is from their gravity* it is a natural and efficient first step to suppose their properties are as simple as allowed by the phenomenology. But it makes sense to watch for hints of more complex physics within the dark sector.

The past *eight decades* have seen steady advances in the technology of application of the cosmological tests, from telescopes to computers; advances in the theoretical concepts underlying the tests; and progress through the learning curves on how to apply the concepts and technology. We see the results: the basis for cosmology is much firmer than a decade ago. And the basis surely will be a lot more solid *a decade from now*.

…

Though this much is clear, we see no basis for a prediction of whether the standard cosmology *a decade from now* will be a straightforward elaboration of ΛCDM, or whether there will be more substantial changes of direction.

VI. APPENDIX: RECENT DARK ENERGY SCALAR FIELD RESEARCH

…

At the time of writing, while there has been much work, *it appears that the dark energy scalar field scenario still lacks a firm, high energy physics based, theoretical foundation*. While this is a significant drawback, the recent flurry of activity prompted by developments in superstring/M and brane theories appears to hold significant promise for shedding light on *dark energy*. Whether this happens before the observations rule out or "confirm" *dark energy* is an intriguing question.

[We are now two decades further on. There is still no progress in understanding the *physics of gravity*, what it is, as opposed to describing its effects.]

[Richard Feynman. (1963). *The Feynman Lectures on Physics*, Volume I, p. 7-9: "*But is this such a simple law? What about the machinery of it? All we have done is to describe how the Earth moves around the Sun, but we have not said what makes it go. Newton made no hypothesis about this; was satisfied to find what it did without getting into the machinery of it. No one has since given any machinery. ... Why can we use mathematics to describe nature without a mechanism behind it? No one knows. We have to keep going because we find out more that way.*']

Gravitational energy

Gravitational energy, or *gravitational potential energy*, is the potential energy a massive object has because it is within a *gravitational field*. In classical mechanics, *two or more masses always have a gravitational potential. Conservation of energy requires that this gravitational field energy is always negative*, so that it is zero when the objects are infinitely far apart. As two objects move apart and the distance between them approaches infinity, the *gravitational force* between them approaches zero from the positive side of the real number line and the *gravitational potential* approaches zero from the negative side. Conversely, *as two massive objects move towards each other, the motion accelerates under gravity causing an increase in the (positive) kinetic energy of the system and, in order to conserve the total sum of energy, the increase of the same amount in the gravitational potential energy of the object is treated as negative*.

A universe in which positive energy dominates will eventually collapse in a "Big Crunch", while an "open" universe in which negative energy dominates will either expand indefinitely or eventually disintegrate in a "big rip". In the *zero-energy universe model* ("flat" or "Euclidean"), the *total amount of energy in the universe is exactly zero*: its amount of *positive energy in the form of matter* is exactly cancelled out by its *negative energy in the form of gravity*. It is unclear which, if any, of these models accurately describes the real universe.

Why does matter attract matter?

Facts related to gravity and other forces of nature are listed and reviewed below in an attempt to answer this question.

(1) *The gravitational attractive force between two bodies obeys an inverse square law as if it were a radiated force in three-dimensional space*, similar to the electrical and magnetic attractive and repulsive forces. The gravitational force is given by

$$F = G\, m_1 m_2 / r^2,$$

where G is *Newton's universal gravitational constant*, m_1 and m_2 are the *masses* of the bodies being attracted, and r is the distance between them.

In *SI units*,

$$F = 6.674 \times 10^{-11}\, m_1 m_2 / r^2 \text{ N}$$

where F represents the *force* in Newtons, m_1 and m_2 represent the two *masses* in kilograms, r represents the separation in meters, and G represents the *gravitational constant,* which has a value of $6.674 \times 10^{-11}\ N \cdot m^2 \cdot kg^{-2}$.

[*Force* is a fundamental concept that describes the interaction between objects. It is a vector quantity, meaning it has both magnitude and direction. *Force* is responsible for causing changes in the motion or shape of an object. *Newton's laws of motion* provide a framework for understanding the relationship between force and the resulting motion.

According to *current theory* all of the known *forces* of the universe are classified into *four fundamental interactions*. The *strong and the weak forces* act only at very short distances, and are responsible for the interactions between subatomic particles, including nucleons and compound nuclei. The *electromagnetic force* acts between electric charges, and the *gravitational force* acts between masses. All other forces in nature derive from these four fundamental interactions operating within quantum mechanics, including the constraints introduced by the Schrödinger equation and the Pauli exclusion principle.

The development of *quantum mechanics* led to a modern understanding that the first three fundamental forces (*all except gravity*) are *manifestations of matter (fermions) interacting by exchanging virtual particles called gauge bosons*. This Standard Model of particle physics assumes a similarity between the forces and led scientists to predict the unification of the weak and electromagnetic forces in electroweak theory, which was subsequently confirmed by observation.]

(2) *The gravitational attractive force between two bodies is proportional to the product of the masses, m_1 and m_2, of the bodies being attracted*, and is otherwise independent of their form:

$$F = G\, m_1 m_2/r^2.$$

(3) *The gravitational attractive force is additive in the sense that each of the constituents (one atom or one molecule) of one body are attracted by the each of the constituents of the other body in proportion to the product of each of their masses,* a_1 and a_2,

$$F_{ij} = G\, (a_1)_i \,.\, (a_2)_j/r^2;$$

resulting in the sum

$$\Sigma_{ij} F_{ij} = G\, \{\Sigma_i\, (a_1)_i \,.\, \Sigma_j\, (a_2)_j\}/\, r^2;$$
$$F = Gm_1m_2/r^2$$

where $m_1 = \Sigma_i\, (a_1)_I$ and $m_2 = \Sigma_j\, (a_2)_j$.

(4) *The force of attraction between two objects with opposite charges also obeys an inverse square law*. It is given by

$$F = 1/4\pi\varepsilon_0\, q_1q_2/r^2, \text{ or } F = k_e\, q_1q_2/r^2.$$

In *SI units*

$$F = 8.99 \text{ x } 10^9\, q_1q_2/r^2 \text{ N}$$

where F represents the *force* in Newtons, q_1 and q_2 represents the two *charges* in coulombs, r represents the separation in meters, ε_0 represents the *permittivity of a vacuum*, which has a value of $8.85 \text{ x } 10^{-12}\, N^{-1} . C^2 . m^{-2}$, and $k_e = 1/4\pi\varepsilon_0$ represents *Coulomb's constant*, which has a value of $8.99 \text{ x } 10^9\, N . m^2 . C^{-2}$.

The force of attraction between an electron and proton due to their charges is given by

$$F = 1/4\pi\varepsilon_0\, |\, e\, |^2/r^2, \text{ or } F = k_e\, |\, e\, |^2/r^2,$$

or in SI units,

$$F = 8.99 \times 10^{9} \times 1.6022 \times 10^{-38}/r^2 = 2.307 \times 10^{-28}/r^2 \text{ N},$$

where e represents the *charge on the electron*, which in *SI units* is equal to 1.602 x 10^{-19} coulombs.

In the *ground state of the hydrogen atom*, a single electron "orbits" a proton at a most probable distance of 5.29 x 10^{-11} meters. The *electric attraction between an electron and proton* is given by

$$F = 2.307 \times 10^{-28}/(5.29 \times 10^{-11})^2 = 8.243 \times 10^{-8} \text{ N}.$$

At this distance, the *gravitational attraction between two hydrogen atoms* is

$$F = 6.674 \times 10^{-11}\ m_1 m_2/r^2 = 7.949 \times 10^{-44} \text{ N}.$$

(5) *The gravitational force is about* 10^{36} *times weaker than the electric force. As both forces are subject to inverse square laws, this relationship will apply at all distances.* It is difficult to imagine that these forces are in any way similar. *Gravity appears to stand on its own.* Einstein's attempt at a unification was futile.

(6) *The net inward gravitational force on 1 kg of matter at a distance s from the center of a planet or a star with radius r and density ρ is given by* $F_{net} = 6.674 \times 10^{-11}\ 4/3\ \pi(2s - r)\ \rho$ *N.*

Within a high-density inner core of a planet or star the *inward gravitational force* on 1 kg of matter at a distance s from the center due to matter closer to the center, *assuming constant density ρ*,

$$F_{in} = 6.674 \times 10^{-11}\ 4/3\ \pi s^3 \times \rho/s^2 = 6.674 \times 10^{-11}\ 4/3\ \pi s \rho \text{ N (Newtons).}$$

The *outward gravitational force* due to the matter further from the center,

$$F_{out} = 6.674 \times 10^{-11}\ 4/3\ \pi(r - s)\rho \text{ N}.$$

So, the *net inward gravitational force* on the 1 Kg of matter at a distance s from the center of a planet or a star is given by

$$F_{net} = 6.674 \times 10^{-11}\ 4/3\ \{\pi s \rho - \pi(r - s)\rho\}, \text{ or}$$
$$F_{net} = 6.674 \times 10^{-11}\ 4/3\ \pi(2s - r)\ \rho \text{ N}.$$

At the center of the planet or star (s = 0),

$F_{net} = -\,6.674 \times 10^{-11}\ 4/3\ \pi r\rho\ N.$

Substituting $M = 4/3\ \pi r^3 \times \rho$, or $4/3\ \pi r\rho = M/r^2$

$F_{net,0} = -\,6.674 \times 10^{-11}\ M/r^2\ N.$

At the surface of the planet or star (s = r),

$F_{net} = 6.674 \times 10^{-11}\ 4/3\ \pi r\rho\ N.$

or $F_{net,r} = 6.674 \times 10^{-11}\ M/r^2\ N.$

So, *the force on 1 kg at the center of a sphere due to gravity is outward and equal to the inward force on 1 kg of matter at the surface of the sphere; and the net force is zero when s = r/2, i.e. at half of the radius of the sphere.*

The inward force on 1 kg outside the sphere is, at a distance s

$F_{ext} = 6.674 \times 10^{-11}\ M/s^2\ N,$

in accordance with Newton's law.

In the case of the Earth, with $M_E = 5.9722 \times 10^{24}$ kg and the average radius, $r_E = 6{,}371 \times 10^3$ m, the *outward force on 1 kg of matter at the center of the Earth*, and *inward force on 1 kg of matter at the surface of the Earth*, is given by

$F_{Earth} = 6.674 \times 10^{-11}\ M_E/r_E^2$
$F_{Earth} = 6.674 \times 10^{-11} \times 5.9722 \times 10^{24}/(6{,}371 \times 10^3)^2,$
$F_{Earth} = 9.82$ N

At average gravity on Earth (conventionally, g = 9.80665 m/s^2), the force on a kilogram mass is about 9.8 Newtons. This force at the surface of the Earth depends on the high average density of the Earth, which is equal to about 5,515 kg/m^3.

At a distance of ten times the Earth's radius from the surface of the Earth, at an altitude of 63,000 Km, where $s = 10r_E$, the inward gravitational force on a 1 kg of matter is given by

$F_{Earth} = 6.674 \times 10^{-11}\ M_E/(10r_E)^2$
$F_{Earth} = 6.674 \times 10^{-11} \times 5.9722 \times 10^{24}/(6{,}371 \times 10^4)^2,$
$F_{Earth} = 0.0982$ N.

The force of gravity is reduced by a factor of 100. With a very small force, *humans would fly off.*

In the case of the Sun, with $M_S = 1.9885 \times 10^{30}$ kg and the average radius, $r_S = 695{,}700 \times 10^3$ m, the outward force on 1 kg of matter at the center of the Sun, and inward force on 1 kg of matter at the surface of the Sun,

$$F_{Sun} = 6.674 \text{ x } 10^{-11}\ M_S/r_S^2$$
$$F_{Sun} = 6.674 \text{ x } 10^{-11} \text{ x } 1.9885 \times 10^{30}/(695{,}700 \times 10^3)^2,$$
$$F_{Sun} = 1.91 \times 10^3 \text{ N}.$$

The *gravitational force of the Sun on the Earth* is given by

$$F_{Sun,Earth} = = 6.674 \text{ x } 10^{-11}\ M_E M_S/r_{ES}^2,$$

where the average distance of the Earth from the Sun, r_{ES} is 149.6 x 10^9 m.

[Using *Newton's law of gravitation*, *Kepler's third law* (a^3/T^2 = const.) can be derived in the case of a circular orbit *by setting the centripetal force equal to the gravitational force*:

$$M_E r\omega^2 = GM_E M_S/r_{ES}^2.$$

Then, expressing the angular velocity ω in terms of the orbital period T and rearranging,

$$M_E r_{ES}(2\pi/T)^2 = GM_E M_S/r_{ES}^2,$$

or $\quad T^2 = 4\pi^2 r_{ES}^3/GM_S,$

so $\quad T = 2\pi\ \sqrt{(r_{ES}^3/GM_S)}.$

Substituting for r_{ES} and M_S,

$$T = 2\pi\ \sqrt{(r_{ES}^3/GM_S)}$$

or $\quad T = 2 \text{ x } 3.14159 \text{ x } \sqrt{\{(149.6 \text{ x } 10^9)^3/(6.674 \text{ x } 10^{-11} \text{ x } 1.9885 \times 10^{30})\}},$

or $\quad T = 2 \text{ x } 3.14159 \text{ x } 502.3 \text{ x } 10^4 = 3.156 \text{ x } 10^7 \text{ s} = 365.28 \text{ days}$

The Earth orbits the Sun every 365.28 days.]

The point at which the gravitational force of the Earth and Sun on 1 kg of matter are equal is given by

$$F_{Earth} = 6.674 \text{ x } 10^{-11} M_E/s_E^2 = F_{Sun} = 6.674 \text{ x } 10^{-11} M_S/s_S^2,$$

where $s_E + s_S = r_{ES} = 149.6 \text{ x } 10^9$ m.

$$M_E/s_E^2 = M_S/s_S^2, \text{ so } M_E s_S^2 = M_S s_E^2, \text{ and}$$
$$M_S s_E^2 = M_E(r_{ES} - s_E)^2.$$

With s_E small compared with r_{ES},

$$M_S s_E^2 = M_E r_{ES}^2,$$
$$s_E = \sqrt{}(M_E/M_S)\, r_{ES},$$

so substituting values, the *distance from the Earth where the gravitational forces of the Earth and Sun are equal*,

$$s_{ES,0} = \sqrt{}\{(5.9722 \text{ x } 10^{24})/1.9885 \times 10^{30})\}\text{x } 149.6 \text{ x } 10^9, \text{ or}$$
$$s_{ES,0} = 1.733 \text{ x } 10^{-3} \text{ x } 149.6 \text{ x } 10^9 = 259 \text{ x } 10^6 \text{ m} = 259{,}000 \text{ km},$$

which is at about 40.65 times the radius of the Earth or around 1/1731 of the distance to the Sun. *Beyond this distance matter is attracted by gravity towards the Sun.*

However, *at this distance, the force of gravity, is very small*; the gravitational force exerted by the Sun or Earth on 1 kg of matter is equal to

$$F_{EarthSun,0} = 6.674 \text{ x } 10^{-11} M_E/(40r_E)^2,$$
$$F_{EarthSun,0} = 6.674 \text{ x } 10^{-11} \text{ x } 5.9722 \text{ x } 10^{24}/(40 \text{ x } 6{,}371 \text{ x } 10^3)^2, \text{ or}$$
$$F_{EarthSun,0} = 6.674 \text{ x } 10^{-11} \text{ x } 5.9722 \text{ x } 10^{24}/(6.4943 \text{ x } 10^{16})$$
$$F_{EarthSun,0} = 6.14 \text{ x } 10^{-3} \text{ N},$$

less than one thousandth of the force, F_{Earth} = 9.82 N, at the surface of the Earth.

(7) *There appears to be no equivalent gravitational repulsive force*, unless this is what is referred to as *dark energy*, which it is claimed is driving matter apart, or equivalently driving the expansion of the universe.

(8) *The gravitational field of each body is present at every other body.*

[*Field theories* are a powerful framework used in physics to *describe the interactions between objects without direct contact*. They provide a way to understand *how forces act at a distance*. Field theories involve the concept of *fields*, which are *regions of influence that extend throughout space*.

In *classical field theories*, such as electromagnetism and gravity, fields are represented by mathematical models. For example, *electromagnetic fields are described by Maxwell's equations*. These mathematical models allow us to predict and understand the behavior of forces in these fields.]

(9) *According to Big Bang theory, the force of gravity separated from the other forces as the universe's temperature fell, immediately prior to the cosmic inflation during which the universe grew exponentially by a factor of at least 10^{78}.*

[In the most common *Big Bang models*, "the universe was filled homogeneously and isotropically with a *very high energy density and huge temperatures and pressures*, and was very rapidly expanding and cooling. The period up to 10^{-43} seconds into the expansion (Planck time), the Planck epoch, was a phase in which the four fundamental forces—the *electromagnetic force*, the *strong nuclear force*, the *weak nuclear force*, and the *gravitational force*, were unified as one.[4]

[4] Unruh, W. G., Semenoff, G. W., eds. (1988). *The Early Universe*.

In this stage, the characteristic scale length of the universe was the Planck length, 1.6×10^{-35} m, and consequently had a temperature of approximately 10^{32} degrees Celsius. Even the very concept of a particle breaks down in these conditions. *A proper understanding of this period awaits the development of a theory of quantum gravity. The Planck epoch was succeeded by the grand unification epoch beginning at 10^{-43} seconds, where gravitation separated from the other forces as the universe's temperature fell.*

At approximately 10^{-37} seconds into the expansion, a phase transition caused a *cosmic inflation*, during which the universe grew exponentially, unconstrained by the light speed invariance, and *temperatures dropped by a factor of 100,000*.

As the universe cooled, the *rest energy density of matter came to gravitationally dominate that of the photon radiation*. This concept is motivated by the flatness problem, where the density of matter and energy is very close to the critical density needed to produce a flat universe. That is, the shape of the universe has no overall geometric curvature due to gravitational influence. Microscopic quantum fluctuations that occurred because of Heisenberg's uncertainty principle were "frozen in" by inflation, becoming amplified into the seeds that would later form the large-scale structure of the universe. At a time around 10^{-36} seconds, the electroweak epoch begins when the strong nuclear force separates from the other forces, with only the electromagnetic force and weak nuclear force remaining unified.

Inflation stopped locally at around 10^{-33} to 10^{-32} seconds, with the observable universe's volume having increased by a factor of at least 10^{78}. Reheating occurred until the universe obtained the temperatures required for the production of a quark–gluon plasma as well as all other elementary particles. Temperatures were so high that the random motions of particles were at relativistic speeds, and particle–antiparticle pairs of all kinds were being continuously created and destroyed in collisions. At some point, an unknown reaction called baryogenesis violated the conservation of baryon number, leading to a very small excess of quarks and leptons over antiquarks and antileptons—of the order of one part in 30 million. This resulted in the predominance of matter over antimatter in the present universe.]

(10) *According to Newton's universal law, the inward gravitational force F on a mass m at the at the surface of the universe,*

$$F = GmM/r^2,$$

and the change in the inward gravitational force F on a mass m at the at the surface of the universe with the increase of the radius, due to the expansion of the universe, is inversely proportional to the cube of the radius of the universe.

$$dF/dr = -2\,GmM/r^3.$$

Substituting $M = 1.5 \times 10^{53}$ kg and $r = 4.4 \times 10^{26}$ m,

the *inward gravitational force F on a mass m at the at the surface of the universe,*

$$F = Gm \times 1.5 \times 10^{53}/(4.4 \times 10^{26})^2 = 0.7748\ Gm\ N;$$

and the change in the *inward gravitational force F on a mass m at the at the surface of the universe with the increase of the radius, due to the expansion of the universe*

$$dF/dr = -2\ Gm \times 1.5 \times 10^{53} / (4.4 \times 10^{26})^3$$
$$dF/dr = -3.3522 \times 10^{-27}\ Gm\ N.\ m^{-1}.$$

With $G = 6.674 \times 10^{-11}$,

$$F = GmM/r^2 = = 0.7748 \times 6.674 \times 10^{-11}\ m = 5.171 \times 10^{-11}\ m\ N;$$
$$dF/dr = -3.3522 \times 10^{-27} \times 6.674 \times 10^{-11}\ m = -2.237 \times 10^{-37}\ m\ N.\ m^{-1}.$$

(11) *One way of looking at the expansion of the universe is as the expansion of the metric which determines the size of the universe as we observe it, in which all distances appear to have expanded and to continue to expand at the same rate, separating matter without any apparent force.*

Gravitational attraction of matter may then simply be a reflection of the resistance of matter to this separation.

(12) *On the other hand, the universe may have begun literally with a big bang which resulted in a uniform increase in the distance between matter as it evolved.* There appears to be plenty of indirect evidence of a *huge quantity of energy* in the first few seconds of the universe, with a temperature of approximately 10^{32} degrees Celsius after 10^{-43} seconds, when *gravitation* first separated from the other forces.

(13) *The existence of a large amount of energy at the origin of the universe helps explain the subsequent expansion without the need to refer to it as dark energy* or to evoke some form of expansion of the metric of space.

(14) *The fact that matter, comprising protons, neutrons and electrons, were formed whilst the distance between them was increasing may have something to do with the emergence of gravity.*

(15) *If the inward attraction of matter is a reflection of the resistance to the outward expansion of the universe, it may be possible to relate the gravitational constant G to the current rate of expansion of the universe.*

(16) *Gravity appears to be some sort of entanglement of the protons, neutrons and electrons comprising matter that has existed since soon after the Big Bang*, when the observable universe's volume had increased by a factor of 10^{78} (an expansion of distance by a factor of at least 10^{26} in each of the three dimensions).

(17) *Gravity appears to be similar to (non-relativistic) quantum entanglement between two particles that interact and then separate in such a way that the quantum state of each particle cannot be described independently of the state of the others?*

It is also possible that the *electromagnetic force* is also a form of *entanglement* that occurred as the density and temperature of the universe fell, resulting in the production of *protons and neutrons* at around 10^{-6} seconds, and *electrons and positrons* at around 1 second.

> ["The universe continued to decrease in density and fall in temperature, hence the typical energy of each particle was decreasing. Symmetry-breaking phase transitions put the fundamental forces of physics and the parameters of elementary particles into their present form, with the *electromagnetic force* and *weak nuclear force* separating at about 10^{-12} seconds."]

(18) *The mass of the observable universe (including ordinary matter, the interstellar medium (ISM), and the intergalactic medium (IGM), but excluding dark matter and dark energy) is around 1.5 x* 10^{53} *kg.*

(19) *The actual density of atoms in the universe is equivalent to roughly 1 proton per 4 cubic meters =* $1.6726 \times 10^{-27}/4 = 4.18 \times 10^{-28}$ *kg* m^{-3}.

The current *mass density* of the universe is 9.9×10^{-30} g/cm^3 (9.9×10^{-27} kg/m^3) which is equivalent to 5.9 protons per cubic meter. Of this *total density*, we know the breakdown to be: 4.6% Atoms; (The *actual density* of atoms is equivalent to roughly 1 proton per 4 cubic meters = $1.6726 \times 10^{-27}/4 = 4.18 \times 10^{-28}$ kg m^{-3}); 24% *Cold Dark Matter*; (*Dark Matter* is likely to be composed of one or more species of sub-atomic particles that interact very weakly with ordinary matter.) 71.4% *Dark Energy*.

(20) *This implies that the current volume of the observable universe is* $(1.5 \times 10^{53})/(4.18 \times 10^{-28}) = 3.6 \times 10^{80}$ m^3*, and its radius is* 4.4×10^{26} *m.*

(21) *The rate of expansion of the universe is estimated to be 73.3 kilometers per second per megaparsec (1 megaparsec = 3,260,000 light years =* 3.0857×10^{19} *km) or for every km from the Earth* $73.3/3.0857 \times 10^{19} = 2.375 \times 10^{-18}$ *km/sec.*

Using a relatively new and potentially more precise technique for measuring cosmic distances, astronomers calculate a rate of expansion of 73.3 kilometers per second per megaparsec (1 megaparsec = 3,260,000 light years = 3.0857×10^{19} km). This means that for every 3.0857×10^{19} km from Earth (or presumably from any point in the universe), the universe is expanding an extra 73.3 ± 2.5 kilometers per second (or for every km from the Earth $73.3/3.0857 \times 10^{19} = 2.375 \times 10^{-18}$ km/sec).

[Estimates of the local expansion rate based on measured fluctuations in the cosmic microwave background and, independently, fluctuations in the density of normal matter in the early universe (baryon acoustic oscillations), give a different answer: 67.4 ± 0.5 km/sec/Mpc.]

This implies that at the limit of the observable universe, 4.4 x 10^{23} km from the Earth, the universe is expanding by 4.4 x 10^{23} x 2.375 x 10^{-18} = 1.045 x 10^{6} km/sec, about 3.48 times the speed of light.

[The speed of light is 300,000 km/sec. It is argued that although this would be a problem at the local level under the theory of special relativity, it is not a problem according to the theory of general relativity. An alternative argument is that it does not matter that is moving greater than the speed of light, it is the space between matter that is expanding, that is the metric that is expanding, and there is no limit on that. From a non-relativistic point of view, this is not be a problem, but this raises questions about how the observed redshift is used to measure intergalactic distances.]

(22) *According to mass-energy equivalence,* $E = mc^2$*, the energy equivalent of one kilogram of mass is 8.99 x* 10^{16} *joules, so the energy equivalent of the mass of the observable universe is around 1.5 x* 10^{53} *x 8.99 x* 10^{16} *= 1.35 x* 10^{70} *joules.*

See Underwood, T. G., (2023). *General Relativity*; pp. 100-32.

(23) *The zero-energy universe hypothesis proposes that the total amount of energy in the universe is exactly zero: its amount of positive energy in the form of matter is exactly canceled out by its negative energy in the form of gravity*. This is largely semantics. It is a statement that can be applied to any force and the associated field.

[Some physicists, such as Lawrence Krauss, Stephen Hawking or Alexander Vilenkin, called this state "a universe from nothingness", although the zero-energy universe model requires both a *matter field* with positive energy and a *gravitational field* with negative energy to exist. According to Hawkin, "*When the Big Bang produced a massive amount of positive energy, it simultaneously produced the same amount of negative energy*. In this way, the positive and the negative add up to zero, always". *The argument was that gravitational energy, or gravitational potential energy, is the potential energy that a massive object has because it is within a gravitational field.* In classical mechanics, two or more masses always have a *gravitational potential. Conservation of energy requires that this gravitational field energy is always negative*, so that it is zero when the objects are infinitely far apart. As two objects move apart and the distance between them approaches infinity, the *gravitational force* between them approaches zero from the positive side of the real number line and the *gravitational potential* approaches zero from the negative side. Conversely, as two massive objects move towards each other, the motion accelerates under gravity causing an increase in the (positive) kinetic energy of the system and, in order to conserve the total sum of energy, the increase of the same amount in the gravitational potential energy of the object is treated as negative.]

In this *zero-energy universe model* ("flat" or "Euclidean"), the total amount of energy in the universe is exactly zero: its amount of *positive energy* in the form of *matter* is exactly cancelled out by its *negative energy* in the form of *gravity*. Experimental proof for the observable universe being a "*zero-energy universe*" in this sense is currently inconclusive. *Gravitational energy* from visible matter accounts for 26–37% of the observed total *mass–energy density*. Therefore, to fit the concept of a "*zero-energy universe*" to the observed universe, other negative energy reservoirs besides gravity from baryonic matter are necessary. These reservoirs are frequently assumed to be *dark matter*.

(24) *Alternatively, a "closed" universe, where the density parameter $\Omega > 1$, and Ω is defined as the average matter density of the universe divided by a critical value of that density, in which positive energy dominates, will eventually collapse in a "Big Crunch"; while an "open" universe, where $\Omega < 1$, in which negative energy dominates, will either expand indefinitely or eventually disintegrate in a "Big Rip".*

In a *closed universe*, gravity eventually stops the expansion of the universe, after which it starts to contract until all matter in the universe collapses to a point, a final singularity termed the "Big Crunch", the opposite of the Big Bang. In an *open universe*, the universe continues to expand.

[However, it is also argued, that *if the universe contains dark energy*, then the resulting repulsive force may be sufficient to cause the expansion of the universe to continue forever—even in a *closed universe*, where the *density parameter*, $\Omega > 1$. This is the case in the currently accepted Lambda-CDM model, where *dark energy* is found through observations to account for roughly 68% of the total energy content of the universe. According to the Lambda-CDM model, the universe would need to have an average *matter density* roughly seventeen times greater than its measured value today in order for the effects of *dark energy* to be overcome and the universe to eventually collapse. This is in spite of the fact that, according to the Lambda-CDM model, any increase in *matter density* would result in $\Omega > 1$.

Conversely, the ultimate fate of an *open universe* where $\Omega < 1$ *with dark energy* is either universal heat death or a "Big Rip", where the acceleration caused by *dark energy* eventually becomes so strong that it completely overwhelms the effects of the *gravitational, electromagnetic and strong binding forces*.]

[The second *Friedmann* equation

$$\ddot{a}/a = -\,4\pi G/3\ (\rho + 3p/c^2) + \Lambda c^2/3,$$
$$[2\ddot{a}/a + (\dot{a}/a)^2 + kc^2/a^2 - \Lambda c^2 = -\,8\pi G/c^2\ p \text{ below.}]$$

(where *a* is the scale factor, G, Λ, and c are universal constants (G is the *Newtonian constant of gravitation*, Λ is the *cosmological constant* with dimension length^{-2}, and c is the speed of light in vacuum), ρ and p are the *volumetric mass density* (and not the volumetric energy density) and the *pressure*, respectively, and k is constant throughout a particular solution, but may vary from one solution to another), *states that both the (negative) energy density and the (positive) pressure cause the expansion rate of the universe* $\dot{a}$ *to decrease*, i.e., *both cause a deceleration in the expansion of the universe. This is a consequence of gravitation,* with pressure playing a similar role to that of energy (or mass) density, *according to the principles of general relativity*. The *cosmological constant*, on the other hand, *causes an acceleration in the expansion of the universe*.]

[The *cosmological constant* can be interpreted as arising from *a form of energy which has negative pressure, equal in magnitude to its (positive) energy density*: $p = -\rho c^2$.]

(25) *According to the "closed" universe model, the universe might have started, after the initial Big Bang expansion, as a large sphere in space containing uniformly distributed matter, largely in the form of atoms and molecules, comprised of protons, neutrons and electrons, which continued to expand after the force causing the initial expansion ceased, based on the outward momentum of the matter. The force of gravity*, acting in the opposite direction, *was assumed to be strong enough* to compress molecules of hydrogen, eventually with such pressure as the accreting mass and density increased, as to create nuclear fusion at the center and form stars. *The force of gravity* was also assumed strong enough to compress other molecules into larger particles of matter, known as *cosmic dust*, and these into planets, even

at the most remote regions of the observable universe. This was expected to gradually slow down the expansion and eventually bring the matter together in a Big Crunch. However, as describe in the next section, this argument fails as *the force of gravity is not strong enough.*

(26) *However, there appears to be a problem. Based on the actual density of atoms in the universe, the time taken for molecules of hydrogen or cosmic dust to accrete due to the force of gravitation is far too long*; and if the *gravitational force* on matter at the outer regions of the universe is calculated, as Newton observed, by assuming that the other matter is concentrated at the center of the sphere, *the gravitational force would be far too weak to stop these regions from continuing to expand in any reasonable timeframe.*

If the *actual density* of atoms, as noted above, is equivalent to roughly 1 proton per 4 cubic meters = $1.6726 \times 10^{-27}/4 = 4.18 \times 10^{-28}$ kg m^{-3}, or 2 protons (or 2 neutrons, to avoid the charge force) per 8 cubic meters (a volume 2 m x 2 m x 2 m). If 2 neutrons are placed at the center of a 2 m cube, 1 m apart, the distance between them in this direction when the cubes are stacked is 1 m, but the distance between them in the other two directions is 2 m, so the average distance between the neutrons is 1 2/3 m.) The time taken for two neutrons to come together due to the force of gravitation between them is given by

$$F = ma = 6.674 \times 10^{-11}\ m^2/s^2 \text{ or}$$
$$a = 6.674 \times 10^{-11} \times m/s^2 = 1.1163\ 10^{-37}/s^2$$

where s(t) is the distance between them at time t, and $a = dv/dt = d^2s/dt^2$.

Substituting $a = 6.674 \times 10^{-11} \times m/s^2$,

$$d^2s/dt^2 = 6.674 \times 10^{-11} \times m/s^2;$$
$$s^2\ d^2s = 6.674 \times 10^{-11} \times m/s^2 = 1.1163\ 10^{-37}\ dt^2.$$

The time taken for two neutrons to come together due to the force of gravitation between them is given by

$$\iint_0^s s^2\ d^2s = 1.1163\ 10^{-37} \iint_0^t dt^2;$$
$$\int_0^s 1/3\ s^3\ d^2 = 1.1163\ 10^{-37} \int_0^t t\ dt;$$
$$1/12\ s^4 = 1.1163\ 10^{-37}\ t^2/2; \text{ or}$$
$$t^2 = s^4/6 \times 1.1163\ 10^{-37} = 1.493 \times 10^{36} \times s^4; \text{ or}$$
$$t = 3.864 \times 10^{17}\ s^2 \text{ or with } s = 1.67 \text{ m};$$
$$t = 3.864 \times 10^{17} \times 1.67^2 = 1.08 \times 10^{18}\ s = \textit{34.2 billion years.}$$

It is thus implausible that gravity at its current level contributed to the accretion of cosmic dust to form planets.

Whilst gravity has no role in the accretion of cosmic dust into planets or stars, nor in creating fusion in the center of stars, such as the Sun, with or without taking into account the expansion of the universe, *it is sufficiently strong once matter is condensed into massive dense bodies to keep humans and other matter attached to the surface of the Earth,* despite the Earth's violent motion around the Sun, and the Solar system's motion through the galaxy, and *to keep celestial bodies in their orbits, even at great distances.*

(27) *The existence of a very large amount of energy at the time of the origin of the current universe, makes the idea of a universe in which gravitational attractive forces eventually overcome the forces causing the expansion of the universe particularly attractive, in that it provides an explanation for this energy and for the Big Bang without invoking dark matter or dark energy.*

If the current universe began with a large amount of *positive energy*, for example from the Big Crunch of the previous universe, the diffuse outward kinetic energy would end up as the outward kinetic energy of the large masses, such as the galaxies, stars and planets; and in an expanding universe the *matter density* reduces in proportion to the volume (third power of the radius) of the universe, the *gravitational forc*e on matter decreases according surface (second power of the radius) of the universe, so as the universe expands, the force of expansion is eventually overcome by the force of gravitation. Then the Big Bang is nothing more than the Big Crunch of the previous universe. This *is consistent with the known laws of (non-relativistic) physics* without the introduction of unobservable *dark energy* or *dark matter.* It is also consistent with the expected end of our planet and solar system as the planets are engulfed by the Sun.

(28) *Under this theory, the current universe, which originated about 13.8 billion years ago, evolved for about 9.6 billion years before a primitive form of life originated under the particular conditions of a small rocky planet.* After a further period of about 4.2 billion years, around 30 million years ago, an *intelligent form of life*, referred to as *humans* (*homo sapiens*), evolved from this primitive form of life on this planet, in accordance with the reproductive laws of the survival of the fittest. Members of this species could communicate in a number of languages in different parts of the planet, which they called the Earth. They were relatively small, presumably as a result of the need to ensure arboreal cover and food for their early ancestors, putting size and energy constraints on the size of their brains. I am a

member of this species. Over the last 5,000 years members of this species learned to communicate in written form, leading to a sequence of civilizations in different parts of the planet. In the last few hundred years, some of the more intelligent of its members were able to develop electronic communications and computation, airborne transportation and to visit nearby planets; and even to learn about their own origins, and the origin of the planet, solar system and universe in which they live; of which this book is an example.

(29) Although this helps to clarify what gravity does and does not do, *this is still about the effects of gravity. The underlying cause remains elusive*. From this analysis it is most likely to be found in the *initial creation of matter and expansion of the universe*, not in quantum theory, nor in a unified theory of electromagnetism and gravity.

Gravitational energy or gravitational potential energy U is the potential energy a massive object m has in relation to another massive object M due to gravity; U = GmM/R, where R is the distance between the centers. It is the potential energy associated with the gravitational field, which is released (converted into kinetic energy) when the objects fall towards each other. Gravitational potential energy increases when two objects are brought further apart.

In classical mechanics, two or more masses always have a *gravitational potential*. Conservation of energy requires that this gravitational field energy is always negative, so that it is zero when the objects are infinitely far apart. The gravitational potential energy is the potential energy an object has because it is within a gravitational field.

The most common use of *gravitational potential energy* is for an object near the surface of the Earth where the *gravitational acceleration* can be assumed to be constant at about 9.8 m/s^2. Since the zero of gravitational potential energy can be chosen at any point (like the choice of the zero of a coordinate system), the potential energy at a height h above that point is equal to the work which would be required to lift the object to that height with no net change in kinetic energy. Since the force required to lift it is equal to its weight, it follows that *the gravitational potential energy is equal to its weight times the height to which it is lifted.*

$$U = m \times g = \text{weight} \times \text{height} \; [\text{kg} \times 9.8\ \text{m/s}^2 \times \text{m} = \text{joules}].$$

The force between a point mass, M, and another point mass, m, is given by Newton's law of gravitation:

$F = GmM/r^2$

To get the total work done by an external force to bring point mass m from infinity to the final distance R (for example the radius of Earth) of the two mass points, the force is integrated with respect to displacement:

$$W = \int_{\infty}^{R} GmM/r^2\, dr = -\, GmM/r\, |_{\infty}^{R}.$$

Because $\lim_{r\to\infty} 1/r = 0$, the total work done on the object can be written as:

Gravitational Potential Energy, $U = GmM/R$.

In the common situation where a much smaller mass m is moving near the surface of a much larger object with mass M, the gravitational field is nearly constant and so the expression for gravitational energy can be considerably simplified. The change in potential energy moving from the surface (a distance R from the center) to a height h above the surface is

$$\begin{aligned} \Delta U &= GmM/R - GmM/(R + h) \\ &= GmM/R\ \{1 - 1/(1 + h/R)\}. \end{aligned}$$

If h/R is small, as it must be close to the surface where g is constant, then this expression can be simplified using the binomial approximation

$$1/(1 + h/R) \approx 1 - h/R$$

to

$$\begin{aligned} \Delta U &\approx GmM/R\ \{1 - (1 - h/R)\}, \\ \Delta U &\approx GmM/R^2, \\ \Delta U &\approx m(GM/R^2)\ h. \end{aligned}$$

As the *gravitational field* is $g = GM/R^2$, this reduces to

$$\Delta U \approx mgh.$$

Taking $U = 0$ at the surface (instead of at infinity), the familiar expression for *gravitational potential energy* emerges:

$$U= mgh.$$

(30) *This returns us to the question of whether it may be possible to relate the gravitational constant G to the current rate of expansion of the universe.*

The gravitational field at the at the surface of the universe, is approximately equal to the gravitational constant.

The gravitational field $g = GM/R^2$. Substituting $M = 1.5 \times 10^{53}$ kg and $R = 4.4 \times 10^{26}$ m, *the gravitational field at the at the surface of the universe, is approximately equal to the gravitational constant.*

$$g = GM/R^2 = G \times 1.5 \times 10^{53}/(4.4 \times 10^{26})^2 = 0.7748\ G;$$

and substituting $G = 6.674 \times 10^{-11}$ N . m^2 . kg^{-2}, *the gravitational field at the at the surface of the universe,*

$$g = 5.171 \times 10^{-11}\ N\ .\ kg^{-1}$$

g = G when

$$R^2 = M = 1.5 \times 10^{53};$$

or $$R = 3.873 \times 10^{26}\ m.$$

This implies that *the volume of the universe,*

$$V = 4/3\ \pi R^3 = 4/3 \times 3.14159 \times (3.873 \times 10^{26})^3,$$

or $$V = 2.4335 \times 10^{80}\ m^3.$$

This is in line with other estimates.

This is probably the closest that I can get.

[The *actual density* of atoms is equivalent to roughly 1 proton per 4 cubic meters = $1.6726 \times 10^{-27}/4 = 4.18 \times 10^{-28}$ kg m^{-3}). The *mass* of the *observable universe* (including ordinary matter, the interstellar medium (ISM), and the intergalactic medium (IGM), but excluding dark matter and dark energy) is around 1.5×10^{53} kg. This implies that the *current volume* of the *observable universe* is $(1.5 \times 10^{53})/(4.18 \times 10^{-28}) = 3.6 \times 10^{80}$ m^3.]

[In general relativity gravitational energy is extremely complex, and there is no single agreed upon definition of the concept. It is sometimes modelled via the Landau–Lifshitz pseudotensor that allows retention for the energy–momentum conservation laws of classical mechanics. Addition of the matter stress–energy tensor to the Landau–Lifshitz pseudotensor results in a combined matter plus gravitational energy pseudotensor that has a vanishing 4-divergence in all frames—ensuring the conservation law. Some people object to this derivation on the grounds that pseudotensors are inappropriate in general relativity, but the divergence of the combined matter plus gravitational energy pseudotensor is a tensor.]

Conclusion.

The underlying cause of gravity remains elusive. From this analysis it is most likely to be found in the initial creation of matter and expansion of the universe, not in quantum theory, nor in a unified theory of electromagnetism and gravity. This appears to suggest that gravity resulted from some form of entanglement between matter after it was driven apart in the expansion of the universe following the Big Bang.

Gravity has no role in the accretion of cosmic dust into planets or stars, nor in creating fusion in the center of stars, such as the Sun. Based on the actual density of atoms in the universe, the time taken for molecules of hydrogen or cosmic dust to accrete due to the force of gravitation is far too long. But it is sufficiently strong once matter is condensed into massive dense bodies to keep humans and other matter attached to the surface of the Earth, despite the Earth's violent motion around the Sun, and the Solar system's motion through the galaxy, and to keep celestial bodies in their orbits, even at great distances.

The existence of a very large amount of energy at the time of the origin of the current universe, makes the idea of a universe in which gravitational attractive forces eventually overcome the forces causing the expansion of the universe particularly attractive, in that it provides an explanation for this energy and for the Big Bang without invoking dark matter or dark energy.

The gravitational field at the at the surface of the universe, is approximately equal to the gravitational constant.

www.ingramcontent.com/pod-product-compliance
Lightning Source LLC
Chambersburg PA
CBHW061323190326
41458CB00011B/3880
* 9 7 9 8 2 1 8 3 9 6 5 4 1 *